Voice Writing Method
FIFTH EDITION

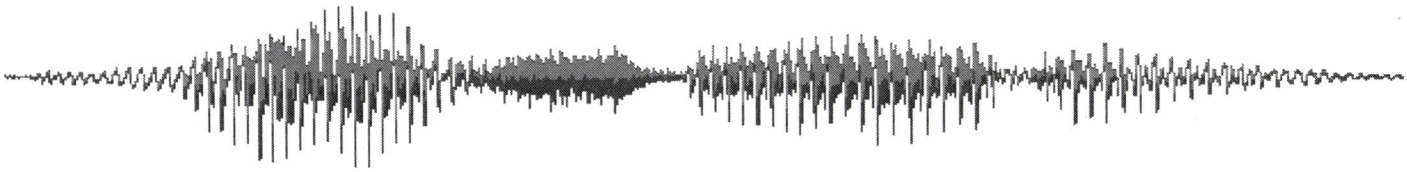

"It is important that an aim never be defined in terms of activity or methods. It must always relate directly to how life is better for everyone."

—W. Edwards Deming

Voice Writing Method
FIFTH EDITION

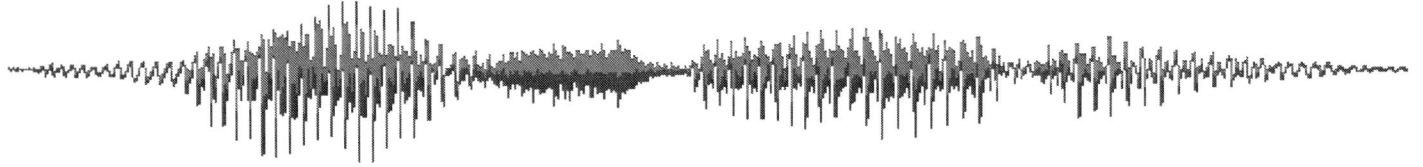

by Bettye Keyes, CCR, CSR, CVR-M, RVR

Voice Writing Method • *Fifth Edition*
(Dragon NaturallySpeaking® 12 Premium and Professional)

(Updated July 22, 2013)

Written by Bettye Keyes, CCR, CSR, CVR-M, RVR

Copyright © 2005-2013 by Bettye Keyes. All rights reserved.

No part of this publication, including the format or outline in which the information is presented, may be reproduced, stored in a retrieval system, or transmitted in any form or by any means, electronic, mechanical, photocopying, recording, scanning, or otherwise, except as permitted under Sections 107 or 108 of the 1976 United States Copyright Act, without either the prior written permission of the author or authorization through payment of the appropriate per-copy fee to the author, Bettye Keyes. Requests to the author for permission should be sent via e-mail to Info@RealtimeUniversity.com.

Trademarks: Dragon NaturallySpeaking® is a registered trademark of Nuance Communications, Inc. All other trademarks are the property of their respective owners.

LIMIT OF LIABILITY/DISCLAIMER OF WARRANTY: THE VOICE WRITING TECHNIQUES HEREIN ORIGINATED FROM A VARIETY OF PUBLIC SOURCES. PROPER DICTATION STYLES WERE INCORPORATED FROM THE FIRST STENOMASK COURSE WRITTEN BY HORACE WEBB, CAT VENDORS, INDIVIDUAL VOICE WRITERS, AND THE AUTHOR HERSELF. ALL OTHER INFORMATION CONTAINED IN THIS PUBLICATION IS A COMBINATION OF THE AUTHOR'S THEORIES AND IDEAS BASED ON THE AUTHOR'S OWN PROFESSIONAL EXPERIENCE, ALONG WITH INFORMATION SHARED AMONG EXPERTS IN THE FIELD OF SPEECH RECOGNITION TECHNOLOGY, INCLUDING TECHNIQUES DEVELOPED AND PUT INTO PRACTICE BY THE AUTHOR PRIOR TO THEIR INCLUSION IN ANY PUBLICATION, COPYRIGHTED MATERIALS, AND THAT MAY OR MAY NOT BE CURRENTLY INCLUDED AS PART OF ANY PREVIOUS OR SUCCESSOR WORKS. WHILE THE AUTHOR AND PUBLISHER HAVE USED THEIR BEST EFFORTS IN PREPARING THIS BOOK, NO REPRESENTATIONS OR WARRANTIES ARE MADE WITH RESPECT TO THE ACCURACY OR COMPLETENESS OF THE CONTENTS OF THIS BOOK AND SPECIFICALLY DISCLAIMS ANY IMPLIED WARRANTIES OF MERCHANTABILITY OR FITNESS FOR A PARTICULAR PURPOSE. THIS TEXT IS NOT INTENDED TO BE USED AS A SOLE SOURCE OF INFORMATION FOR EDUCATION AND LEARNING; OTHER ACADEMIC MATERIALS AND/OR TRAINING MAY BE NECESSARY TO MEETING THE PROFESSIONAL STANDARDS REQUIRED IN CERTAIN CAREER SPECIALTIES. NO WARRANTY MAY BE CREATED OR EXTENDED BY SALES REPRESENTATIVES OR WRITTEN SALES MATERIALS. THE ADVICE AND STRATEGIES CONTAINED HEREIN MAY NOT BE SUITABLE FOR YOUR COMPUTER OR SOFTWARE CONFIGURATION. YOU SHOULD CONSULT WITH A PROFESSIONAL WHERE APPROPRIATE. THE AUTHOR AND PUBLISHER SHALL NOT BE LIABLE TO ANY PERSON OR ENTITY WITH RESPECT TO ANY LOSS OR DAMAGE, INCLUDING, BUT NOT LIMITED TO, SPECIAL, INCIDENTAL, OR CONSEQUENTIAL DAMAGES CAUSED OR ALLEGED TO BE CAUSED DIRECTLY OR INDIRECTLY BY THE INSTRUCTIONS CONTAINED IN THIS BOOK OR BY THE COMPUTER SOFTWARE AND HARDWARE PRODUCTS DESCRIBED HEREIN. NAMES AND TERMS MENTIONED IN THIS BOOK WHICH ARE KNOWN TRADEMARKS OR SERVICE MARKS HAVE BEEN APPROPRIATELY CAPITALIZED, AND THE USE OF ANY NAME OR TERM IN THIS BOOK SHOULD NOT BE REGARDED TO AFFECT THE VALIDITY OF ANY TRADEMARK OR SERVICE MARK.

Cover art and layout by Bettye Keyes
Library of Congress, Registered July 2005
Library of Congress, Registered August 2006
Library of Congress, Registered October 2009
Library of Congress, Registered June 2011
Library of Congress, Registered March 2013
Library of Congress Control Number: 2006925613
(First edition released July 2005. Second edition released February 2007. Third edition released October 2009. Fourth edition released June 2011. Fifth edition released March 2013.)
ISBN-10: 0-9891134-0-X
ISBN-13: 978-0-9891134-0-3
Manufactured and Distributed in the United States of America

In Loving Memory of

Dave Rogala, CCR, CSR, CVR-CM

(www.DaveRogala.com)

This book is dedicated to my dearest friend, Dave Rogala, who nobly donated most of his personal time to supporting all causes related to the voice writing profession. He traveled far and wide to meet political challenges, as a gift to us all, so that we may see a stable future for the reporting industry. Although he passed away in April of 2009, the efforts of his "mission" to spread voice writing across the U.S. and abroad will continue to bear fruit well after his mission has been accomplished. Personally, I will regard him not merely as a catalyst, but as the most influential impetus behind these important changes.

Much honor and love to you always, my friend.

About The Author

Bettye Keyes is the reporting industry's first certified realtime voice writer to develop and author works related to speech-recognition-based realtime voice writing techniques and theories.

In 1999, she took a self-study approach to learning realtime voice writing, since, at that time, there were no schools teaching the method. In 2000, she became a Certified Court Reporter through the State of Louisiana and Certified Verbatim Reporter through the National Verbatim Reporters Association (NVRA). In 2001, she obtained her Realtime Verbatim Reporter certification through the NVRA, which required a passing score of over 96% accuracy for live transcription at speeds ranging between 180 to 200 words per minute.

Bettye has used speech recognition technology from the inception of her career. Her realtime reporting experience, which began in the court reporting industry, led her to expand into the communications access realtime translation (CART) services and captioning arenas. She currently teaches online realtime training courses and works as a private consultant to corporations and firms throughout the United States and abroad, building realtime training programs across all voice writing disciplines. Bettye is also the founder of Realtime University, and, in addition, has produced training videos, written six other books on the topic of realtime voice writing and contributes her ideas and methods on an ongoing basis to cultivate the future of the realtime reporting profession through the advancement of realtime voice writing practices taught by educational institutions.

Acknowledgments

Many thanks go to the innumerable voice writers who have played leading roles in advancing our industry over the years and for continuing to inspire us as we prepare our next generation of professionals for careers in voice-based reporting.

Special acknowledgment goes to Dave Rogala, CCR, CSR, CVR-CM, for his unrelenting support to our success through our voice writing journey.

Contents

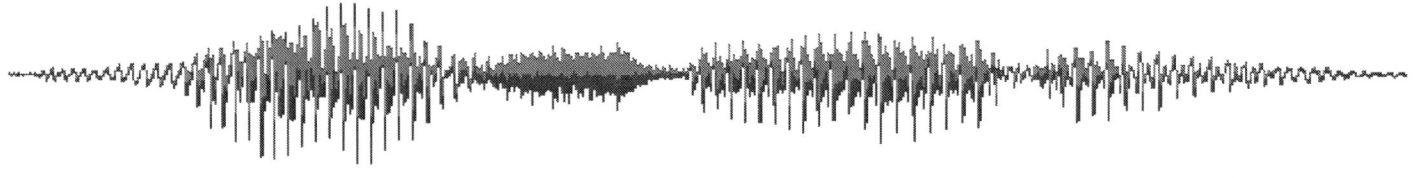

Introduction xiii

About This Book xiii
Speech Recognition Software in This Book xiv
About SpeedMaster™ xiv
Common Terms, Abbreviations, and Acronyms Used in This Book xiv
About Voice Writing.. xv
About Voice Writing Theory xv
What Will be Covered xvi
What You Will Learn... xvii
    Dictation Techniques.. xvii
    Voice Writing Theory xvii
    Realtime Setup and Performance... xvii
What You Will Do... xviii
    Create a Voice Model xviii
    Improve Accuracy... xviii
    Do Realtime.. xviii

vii

Voice Writing Method • FIFTH EDITION
Dragon NaturallySpeaking® 12

—Part 1—
Careers, Equipment, and How It All Works

Chapter 1: The Basics 3

Careers for Realtime Voice Writers 3
    Court Reporting 4
    CART 5
    Captioning 5
    Financial Call Reporting 6
    Other Transcription Services 6
Realtime Voice Writing Software 6
    Speech Recognition Software 7
        How Speech Recognition Works 7
            Sound Parts 8
            Speech-to-Text Conversion 8
        Speech Recognition Tools, Terms, and Functions 10
    CAT Software 11
        The Purpose of a CAT Program 11
        How a CAT Program Works with Speech Recognition 12
            CAT Program with a Dictionary 13
            CAT Program without a Dictionary 14
Realtime Voice Writing Equipment 15
    Computer 15
    Dictation Input Devices 17
        USB Sound Card 17
        Open-Mic Headset or Speech Silencer 18
    High-Gain Microphone (for Recording External Environment) 19
    Foot Pedal 19
    Surge-Protection Power Strip 20

—Part 2—
Realtime Dictation and Voice Theory Development

Chapter 2: Dictation Techniques 23

    Proper Breathing and Volume with Dictation Devices 24
        Dictating into an Open-Mic Headset 24
        Dictating into a Speech Silencer......................................24
    Tone and Modulation 26
    Enunciation 29
    Pace..32
        Fast Speakers 33
        Slow Speakers 34
    Punctuating..34
    Verbatim Dictation versus Paraphrasing and Summarization........35
    Dictation Patterns..36

Chapter 3: Voice Writing Theory 39

    Punctuation...41
    Speaker Identification..43
    Examination Lines 49
    Q&A Markers..50
    Parentheticals..52
    Numbers 61
    Spelling Technique..63
    Conflict Resolution..64
        Homophones...65
        Words with Soft Pronunciations 67
        Undesirable Words 68
        Small Words 68
        Similar-Sounding Words...70

Words versus Acronyms...71
All Other Words..72
Phrases 73
Brief Forms 76
On-the-Fly Translations 79
Delete Command..81

—Part 3—
Getting Set Up and Doing Realtime

Chapter 4: Getting Started.. 85

Computer and Equipment Setup..85
USB Connectivity Settings..88
Turning Off Automatic Updates and Disabling Anti-Virus Software....89
Dragon NaturallySpeaking® Setup 90
Creating a User..93
Setting User Options 109
Setting Formatting Preferences 116
Performing a Dictation Session 117
Saving File of Dictation Session...120

Chapter 5: Vocabulary Setup and Formatting...................... 121

Working with the Vocabulary...121
Written- and Spoken-Form Spelling Rules 122
Accessing the Vocabulary 124
Adding an Entry..125
Deleting an Entry..127
Modifying Formatting Properties of an Entry 128

Vocabulary Setup and Formatting Using Dragon Alone................................129
    Punctuation Marks..130
    Speaker IDs..133
    Q&A Markers..136
    All Other Voice Writing Entries..138
Vocabulary Setup and Formatting Using Dragon with Other Applications....138
    Microsoft Word..138
    CAT Program with a Dictionary..141
    CAT Program without a Dictionary.....................................142
Practice Dictation to Test Formatted Results................................143

Chapter 6: Improving Accuracy 145

Adjusting Audio Levels..146
Correcting Misrecognitions..150
Adding New Words..158
Adding Phrases...160
Vocabulary Building..162
    Document Preparation...163
        Reformatting Documents for Dragon-Alone Use 167
        Reformatting Documents for CAT Software Use..........169
            CAT Software Scenario 1 (with CAT Dictionary)............169
            CAT Software Scenario 2 (without CAT Dictionary) 172
        Programming Macros to Automate Search-and-Replace
            Functions...174
        Saving and Storing Documents................................174
        Preparing Documents Before a Realtime Job.............174
    Vocabulary Customization...175
    Document Analysis..176
Resolving Recognition Problems...183
    Removing Unnecessary Words That Conflict with Necessary Words ..183
    Making Additional Vocal Recordings to Distinguish Between Words
        with Similar Pronunciations 183
    Making Multiple Vocabulary Entries to Obtain Correct Recognition
        for a Single Result...184
    Using Misrecognitions to Produce Recognitions.................184
    Accuracy Reinforcement...184

Chapter 7: Speed Building 187

Speed-Building Plan..188
SpeedMaster™ Software...188
    How to Install..188
    Obtaining Files From the SpeedMaster™ CD..........................189
    How to Operate 189
Dictation Material 191
Determining Accuracy Performance..191

Chapter 8: Computer Maintenance and File Management.193

Computer Maintenance..193
    Archiving Data..193
    Disk Defragmentation...194
    Anti-Virus Scans 194
User File Management...195
    Creating a New User..195
    Opening/Closing Users 198
    Renaming/Deleting Users 198
    Backing Up/Restoring User Files 199
    Exporting/Importing User Files..201
Vocabulary Management..205
    Creating a New Vocabulary 206
    Opening, Renaming, and Deleting Vocabularies 208
    Exporting/Importing a Vocabulary 209
    Exporting/Importing a Word List...212

Comments About CAT Software 215
GLOSSARY 217
INDEX 223

Introduction

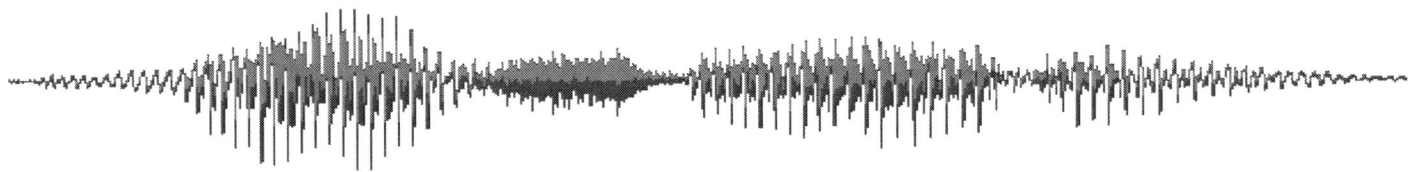

About This Book

This guide offers a fundamental approach to learning how to provide accurate realtime translation services with speaker-dependent speech recognition. All information about speech recognition software in this book is based on the Premium and Professional editions of Dragon NaturallySpeaking® 12 and is intended for use with this version and higher. The how-to steps for accessing various tools, implementing formatting functions, and performing many other software-related tasks will differ if you are using additional programs or modules such as computer-aided transcription (CAT) software.

The techniques presented in this book involve the application of special utterances, or "**voice codes**," which form the basis of dictation shorthand methods for voice-to-text translations with speaker-dependent speech recognition. The culmination of these voice codes and their conceptual usages together comprise a special dictation language known as a "**voice writing theory**."

The careers of focus are court reporting, communications access realtime translation (CART), broadcast captioning, and financial call reporting. The techniques presented herein may also be applied in any other career specialty involving realtime transcription. Whether you are interested in providing realtime services as a court reporter, CART provider, broadcast captioner, or in any other capacity that requires live voice-to-text production, you can use this method of voice writing to keep track of speakers, resolve word conflicts, and produce text in myriad scenarios that are necessary to meet the realtime transcription performance standards in your specific field.

Speech Recognition Software in This Book

All information about speech recognition software in this book is based on Dragon NaturallySpeaking® version 12 (Premium and Professional editions) and is strictly intended for use with these specific editions. All screen captures were taken using the Dragon NaturallySpeaking® Professional edition; if you are a Dragon NaturallySpeaking® Premium edition user, there will be some minor differences in features and functionality, but both editions of Dragon work basically the same way.

The how-to steps for accessing various tools, implementing formatting functions, and performing many other operations will differ if you are using additional programs or modules such as CAT software.

About SpeedMaster™

Accompanying this book on CD is SpeedMaster™ software, a speed-building program designed to help you rapidly build dictation speed as you work to increase speech recognition accuracy. SpeedMaster™ allows you to perform dictation sessions at your own comfortable pace—with unlimited practice material—and takes the guesswork out of measuring speed-building progress. SpeedMaster™ works only with plain text (.txt) files. When these files are opened through SpeedMaster™, text is displayed on the SpeedMaster™ screen. You can set the text-reading display to any speed you want, ranging from 100 to 350 words per minute, and dictate that text at your set word-per-minute rate.

Text files of Dragon's general training stories are also provided with SpeedMaster™ so you can create a voice model at realtime dictation speeds. Once SpeedMaster™ is installed, you may find these files in a folder called *SpeedMaster* on your computer's hard drive.

Common Terms, Abbreviations, and Acronyms Used in This Book

Please note the following abbreviations and acronyms for terms frequently used in this book.

- Dragon NaturallySpeaking® (Dragon)
- artificial intelligence (AI)
- computer-aided transcription (CAT)
- communications access realtime translation (CART)
- speech recognition (SR)
- speech recognition engine (SRE)

Introduction

About Voice Writing

The method of transcription based on human dictation began in the court reporting industry in the 1940s. It originated with a man named Horace Webb who manufactured the Stenomask®, a dictation device designed to allow a court reporter to speak, repeating the words of others, into an enclosed barrel with a built-in microphone so a recording of the dictation could be made. This device also prevented the sound of the court reporter's voice from distracting others in the environment. The court reporter would then listen to a playback of his or her dictation to manually type every word of the transcript.

Court reporters who used the Stenomask® to perform their jobs were known as "**stenomask reporters**" (or "**mask reporters**" for short). Today, other forms of dictation masks are available, now generically referred to as "**speech silencers**," and the new term being used to describe a stenomask reporter is "**voice writer**."

Now that SR technology is available, voice writers do not need to manually type every word of their dictation from an audio recording since their voices can automatically transcribe text in realtime, and the new term being used to describe a voice writer who provides realtime services is a "**realtime voice writer**."

Realtime voice writers have grown in their understanding of how to attain accuracy results well past the marks of what SR programs were originally designed to achieve. It has been proven that proficient realtime voice writers can pass realtime certification tests at rates of up to 200 words per minute with a minimum of 96% accuracy.

Today, realtime voice writers have additional opportunities extending to them in professions besides court reporting. This includes CART for hearing-impaired persons, closed-captioning services, and wherever else a market exists for live voice-to-text production.

About Voice Writing Theory

The techniques presented in this book involve the application of special utterances, or voice codes, which form the basis of dictation shorthand methods for voice-to-text translations with speaker-dependent speech recognition software. The culmination of these voice codes and their conceptual usages together comprise a special dictation language known as a **voice writing theory**.

Whether you are interested in providing realtime services as a court reporter, CART provider, broadcast captioner, financial call reporter, or in any other career specialty that requires live voice-to-text production, you may use this voice theory to keep track of speakers, resolve word conflicts, and produce text in myriad scenarios that are necessary to meet the realtime transcription performance standards in your specific field.

What Will Be Covered

This book teaches the basics of using SR software, through step-by-step guidance, to meet the reporting profession's established accuracy standards for providing realtime services. Here is an overview of what is covered.

- **Chapter 1** introduces the main career specialties for realtime voice writers as well as voice writing hardware and software to understand how it all works.

- **Chapter 2** covers dictation techniques.

- **Chapter 3** presents a realtime dictation method known as a "voice writing theory."

- **Chapter 4** explains computer and equipment setup, how to create a voice model and set user options, and do realtime dictation.

- **Chapter 5** details how to set up your vocabulary to work with your voice writing theory and produce formatted results where necessary.

- **Chapter 6** describes how to effectively increase SR accuracy performance using various accuracy improvement techniques, including correcting recognition errors and adapting your vocabulary to reflect the grammar style of your realtime voice writing dictation language.

- **Chapter 7** provides a speed-building plan for you to follow to effectively build dictation speed as you work to increase speech recognition accuracy.

- **Chapter 8** outlines how to maintain your computer and voice files.

What You Will Learn

Dictation Techniques

You will learn special dictation techniques involving volume, pitch, enunciation, pace, and punctuation. A conclusive working dictation theory is presented for practical SR application.

Voice Writing Theory

You will learn about the special dictation language used by realtime voice writers to produce text in myriad reporting scenarios.

Realtime Setup and Performance

You will learn a step-by-step approach for creating a voice model and how to translate speech into text to meet the reporting industry's realtime accuracy and speed performance standards.

What You Will Do

Create a Voice Model

You will set up a user within Dragon and perform voice training to create a voice model. Then you will do the required vocabulary setup work for using your voice writing theory.

Improve Accuracy

You will do practice dictation sessions and make corrections to SR errors. Then you will build the grammatical model to reflect the style of your voice writing language and improve Dragon's ability to determine proper word selections based on the syntax of the words you commonly dictate. You will make the remainder of accuracy improvements by continuously dictating and correcting misrecognitions until your speed and accuracy performance meets or exceeds the reporting industry's realtime standards.

Do Realtime

Once you have met the standard realtime speed and accuracy requirements set by the reporting industry, you are ready to seek a fulfilling, profitable place among other practicing realtime professionals by contacting NVRA to learn about certification requirements. If your chosen field requires the additional use of specialty realtime software, a list of CAT vendors is provided on NVRA's website (www.nvra.org).

Part 1

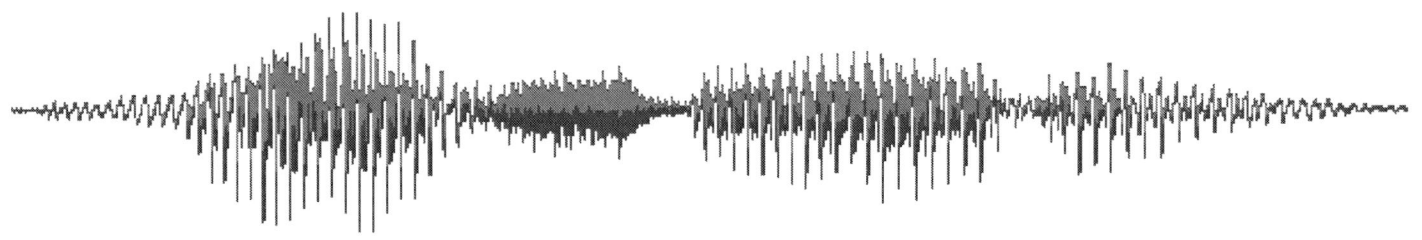

Careers, Equipment, and How It All Works

Chapter 1: The Basics

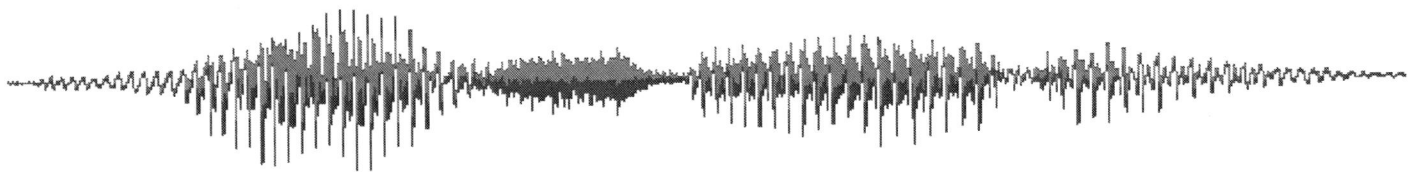

Careers for Realtime Voice Writers

The advent of realtime communications in the business world, consumer electronics marketplace, government sector, and homes across the world has fueled a customer expectation that realtime reporting be part of daily life. This has trickling effects on transcription-based careers, where the availability of realtime technology is becoming a prominent influence. However, we are currently facing a shortage of skilled realtime stenographers available to fill many of the jobs throughout the United States and international markets which require realtime transcription. Court reporting schools teaching stenographic-based reporting methods are not churning out certified professionals fast enough to meet the needs of court systems and attorneys' offices requiring court reporting services. Meanwhile, since the U.S. Congress has mandated that all new television programming be closed-captioned, SR is on the rise in broadcast captioning, and there is an ever-increasing demand for provision of CART services to hard-of-hearing citizens. Also, the financial services market is now providing live transcription services of financial calls to keep their clients informed with the most up-to-date investment information, while the medical field is backlogged with the untimely reporting of patient records due to the length of time it takes medical transcriptionists to manually type medical reports. This all means a growing number of career opportunities for realtime voice writers.

Successful campaigns to present realtime transcriptionists as leading members of the information communications technology sector, and to grow realtime reporting into new areas, have only increased pressure to keep filled the ranks of all reporting occupations nationwide. Realtime reporting will only continue to grow. It is now and will be for the foreseeable future a realtime reporter's market for providing realtime transcription services.

The following sections provide basic information about careers available to realtime voice writers.

Court Reporting

In standard **court reporting**, the duty of a voice writer is to use **verbatim dictation**—where a voice writer clearly dictates every word spoken by others—as a means from which to produce a verbatim record of a legal proceeding. A **verbatim record** is a transcript that has been edited for accuracy to reflect the exact words of speakers (i.e., verbatim content), including correct spellings of all names and words. The transcript is proofread to ensure that it is a true account of the words spoken in a legal proceeding and is then delivered as the certified legal document.

Many court reporters choose to provide their services in realtime, as this normally means a healthy increase in marketability and income. In **realtime court reporting**, the duty of a voice writer is to produce live transcription of content in the form of a rough draft in addition to following up with a verbatim transcript. A **rough draft** is the original version of a realtime transcript that is not considered a verbatim account of spoken content since it has not yet been edited or proofread. The live text may be fed directly to other computers via cables or an Internet connection for simultaneous view by attorneys and judges. The rough version of the transcript is then edited and proofread later in order to be certified and delivered as the final verbatim record.

Many states require court reporting certifications for voice writers, but some do not. If your state does not have a means to certify court reporters or does not address court reporters who use the voice writing method, a certificate may not be required in order for you to work as a court reporter in that state.

If your state does require certification, you must pass a proficiency test before you can work as a court reporter. You may also enter a state's reporting workforce by reciprocity or endorsement of license through a different state, where a certifying authority in one state may recognize passing examination scores from another state to award a court reporting certification.

Obtaining certification in your field demonstrates that you are able to perform at the expected competency levels within your industry. So even if you are not required to obtain certification in order to work in your state, you still should consider obtaining a national certification through the National Verbatim Reporters Association (NVRA) to qualify your professional expertise.

Court reporting certifications offered through the NVRA are Certified Verbatim Reporter (CVR), Certificate of Merit (CM), and Realtime Verbatim Reporter (RVR). To learn about certification requirements and test registration, visit www.nvra.org.

For more information about individual state certification standards and requirements, visit www.ncsc.org and click on the *State Court Websites* link.

The United States Court Reporters Association (USCRA) is another certifying body which recognizes voice writers. To obtain certification information, visit their website at www.uscra.org.

Chapter 1: The Basics

CART

CART is an acronym for the term "communications access realtime translation" and is often used synonymously with the term "computer-aided realtime translation."

When providing **CART services**, the duty of a voice writer is to accurately translate the spoken word into visual communications for people with hearing disabilities in realtime. Realtime voice writing skills are used to produce text on computer screens or onto a large screen through a projector. A text feed may also be sent via cables or the Internet. Although it is preferred that realtime translations be verbatim, it is not a requirement. Paraphrasing techniques may be used, where necessary, as long as the captured content accurately reflects its meaning. Common applications for CART include providing realtime translations within a classroom setting, church, convention, business meeting, and miscellaneous functions (such as plays or musicals, etc.) that do not involve television broadcasting.

It is usually not necessary to attain certification in order to provide CART services. However, holding an NVRA certification will increase your marketability if you are seeking employment in this field. A Registered CART Provider (RCP) certification is offered through the NVRA. To learn about certification requirements and test registration, visit www.nvra.org.

To access general information and obtain details about providing CART services for the hearing impaired community, visit www.cartwheel.cc.

Captioning

The term "**captioning**" is commonly used to mean "**broadcast captioning**," where text is superimposed on the bottom of a television screen or combined within motion picture frames for the purpose of communicating dialogue to the hearing-impaired or for translating foreign dialogue.

Realtime voice writers who work as **broadcast captioners** are responsible for producing translations that are as verbatim as possible; however, just as with CART, the textual content may be paraphrased as long as the meaning reflected is accurate and readable. A realtime text feed may be sent through a captioning encoder or via the Internet.

Voice writers who work as **off-line captioners** do not send realtime text feeds and have the added responsibility of editing content to reflect a verbatim account of words spoken. Paraphrasing is generally not used unless rewording is necessary for translations involving interpretations from one language to another.

It is not necessary to attain certification in order to provide captioning services, but holding an NVRA certification is likely to increase your employability and earning potential in this market segment as well. A Registered Broadcast Captioner (RBC) certification is offered through the NVRA. To learn about certification requirements and test registration, visit www.nvra.org.

For more information about different types of captioning careers and broadcast captioning guidelines, visit www.fcc.gov and type "closed captioning" for your keyword search.

Financial Call Reporting

In **financial call reporting**, a voice writer listens to audio of financial calls and repeats aloud the words he/she hears in order to produce realtime text. A live feed of the text is either streamed directly to a website via the Internet, sent to a scopist for simultaneous editing, or a combination of both. Regardless of the direction of information flow, a realtime voice writer's role is to output text that accurately matches the verbatim words spoken during a financial call.

As with CART and captioning, it is not necessary to attain certification in order to provide realtime services in the financial call services market, but holding an NVRA certification is likely to increase your employability and earning potential.

For information about employment opportunities in the financial call reporting market, do keyword searches through Google about this career.

Other Transcription Services

Other careers for realtime voice writers include **medical transcription** for doctors' offices and hospitals, **legal transcription** for court documents and briefs, and **corporate transcription** for board meetings and minutes that contain sensitive company information. No certification is required for such fields, but it would still behoove you to obtain a base-level certification from NVRA to increase your marketability for employment.

You can locate general information related to employment opportunities for any of these transcription careers by doing Google searches on the Internet.

Realtime Voice Writing Software

The following sections cover which software applications may be used for providing realtime voice writing services. Speech recognition software is a requirement for all realtime voice writing career specialties. However, additional software may also need to be used in conjunction with speech recognition software for those careers that involve outputting text feeds to other computers, sending text via the Internet, or for combining text within a video picture.

Chapter 1: The Basics

Speech Recognition Software

There are many wonderful capabilities of SR software, and the functions are widespread. Not only can you convert speech into text, but you can use your voice as a tool to open and close programs, navigate on the Internet, and replace virtually every keyboard-and mouse-controlled operation. Many of the bell-and-whistle features available in an SR program are not necessary to performing your job as a realtime voice writer since your focus will be on one aspect of using it, for transcription only.

Of all the SR programs that exist in today's marketplace, Dragon NaturallySpeaking® is by far the most technologically advanced speaker-dependent speech recognition product available, which is the reason this book focuses on Dragon. Dragon provides a base vocabulary of around 300,000 of the most common English words, which means users enjoy the benefit of having most of their spoken words recognized without ever having to personally add them, plus are given the capability to add thousands more if desired. Current versions of Dragon can process speech into text at up to 99% accuracy within two weeks, or even sooner, for a general user who speaks at an average of 150 words per minute. Accuracy results can even realistically be around 95% at that same rate right out of the box if a person dictates properly. Although these are excellent initial results, voice writers are expected to achieve this level of accuracy at a much faster rate to meet realtime transcription accuracy standards, which can average between 180 to 200 words per minute. This means you will be pushing Dragon past its reported performance measures. However, as a voice writer whose job is to dictate properly and whose occupation focuses on transcription accuracy, you will spend more time dictating and improving your speech recognition accuracy than average users. As daunting as it may sound at first, rest assured that with practice you will attain this degree of speed-and-accuracy competence, as the methods you will be learning about in this guide have been demonstrated as a proven means to get you there.

The speech recognition products covered in this book are Dragon NaturallySpeaking® Premium and Professional, as these are the editions recommended for most voice writing careers. However, the Medical edition would be most suitable for a medical transcription career. To learn more about other available editions of Dragon NaturallySpeaking®, visit the www.nuance.com website.

Dragon NaturallySpeaking® 12 (Premium edition)

Dragon NaturallySpeaking® 12 (Professional edition)

—How Speech Recognition Works—

In everyday conversation, we run words together and drag out certain syllables to emphasize how we feel, and we usually are not accustomed to enunciating sounds in small words, such as the "d" in "and." That's probably because other humans better recognize us when we use more expressive styles of speech to impart meaning. And it isn't necessary for us to distinctively pronounce the ending "d" sound in "and" for another person to understand what we said. Humans and computers process data through different means.

7

We recognize speech through top-down processing, where we distinguish words based on concepts and circumstantial knowledge. Computers, on the other hand, recognize speech through bottom-up processing, which involves analyzing basic sound structures to identify words. So if a child says to us, "I want my blankowit," we understand the word they mean is "blanket"; this is our interpretation based on meaning and circumstantial knowledge. A computer may recognize the child to have said "blank or with" or "banquet"; this is the computer's interpretation based on its analysis of the actual sounds it heard.

When we run words together, such as when we say, "Ya know what I'm sayin'?" yes, I do know what you're saying, but the computer will not unless you speak clearly and at a rate that allows for distinctions of words to be made apart from other words. This does not mean, however, that we must discretely speak, significantly pausing between each word. The SRE actually recognizes small words better when they are spoken in conjunction with other words and in phrase groupings, and pausing at the right times is important for letting the computer know what to process as either a single word or group of words.

Sound Parts

We're all familiar with the fact that words are comprised of syllables, but words contain even smaller parts. The most basic sound unit of speech is called a "phoneme." The word "sit," for example, contains three phonemes: the sounds "s," "i," and "t," and the English language contains approximately 48 of them altogether. Computers recognize speech by paying attention to these sound parts as it breaks them down even further into sub-parts through a process called digitization, where analog information is converted into digital form.

Speech-to-Text Conversion

There is a three-step process for converting speech into a form the computer can use to produce text:

Speech to Analog > Analog to Digital > Digital to Text

Speech produces sound waves. When we speak into a microphone, those sound waves are converted into analog electrical signals in a process similar to the transmission of sound through a telephone. The computer uses a sound card and software algorithms to digitize those analog signals. The process begins in the sound card's ADC (analog-to-digital converter), which breaks signals down into thousands of discrete steps, measured at specific time intervals within a continuous analog signal range, and converts them into a predefined range of distinct number sets.

Once processed into number sets, the digital patterns are then measured against a set of other digital patterns representing prototypes of phonemes stored within the SRE. A suite of algorithms is used to link the digital patterns of phonemic representations to digital patterns of word representations to determine which choice of words have the highest statistical probability of being correct. Then a best-match guess is made and the text is produced.

Factored into speech-to-text conversion is the retrieval and dissemination of information through the acoustic model, grammatical model, and vocabulary within the SRE. The acoustic model holds sound data associated with your voice, which is then shared with the SRE's vocabulary and grammatical model

Chapter 1: The Basics

for word-priority selections. After possible word choices are retrieved from the SRE's vocabulary (or word bank), information from the grammatical model is used to make a best-match guess based upon complex algorithmic analyses that look for syntax of single-word and word-group relationships and occurrences.

Diagram 1 may give you a better grasp in understanding how the components work together in the conversion process, by looking at it in more visual terms.

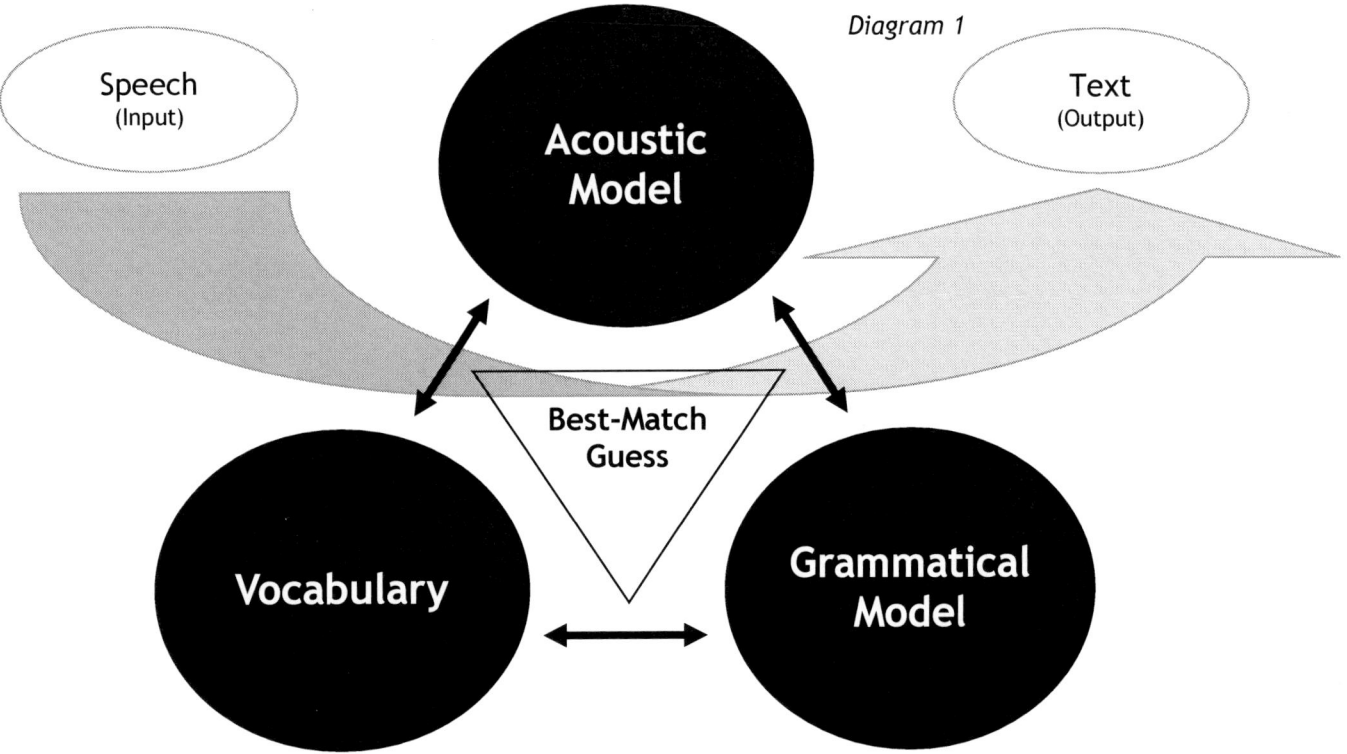

Diagram 1

If you dictate the following sentence while enunciating clearly and speaking at a consistent rate . . .

Mrs. Wright will write a letter right now.

. . . the SRE will use the processes just described to make a best-match guess and probably result in the proper word selections.

With such complex conversions of digital patterns taking place, the selection of words can be affected by minor variations in speech, including enunciation, inflection, and rate of dictation. Common ground must be found and established sets of rules must be followed in the successful merging of human-to-computer interactions. The human must learn the proper method of delivering input to get the computer to produce accurate responses. Through a basic understanding of how SR works, a voice writer gains better insight on how to achieve top levels of SR performance. Being consistent in your dictation style will minimize the risk of error. The dictation methods described in Chapter 2 will give you a firm grasp on how to get higher quality speech-to-text output based on the way you speak.

Speech Recognition Tools, Terms, and Functions

Review the table below to familiarize yourself with the names of Dragon's utilities as they relate to the generic terms being referenced. These are the main tools you will be using in Dragon, and these terms appear commonly throughout this book.

Table 1

Names of Tools	
Generic Term	**Dragon NaturallySpeaking® Screen or Utility Name**
Voice Training	General Training
Vocabulary	Vocabulary Editor
Correction Tool	Spelling Window
Vocabulary-Building Utility	Learn from specific documents
Audio Setup	Check Microphone

To apply the dictation techniques and voice writing theory presented in this guide, a voice model will first be created through the **voice training** process, where Dragon provides stories for you to read so it can analyze your voice and learn the way you speak. You will also input your voice writing theory entries into the **vocabulary** so Dragon will be able to recognize them in your dictation, and then you will use other software—such as Microsoft Word or a CAT program—when required, to handle any special formatting needs that Dragon isn't able to produce. This is the most basic part of the total setup involved in your realtime voice writing preparation work.

The remainder of the setup tasks relate to maximizing Dragon's ability to accurately recognize your speech. In order to achieve the highest levels of SR accuracy, focus is placed on three critical adjustment factors to tune Dragon's performance measures based on its three main design elements: 1.) the acoustic model, 2.) the vocabulary, and 3.) the grammatical model.

Adjustments to the acoustic model are made using the **correction tool**, where you will dictate and correct any speech recognition errors. This improves accuracy by giving Dragon additional layers of acoustic data to properly recognize how you pronounce certain words and to account for sound-pattern inconsistencies associated with the fluctuations and varieties for ways you say them in connection with other words.

Finally, the vocabulary and grammatical model must be customized using the **vocabulary-building utility**. This is where Dragon analyzes documents for the purpose of adding words to its vocabulary that it doesn't already contain and enhance the recognition of them within the context of your voice writing language by establishing relationships between the words you dictate based on syntax and how often they will occur.

Additional measures taken for accuracy improvement involve adding individual words and phrases to the vocabulary and performing an **audio setup** to adjust for acoustic changes related to the level of your speech input and/or extraneous environmental sounds.

Chapter 1: The Basics

CAT Software

In professions requiring you to send a realtime text feed, you will have to use a CAT program in addition to SR software to have that capability. You will not have need for a CAT program if you are using SR software for regular transcription work.

There are several vendors in the market offering CAT software. These programs are all compatible with Dragon NaturallySpeaking®. But for realtime careers requiring text to consistently appear within or under three seconds, such as broadcast captioning or Internet text-streaming of financial calls, be sure to choose CAT software that meets those text delivery speed requirements.

For a list of CAT vendors, visit the www.nvra.org website.

The Purpose of a CAT Program

When using an SR program alone, you run the risk of experiencing instability factors, such as your microphone turning off in the middle of a dictation, programs opening and closing by surprise, or words being deleted that you actually wanted to keep, all because sets of words you dictated may have been mistakenly recognized as commands to perform those operations. Also, let's say you are dictating on page 60 of a transcript, and your cursor jumps to page 1 by mistake. If you are dictating directly into an SR program, your text would begin inserting itself at the beginning of the transcript, not on page 60 where you actually want the text to appear. You do not have to worry about these mishaps if you use CAT software.

Benefits of using CAT software include the ability to send realtime feeds, simultaneously make recordings of your voice and the external room, audio-sync text with a digital recording, play back an audio file while it is still in the process of recording, instantly resolve language conflicts the SRE has trouble with, and handle all of your text-formatting requirements. The SR commands that are unnecessary for realtime voice writing are disabled so they do not interfere with your realtime transcription work, and if you are dictating on page 60 of a transcript and your cursor jumps to page 1 by mistake, your text will continue being inserted at the end of the transcript. You can also do simultaneous editing as you are dictating and even produce words on the fly that are not contained in the SRE's vocabulary.

There are many other wonderful advantages to having CAT software that are not mentioned in this book but are very useful. Additional features may include a user-friendly file-management system for helping you keep track of your work files and archive them. Other utilities may also be available for automatically generating a table of contents, word index lists, and condensing transcripts. You'll also be able to edit your transcript while you are dictating.

These are just a few of the capabilities of CAT software. For information about the functions of the particular CAT program you own, refer to the Help menu or user's manual. If you do not have CAT software and want to explore your options, begin by searching for a CAT vendor list on NVRA's website at www.nvra.org.

How a CAT Program Works with Speech Recognition

Most SR tools are available directly from within a CAT program, but the steps for accessing them and performing certain tasks will vary. The SR application principles are the same, however, no matter which type of CAT program you work with, because the SR program is what is responsible for converting your speech into text in the first place.

As you dictate, the SRE outputs a continuous stream of text to a CAT program. The CAT program filters the information it receives and either allows it to pass through and appear on your screen as is or be further processed to produce different output as an ending realtime result.

For special output requirements, instructions must be given to the CAT program on what output it should produce and how the output should be generated. This means that when voice writing entries are made into your vocabulary, they must be set up according to the operational parameters of your CAT program's design, and every CAT program is different. A CAT program has an established set of guidelines it follows for definitions of output requirements, where it uses combinations of characters and symbols as code (referred to in this book as "**CAT code**") for interpreting the user's instructions on what text to produce and/or formatting functions to perform. When vocabulary entries are made based on those guidelines, the SRE and CAT program communicate together to achieve a more specific result.

The most common example used to explain how a CAT program converts SRE output is in describing how a question marker is generated in a realtime court reporter's transcript. The string of text and other commands associated with the production of a question marker is as follows:

- *insert a period (.) and remove the previous space, unless other punctuation is present*
- *produce a hard return*
- *insert the capital letter Q*
- *produce a tab*
- *capitalize the first letter of the next word*

Say, for example, the voice code used to represent a question marker is "quesco" and the CAT code associated with producing the necessary output requirements is "{Q}". During the court reporter's dictation, the voice code (quesco) invokes the CAT code ({Q}) to produce that string.

To better understand how an SRE and a CAT program work together to produce realtime output, a simple illustration appears on the next page (see Diagram 2).

Chapter 1: The Basics

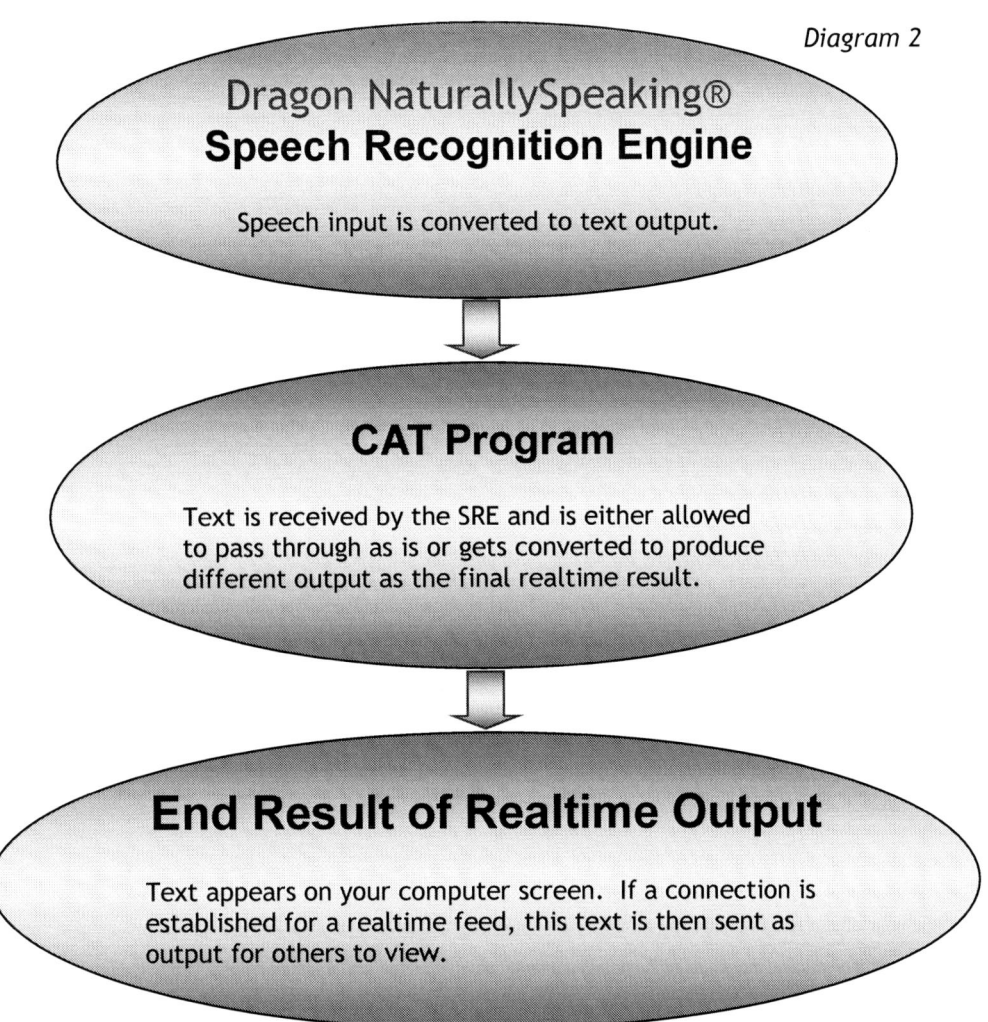

Diagram 2

—CAT Program with a Dictionary—

A CAT program with a dictionary module allows the user to define input and output of if-this-then-that scenarios using either text, codes, and/or a combination of both. For voice codes to become useful in providing realtime translations, definitions must be given to your CAT software to instruct it on what output it should produce and how the output should be generated. Your CAT program's dictionary is the database which stores that information.

Once you have entered a voice code into the SRE's vocabulary, you would also need to have a dictionary entry that includes that voice code along with a corresponding CAT code. As an example, let's use the voice code "quesco" to represent a question marker and assume that the CAT code associated with producing the required text and formatting functions is "{Q}". Table 2 (see the next page) gives you an idea of how that information would be processed.

Table 2

You say this...	Dragon produces this...	CAT program reads dictionary entry...		This is your final output result...
		input field	output field	
quesco	quesco	quesco	{Q}	Inserts period (.), hard return, Q, tab, and capitalizes the first letter of the next word.

If you choose not to use the CAT dictionary at all, you have that option as well. You could enter the CAT code directly into the SRE's vocabulary just as you would in a CAT program that does not have a dictionary module (see Table 3 on the next page).

The purpose of a CAT dictionary is not only to store voice code and CAT code entries, but also to resolve conflicts that can be generated by the SRE and benefit from a wider range of output options through the use of your CAT program's additional AI features. Some of these types of CAT programs are even sophisticated enough to judge the correct form of output from a selection of choices based on input from a single voice code. For example, if you used the brief form "ITC" to represent the phrase "is that correct" (see the section on *Brief Forms* in Chapter 3), depending on where you are in a sentence as you dictated "ITC," those same words could appear in different ways according to context. Here's an example:

If you dictated "ITC". . .

"Is that correct" would appear at the beginning of a sentence.
", is that correct," would appear in the middle of a sentence.
"; is that correct?" would appear at the end of a sentence.

Read your CAT software manual for further details on how your CAT program's AI process works.

—CAT Program without a Dictionary—

If your CAT program does not have a dictionary module, the CAT code that is used to instruct the CAT program on what to produce should be entered into the SRE's vocabulary. Using the same example noted in the previous section, the voice code "quesco" represents a question marker, and the CAT code that is associated with producing the required text and formatting functions is "{Q}". Table 3 gives you an idea of how that information would be processed (see the next page).

Chapter 1: The Basics

Table 3

You say this...	Dragon produces this...	CAT program reads...	This is your final output result...
quesco	{Q}	{Q}	Inserts period (.), hard return, Q, tab, and capitalizes the first letter of the next word.

Speech-to-Text-to-CAT Conversions
(Using a CAT Program without a Dictionary)

Not all voice codes in this book require a CAT code to be used for producing the desired realtime result. For example, if you simply want to produce a certain form of text that does not require formatting, such as the affirmative response of "uh-huh" when you dictate a voice code such as "yay yay," you would set up your vocabulary entry with "uh-huh" as the written form and "yay yay" as the spoken form; "uh-huh" is the text that would appear on your screen during your dictation. (For details about written and spokens forms, see the *Written and Spoken Form Spelling Rules* subsection under the *Working with the Vocabulary* section of Chapter 5.)

This type of CAT program is more simple to operate because most of the entries you input are in one place: the SRE's vocabulary. You can still resolve conflicts generated by the SRE, but this is mostly done using the SRE's tools instead of tools within the CAT program.

Realtime Voice Writing Equipment

The basic equipment to obtain are a computer, SRE and CAT software, a headset microphone or speech silencer, and a USB sound card. An external microphone and foot pedal are additional components that are generally used for court reporting.

Computer

You may use a desktop computer if you will be working in the same physical environment every day, but if you require mobility, as most voice writers do, you will need a laptop.

When choosing a desktop computer, selecting one with a fast processor is important to obtaining the best SR performance results. You will be able to get a desktop computer with a faster processor less expensively than what is available in laptops. A great advantage to buying a desktop computer is that you can upgrade the processor and other internal hardware components very easily as SR system requirements change in the future.

When choosing a laptop, it is important to select one with the fastest processor available on the market, since many internal components of laptops are not upgradable. As developments in SR technology grow, so too do the demands on a computer's resources and power requirements. You do not want to be stuck with a computer that will not be powerful enough a year from now to support newer versions of software.

Fast processors generate lots of heat, so if you're using a laptop with a very fast central processing unit (CPU), maximize the fan cooling system setting within the Control Panel to help keep your computer from overheating if your brand of laptop provides this level of control. The location of this setting varies among laptops, so contact a local computer technician if you cannot find this setting on your own. Your laptop should never be hot to the touch, as too much heat can diminish its life. If your computer still generates too much heat, you can find an external cooling device with built-in fans at your local computer superstore.

If you are using Dragon alone or in conjunction with a regular word-processing program such as Microsoft Word, refer to your Dragon user's manual or visit the www.nuance.com website to make sure your computer meets the minimum specifications. If you are using CAT software in addition to Dragon, your computer will need to have more powerful capabilities than the specs listed for Dragon if you want optimal performance, so rely on your CAT vendor's recommendation for the type of computer that will best function with Dragon and your particular CAT software, including a list of all specifications, leading computer brands, and where to find them. Your CAT vendor may even recommend you purchase a computer from them directly if they do specialty configuration work to ensure optimum voice writing performance.

You may want to double up on the recommended minimum requirements for random access memory (RAM), as more RAM has demonstrated marked improvement in SR performance.

As far as external device plug-in capabilities, make sure there are at least three universal serial bus (USB) ports. You will need one for the USB sound card, one for plugging in a flash drive (or "thumb drive") for transferring files, and another for your CAT software key if you have CAT software. You may even want a fourth USB port in case any other components you will use, such as certain brands of external microphones, are USB-powered.

To prevent your computer from periodically disabling any of your USB connections in an effort to conserve resources during intensive SR processing tasks, you should set your computer's USB root hubs to continuous operation. (See the *USB Connectivity Settings* section of Chapter 4 for instructions.)

If you have CAT software, ask your CAT vendor about turning off unused applications that run in the background. Turning these off will make more RAM available for the compute-intensive task of producing accurate SR. Anti-virus software may also need to be turned off or disabled, as it may interfere with speech recognition processing by generating false virus reports. Anti-virus scans that occur in the middle of your dictation will also negatively affect SR accuracy, as the running of any additional application processes takes away from your computer's resources. You can always re-enable anti-virus applications when you need to access the Internet. (See the section on *Turning Off Automatic Updates and Disabling Anti-Virus Software* in Chapter 4. If you require further help with these settings, contact a local computer technician for assistance.)

Chapter 1: The Basics

Consider the computer you use for realtime reporting as your means of livelihood. Much of the nasty malware found on the Internet can sideline a computer for several days. Therefore, if at all possible, use a different computer when connecting to the Internet for purposes other than sending realtime text feeds.

Dictation Input Devices

As far as which dictation input devices to use for voice writing, you will be working with a USB sound card along with a dictation microphone: either an open-mic headset or a speech silencer.

Any combination of USB sound card and dictation microphone will work successfully with speech recognition as long as that exact combination of components that was used to create a specific voice model is always used while performing actual dictation work within that voice model. It is important to note that this equipment configuration must be unchanged if you want your your accuracy results to be consistent. If you decide to use either a different USB sound card or dictation microphone in the future, even if those devices are the same models of the same brand made by the same manufacturer, you must create a new voice model based on that new configuration.

—USB Sound Card—

A **USB sound card** (also referred to by other names such as a "sound pod" or "USB speech processor") is an external device that converts your voice's analog signals into digital form so the SR program can process your speech into text. By being external, it bypasses the noisy interior of a computer, avoiding electrical interference with other components. This also allows a computer's internal sound card to be assigned the task of recording an external room track when necessary. And since the qualities of internal sound cards vary from computer to computer, using a dedicated USB sound card will maintain speech-to-text processing quality when sharing sets of speech files between computers and when permanently migrating them from one computer to another.

If you have CAT software, ask your CAT vendor which USB sound card best matches your system's requirements.

USB Sound Card

17

—Open-Mic Headset or Speech Silencer—

There are two types of dictation input devices used by voice writers for providing realtime services: an open-mic headset and a speech silencer.

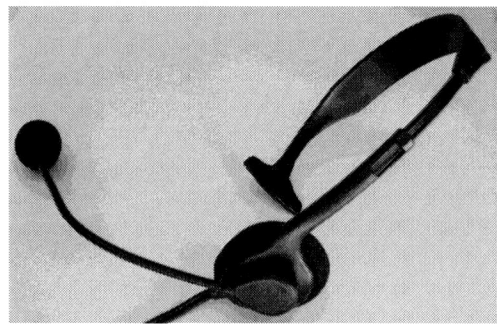
Headset Microphone

A **headset microphone** (also "open-mic headset") is used by voice writers who have careers working in isolated settings. It is designed to filter out extraneous noise, but is ideal for use in quieter environments where there is little background interference. In heavily populated or sound-intensive environments where you must dictate while others are speaking—such as in courtrooms, offices, or classrooms—an open-mic headset should never be used, as the sound of your dictation will be a distraction to the people who are in the same room with you.

Although a Dragon full-version software package includes a headset microphone, voice writers should use a higher quality microphone and headset with better noise-cancelling capabilities when providing professional realtime services remotely or working as a captioner. "**Noise-cancelling**" refers to a design that filters out extraneous sounds and removes interference of background noises. This will help prevent background sounds from affecting speech recognition performance during dictation and also help keep you from becoming distracted by the sound of your own voice as you listen to speakers while dictating. Therefore, it is recommended that the open-mic headset you choose for voice writing be a specialty-item purchase that is higher in quality than Dragon's stock microphone. If you cannot decide which microphone headset to buy, ask your employer when you are ready to begin providing realtime services. Until then, you can work with the headset microphone that you already have.

A **speech silencer** is a dictation device, usually covering your mouth and a portion of your chin area, designed to allow a court reporter or on-site CART provider to dictate while preventing his/her speech from distracting others nearby in the room. A silencer contains a built-in microphone that captures sound for recording dictation and/or for realtime text production with speech recognition and CAT software.

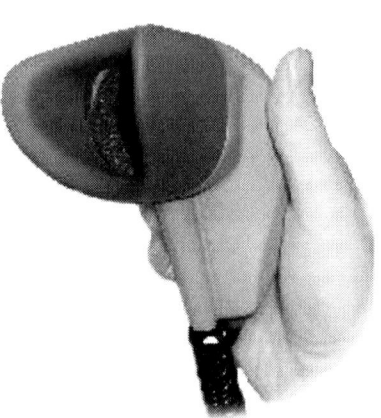
Speech Silencer

Vendors of speech silencers include Talk Technologies Inc. and Martel Electronics. You can find the Sylencer®, manufactured by Talk Technologies, on the Web at www.talktech.com, and the Mini Mask™ by Martel at www.martelelectronics.com. For a complete list of vendors, visit the www.nvra.org website.

Chapter 1: The Basics

High-Gain Microphone
(for Recording External Environment)

In order to make the most clear digital recording of the audio in your external environment, you will need a high-quality microphone.

Many court reporters purchase high-gain microphones from Martel Electronics, and you can visit their website at www.martelelectronics.com. If you have CAT software, talk to your CAT vendor about which type of microphone to buy. When asking around for further microphone suggestions, be sure to visit NVRA's court reporters forum, as it is full of helpful information from folks who freely share the benefits of their experiences.

Also, a microphone extension cord is a very worthwhile accessory which will allow you to place your microphone in areas of a room that are too far for your regular microphone cord to reach.

Microphone with Tripod

Foot Pedal

A USB foot pedal can be plugged into your computer, giving it the functionality of a cassette tape transcriber. With the proper software, you can control playback of your digitally recorded audio files while editing or proofreading transcripts. Most CAT programs allow the optional use of hot keys instead of a foot pedal, but most voice writers prefer to use a foot pedal because it frees their hands for editing. If you have CAT software and your vendor's package did not already include a foot pedal, you may purchase one separately, but first contact your vendor to find out which foot pedal is compatible with your CAT program.

Foot Pedal

Surge-Protection Power Strip

As a primary means of income, your computer is worthy of all the insurance you can provide, and a power strip with a surge-protection feature is among the most cost-effective ways to help protect your investment. Having your computer plugged into this device, as opposed to directly into an electrical socket, will help safeguard it from permanent damage caused by an electrical surge.

Surge-Protection Power Strip

Part 2

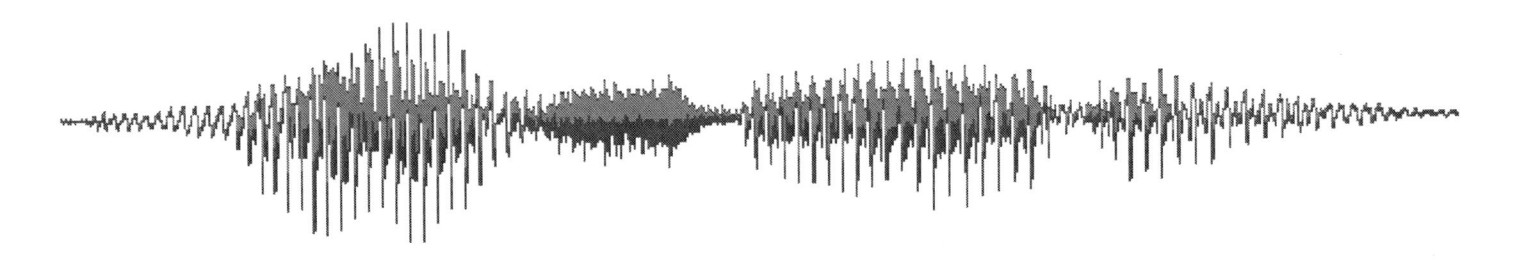

Realtime Dictation and Voice Theory Development

Chapter 2: Dictation Techniques

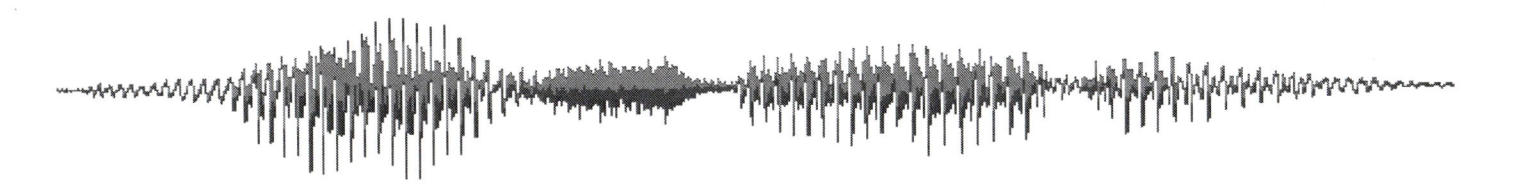

The majority of your success with accurate realtime voice writing will be won by mastering proper dictation style. You can be the fastest-dictating voice writer in the world, but if you do not speak in a way comprehensible to Dragon, you will not produce accurate text.

The average rate of speech is between 140 and 160 words per minute, but the generally accepted standards for realtime textual output ranges from 180 to 200 words per minute—at a minimum of 96% accuracy—so you must train yourself to dictate clearly at these higher-than-average rates. Initially, as you practice careful enunciation at a normal pace, be cautious not to clip apart sounds of words or elongate or exaggerate their pronunciations. You must also avoid the tendency of running words together and over-enunciating them. At first your words will not sound clear when spoken quickly, but over time you will find that you can increase your rate of speaking with no loss of clarity at realtime speeds.

Bear in mind that even as you put these dictation techniques into practice, you will still encounter speech recognition errors that need to be corrected. As you work on building your speed, you will want to continuously correct misrecognitions until you reach a minimum 96% level of accuracy dictating between 180 and 200 words per minute. You will learn how to improve speech recognition accuracy in a later chapter.

For now, let's focus on proper dictation techniques. To illustrate how your speech recognition accuracy can be compromised if careful enunciation and dictation pace are not practiced, Table 4 lists some common conflicts that may arise.

Table 4

Dictation Style Conflicts (Examples)	
Conflicts	Reason for Conflict
"ten issues" vs. "tennis shoes"	pace
"position" vs. "physician"	enunciation
"have you" vs. "heavy"	pace and enunciation

There are times, of course, when perfect pace and enunciation may not always be possible during high-speed dictation, and special techniques for getting around speech recognition errors in those instances are explained in this chapter.

Proper Breathing and Volume with Dictation Devices

The following sections describe proper breathing and volume levels that should be applied when dictating into an open-mic headset or speech silencer.

Dictating into an Open-Mic Headset

Dictate at the same volume you would during regular conversation and breathe normally. Just be sure not to exhale directly onto the microphone or Dragon may interpret the sound of your breath as a word.

Dictating into a Speech Silencer

Breathing can affect many factors in the sound of your voice during dictation, especially when using a speech silencer. It is best to inhale and exhale through your nose primarily, breathing out only as much air as is required through your mouth for controlled volume and pace. Breathing that is too restricted will not allow you to enunciate clearly enough to produce the full sound of your words, and gasping or breathing heavily directly into your microphone may be recognized by Dragon as sound input for text production.

Chapter 2: Dictation Techniques

You should not have trouble breathing with a speech silencer unless its seal obstructs your nasal passages. Your facial structure will determine which speech silencer and seal combination is the best design for you. Consult your dealer or manufacturer's website for recommendations on different silencer models and seal solutions to decide on the most comfortable fit.

To form a proper seal, be sure the mouthpiece of your speech silencer is pressed securely enough against your face to prevent the sound of your voice from being heard outside the silencer. You do not want your dictation to distract others nearby. Also, you do not want the built-in microphone to detect any extraneous sounds, such as background noises or other voices in the room, since Dragon may attempt to produce words from those sounds.

To test the security of your seal, position the speech silencer onto your face as you would when preparing to dictate and exhale through your mouth. If air escapes around the edges of the mouthpiece, that is a clear indication that sound will escape. To ensure formation of a good seal, reposition your speech silencer and continue exhaling until no air escapes around the mouthpiece's edges.

As for dictation volume, you should speak at a level just below that of your normal conversational voice and loudly enough for the microphone to receive a clear signal, but whispering is not recommended. A common misconception voice writers have when they first begin using speech recognition is that if they whisper, even while being careful to properly pronounce words, they can eventually force Dragon to produce accurate text, thinking it will recognize any style of dictation if trained well enough. Though words can be discerned by the human ear from a whisper, this practice routinely results in poor recognition by Dragon. Dragon has a difficult time deciphering whispered speech because it is much less distinctive than normal speech. This is an important factor for facilitating proper sound delivery through a microphone.

With the majority of modern-day microphones, certain audio signals cannot be sent at differential amplification levels sophisticated enough for Dragon to adequately detect and measure. However, the latest Sylencer® design from Talk Technologies Inc. (www.talktech.com) has, in fact, demonstrated the ability to effectively deliver such signals. Accurate speech recognition results have been obtained with whispered speech using this new microphone. But, since sound qualities of normal speech are *most* distinctive to Dragon, it is still best not to whisper if you want to achieve the highest levels of accuracy.

In very quiet environments, you will notice that your voice can be heard even while speaking in the lowest volume humanly possible. There is usually enough background noise in the average room to attenuate the sound of any slight hum generated by your speech from within the silencer, but knowing this may not always the case in every room, you will need to modify your dictation style in a way that does not sacrifice volume. A remedy for this is to use the **hold-dictation technique**, where you dictate only when the speakers in the room are talking and hold your dictation when they pause. During pauses, you can retain in your memory the last sentences or phrases you heard, then dictate the remainder of what you remember when they resume talking. This way, you are never dictating when the room is completely quiet. Waiting to dictate until after speakers have begun talking will turn the focus of their attention to the sounds they themselves are producing and draw their attention away from any slight hum of your dictation.

If you are concerned that your voice may still be heard outside the speech silencer, even while using the lowest possible dictation volume and employing the hold-dictation technique, you may find additional silencing benefit from lining its interior with padding to help with sound absorption. Just be careful not to let such material obstruct the pathway between your voice and the microphone. Also, note that lining the interior of your speech silencer will have an effect on the quality of its internal acoustic environment. So when you begin working within a specific voice model using a certain padding thickness and lining placement, be sure to keep those absorptive materials in that same fixed position throughout the course of all future use of that voice model.

Tone and Modulation

As mentioned previously, speech recognition can be affected by minor variations in speech. The idea is to remove as many variables from speech as possible to achieve the highest attainable accuracy levels. One way to do this is by adjusting your speech to a definitive pitch and vibration and avoiding too much inflection, to keep your speech tone consistent. Removing tone variations helps to minimize the speech recognition error rate by reducing the level of complexity involved in Dragon's sound measurement tasks.

Using a monotone style of dictation should be your steady style of expression during dictation, except when you encounter words that may sound similar to others, in which case accuracy can be improved with the use of inflective accents. This usually relates to smaller words, such as "and," "an," and "in." But this same phenomenon can be found with larger words as well, where an inflective accent placed on certain syllables requiring distinction will help recognition.

For example, when saying "position," emphasizing the first syllable "po" will make a clear enough distinction for the SRE to discern whether you said that specific word or another one that may sound similar to it at a fast rate of speech, like "physician."

Likewise, when you say "physician," make sure emphasis is given to the first syllable, because the first syllable's sound is what distinguishes these words from each other.

Thus, while as a general rule, removing inflection from speech lends consistency to our dictation patterns, sometimes the only way to enunciate certain words clearly is to use inflective accents so Dragon can recognize us properly when dictating at high speeds.

The objective is to make clear sound distinctions without having to sacrifice speed. Perform Experiments 1 and 2 to demonstrate this technique's usefulness for yourself.

Chapter 2: Dictation Techniques

> ### Experiment 1
>
> (Say these sentences aloud as quickly as you can, without pausing, using a monotone speaking style.)
>
> How are you?
>
> How were you?

Did you hear how similar "are" and "were" sounded when quickly saying those sentences? We must pay close attention to how we speak if we want Dragon to successfully recognize our speech, especially with small words.

> ### Experiment 2
>
> (Say these sentences aloud as quickly as you can, without pausing, using an inflective accent with the words "are" and "were.")
>
> How **are** you?
>
> How **were** you?

Did you hear a distinction between the pronunciations of "are" and "were"? If the sounds are more recognizable to you, they will also be more recognizable to Dragon.

An inflective accent is also useful for distinguishing between the ending sound of a word and a small word that should stand alone. Experiment 3 is an excellent example of how an individual word can easily sound like only a word part, and it's no wonder that Dragon has trouble properly recognizing speech in such instances; Experiment 4 demonstrates how that sound conflict may be resolved.

> ### Experiment 3
>
> (Say these phrases aloud as quickly as you can, without pausing, using a monotone speaking style.)
>
> decided now
>
> decide it now

> **Experiment 4**
>
> (Say these phrases aloud as quickly as you can, without pausing, but use an inflective accent on the word "it" in the second phrase.)
>
> decided now
>
> decide <u>it</u> now

When performing Experiment 4, did you hear how the word "it" gained independence from the word "decide"? As you come across other confusing word-blend combinations, distinguish sounds by using inflective accents to avoid word misrecognitions.

Words having totally different pronunciations can even sound similar during high-speed dictation. As another experiment, assume you've been dictating for eight hours, you're tired, and your enunciation of small words is not as clear as it was earlier; all the attention and energy you have left is spent on reporting content.

Now perform Experiment 5 to witness for yourself similarities between words you ordinarily wouldn't expect to have sound conflicts. This will help you gain more understanding as to the reasoning behind Dragon's judgment.

> **Experiment 5**
>
> (Without using any special dictation techniques—*remember, you're tired*—say these phrases aloud quickly, and repeat them over and over again as you listen to the sounds of the words "you" and "he." If you do not hear sound similarities by listening to your own voice, ask someone else to quickly say these phrases aloud for you so you can listen more objectively.)
>
> so you did
>
> so he did

You would resolve this conflict using inflective accents as well, by placing emphasis on the words "you" and "he."

Chapter 2: Dictation Techniques

Enunciation

One of the most important aspects of speech is clear enunciation. This idea is generally understood in terms of making full sound production of words. However, the way to enunciate for Dragon to properly recognize speech requires special attention to the formulation of sound and how it is released from our mouths, which is important for detection by your dictation microphone. Before the reasoning behind this is explained, try Experiment 6.

Experiment 6

(Say these words aloud clearly, but in your normal conversational voice, as if you were speaking to a casual acquaintance.)

couldn't

button

important

Many times, even though you fully produce all sounds in words, Dragon still may not accurately recognize your speech. For example, you could get "could" for "couldn't," "but" for "button," and "import" or "imported" for "important." This is not a problem for everyone, but these types of errors can be common for some and can be perplexing at times. After all, if you think you're saying these words correctly, then you expect them to be recognized correctly.

So why, then, does Dragon produce these types of errors? For many of us, it's because we are closing off sound by forming a seal between our tongue and teeth on certain syllables. Though you can hear the sound coming from your throat, the microphone may not detect sound unless you allow the sound to fully escape from your mouth.

If you experience errors such as these, you may want to play back a recording of your voice to hear how you dictated the words that were misrecognized. Chances are, Dragon didn't recognize a certain syllable because it wasn't detectable to begin with. This is especially important to pay attention to when saying contractions, where a misrecognition not only changes context, but actually conveys the opposite meaning (e.g., "I didn't rob the bank" versus "I did rob the bank"). So be sure not to prevent the ending sounds of contractions from being detected by the microphone by swallowing the last syllable.

Now perform Experiment 7.

Experiment 7

(Say these words aloud clearly, but enunciate the word parts underlined in bold by making sure your tongue taps the top of your teeth instead of creating a seal. Since these are multisyllable words, the ideal way to make full sound production involves tapping your tongue to your teeth twice.)

coul**dn't**

but**ton**

impor**tant**

Was there a difference between the way you enunciated these same words in Experiment 6 versus 7? If not, your speech is clear enough for Dragon to accurately recognize. If there was a difference, you should change your dictation style to reflect the way you enunciated when performing Experiment 7.

Performing Experiment 8 will further assist in understanding the level of attention necessary to adequately produce sounds recognizable to Dragon.

Experiment 8

(Below is an example of words that can pose conflicts in Dragon due to enunciation factors. Say these words aloud clearly, paying special attention to pronouncing the word parts underlined in bold.)

woul**d it**

wha**t it**

woul**dn't**

wan**t it**

Chapter 2: Dictation Techniques

You should also be aware than Dragon recognizes larger words more easily than smaller words. This is because larger words contain more syllables. The more syllables a word contains, the more easily Dragon can distinguish it because it holds a more uniquely defined set of sound characteristics.

When you mistakenly emphasize part of a larger, multisyllable word that you ordinarily wouldn't have placed emphasis on, even if you had never dictated it that way in the past, you will find that Dragon may recognize it correctly anyway.

Table 5 lists some examples of large words Dragon will still properly produce if you were to accent a different syllable by mistake.

Many times, sounds you either add or drop in large words will also result in a proper recognition, despite sloppy or mistaken pronunciations. Table 6 lists some of those examples.

Table 5

Mis-Accented Multisyllable Words
(Examples)

(Accented syllables are underlined in bold.)

If you said. . .	Instead of saying. . .	You will still get. . .
in**for**mational	info**rma**tional	informational
cont**ro**versial	contro**ver**sial	controversial
ana**log**ous	a**na**logous	analogous
contextual	con**tex**tual	contextual
ad**ver**tisement	adver**tise**ment	advertisement

Table 6

Mis-Pronounced Multisyllable Words
(Examples)

(Added or dropped word parts are underlined in bold.)

If you said. . .	Instead of saying. . .	You will still get. . .
muscul**s**keletal	muscul**o**skeletal	musculoskeletal
un**n**erstanding	un**d**erstanding	understanding
f**er**nancial	f**i**nancial	financial
techn**a**lly	techn**i**cally	technically
comp**s**ation	comp**en**sation	compensation

31

After you have created a voice model, you can experiment with dictating multisyllable words to see for yourself how easy it is for Dragon to accurately recognize them, though they may be accented or pronounced in different ways.

Since Dragon has more difficulty recognizing small words, those are the ones we must focus on enunciating most clearly. Small, one-syllable words have higher probabilities of conflict with other small, one-syllable words. Examples of such words include "and/an/in," "where/were/we're," "at/that," "than/then," and "what/with." Placing emphasis on small words and their word parts, such as the ending consonant sounds of the "d" in "and" and "t" in "that," while saying them evenly as whole-word units, will increase accuracy.

Overemphasizing a word ending, however, can lead Dragon to misrecognize it as a separate word. For example, if you pronounced the word "and" as "anduh" because you overemphasized the ending "d" sound, Dragon will probably interpret "uh" as a second syllable and produce the words "and a" or "and the." If you enunciate clearly enough for the consonant sound "d" to be fully produced as a distinct sound part, but without overemphasis, you reduce the risk of Dragon misinterpreting your speech. So when enunciating small words, be sure not to elongate their ending sounds.

For any type of word, whether it be large or small, do not syllabize (sil-uh-bize) by discretely emphasizing each of its sounds, because Dragon will not recognize the word better; it will likely translate your syllables into separate words (sell or buys).

Pace

Because speech recognition accuracy relies heavily on consistency in your speech patterns, it is important to dictate at a consistent rate. You should use the reporting industry's generally accepted realtime speed standards of 180 to 200 words per minute as the guideline for your dictation pace. As you will learn later, you will match this pace when creating your voice model so Dragon knows what rate of speech to expect from you in the future when you provide realtime services. Accordingly, you should do your best not to dictate under 180 words per minute or over 200 words per minute. A little slower or faster will not be a problem, but significant dips or rises in pace can be problematic, as depicted in Table 7.

Table 7

Dictation Pace Errors		
Phrase	*Spoken Too Slowly*	*Spoken Too Quickly*
for the record	for the wreck heard	forth a record
recognize speech	wreck a nice beach	reckon I speak

Chapter 2: Dictation Techniques

Dragon may not recognize "record" as one word if spoken too slowly, because it will digitize and process the first syllable of that word before the second syllable's sound has had a chance to be processed. In the example of speaking too quickly, the words "for the" may not have properly been recognized because they were too closely spoken together. If you pause too long between syllables within words, for example, "rec-og-nize-speech," Dragon may think you said "wreck a nice beach."

The pace we use when speaking groups of words also affects recognition accuracy. Depending on when you pause your dictation with certain words in phrases, Dragon may make different word selections. Dragon's decisions are heavily influenced by contextual probability factors. Let's use the homonyms "to," "two," and "too," for example. For the sentences that appear in Table 8, Dragon should make correct homonym word selections, as long as you do not pause before giving Dragon enough words for it to consider when analyzing context.

Table 8

Contextual Effects of Pace

I have **to** go shopping.

I have **two** children.

I have **too** much work.

Now, suppose you paused after saying "to," "two," or "too" in any one of the sentences from Table 8. Dragon will generate whatever word most closely matches your pronunciation based on the context you give to it at the time. In any one of these cases, the result will most likely be "to," since that is the most statistically probable choice ordinarily following the words "I have" according to its analysis of English context.

To avoid contextual errors, always let phrases queue in your memory before you dictate so you can provide the grammatical model with the information it needs to judge words based on their relationships to other words. This is referred to as the **phrase-dictation technique** because the focus is on dictating words together in phrase groupings.

Fast Speakers

When dictating fast speakers (above 200 words per minute), you cannot match their pace without experiencing poor accuracy results. To stay consistently within the realtime speed range (between 180 to 200 words per minute), you must lag behind them, relying on your word-carrying ability to finish dictating what the speakers have said, and exercise your best judgment when having to drop words (see the *Verbatim Dictation versus Paraphrasing and Summarization* section of this chapter). Of course, this means you will not have the ability to create a verbatim record on the spot, but it will be readable nonetheless. If you have to produce a verbatim record, the editing process will be much easier when preparing the final transcript.

Slow Speakers

With slow speakers, again, do not match their pace. You will be creating your voice model at realtime dictation speeds (between 180 and 200 words per minute), so it is best for you to consistently speak within that range. Considering the digital-conversion process that takes place as Dragon analyzes speech, it is looking for consistent pattern matches. A significantly slower rate of speech will be less recognizable to Dragon. Use the phrase-dictation technique to let words queue in memory before you dictate them. This way, Dragon will hear them being spoken together at an expected rate and have an adequate number of words to process for context. If we want Dragon to make a well-informed best-match guess, we must work within the parameters of its statistical-measurement capabilities to produce high-accuracy results.

Punctuating

Differences in meaning between sentences that sound exactly the same can easily be discerned by humans, but Dragon requires punctuation in order to make correct word selections for proper contextual judgment. In the following table, note how the simple use of a punctuation mark, such as a comma (,), determines different word selections made by Dragon.

Table 9

Contextual Effects of Punctuation (Sound-Alike Sentences)	
No Punctuation Dictated	Punctuation Dictated
I don't know	I don't, no
it is common sense	it is common, since
they're not there	there, not there

Dictating punctuation increases speech recognition accuracy because Dragon's grammatical model considers punctuation-to-word occurrences when determining word selections. At the very minimum, you should begin dictating periods (.) and question marks (?). Once you've mastered the ability to dictate verbatim words along with periods (.) and question marks (?), begin incorporating commas (,) and other punctuation marks into your dictation; this will increase your recognition even further by helping Dragon make proper word choices using the context of that punctuation. The punctuation you should be most concerned with dictating are commas (,), periods (.), question marks (?), and dashes (--).

Chapter 2: Dictation Techniques

Verbatim Dictation versus Paraphrasing and Summarization

In all realtime voice writing careers, the objective is always to ensure that content is accurately transcribed during realtime translations. Accurate transcription requires great attention to detail when it is done by manually typing up each word you hear from an actual audio recording of what those speakers are saying. Accurate transcription that is based on **verbatim dictation**—where a voice writer clearly dictates every word spoken by others—demands linguistic skills be developed to the highest possible degree of human performance, as this means every single word must be captured in your dictation no matter how fast speakers are talking. Accurate content, however, can mean different things in different types of transcription scenarios; in other words, accurate content may not always mean verbatim content.

When speakers are talking much faster than you are able to clearly dictate, in which case your efforts to keep up with verbatim content may actually cause accuracy deterioration, your speech recognition accuracy will be better if you drop certain words. In those situations, paraphrasing and/or verbatim summarization options should be weighted in your decisions as to how to make the live version of transcribed content understandable to the reader, whereby text ends up matching words of speakers closely enough to be readably accurate. You should use your best judgment in determining which words are sufficient to omit and/or transcribe differently without sacrificing meaning. Do note, however, that employing these techniques is a matter of subjectivity, and which techniques are acceptable to utilize depends on the type of voice writing work you do.

The skill of **paraphrasing** requires knowledge of which words to drop, which word-replacement choices can be made, and how words can be rephrased in such a way that concisely presents the main ideas without changing the meaning of the content. Although verbatim content is always preferred, paraphrasing content during realtime translations is considered a generally accepted practice for CART providers and broadcast captioners. However, it is not generally accepted in most other voice writing careers—especially court reporting—since paraphrasing practices often involve reworded interpretations that potentially risk misconstrual of the facts that speakers are trying to convey.

When producing a "verbatim record" in court reporting, accurate content always refers to verbatim content. When producing a "realtime record" in court reporting, accurate content refers to a readably-accurate rough-draft version of the content that is considered near verbatim, meaning the arrangement of spoken content has not been restructured in any way, but insubstantive words may have been omitted and certain non-translatable words may have been replaced with translatable words in an effort to make the transcript more readable. The realtime record is, of course, edited and proofread later before being certified and delivered as the final verbatim record.

When producing a realtime record of a legal proceeding where content must be as verbatim as possible, court reporters rely on verbatim summarization skills when necessary. This is different from paraphrasing sentences by rearranging and altering words simply to convey the essence of what is being said. **Verbatim summarization** is the technique of dropping insubstantive words here and there in order to produce a transcript that is readable, yet maintains the complete integrity of the verbatim content. This requires knowledge of which words are acceptable to leave out while keeping the sequence of all other words in their original verbatim order. It is acceptable to omit nonfluencies such as "you know," "like," and repetitively spoken "wells" and "okays" when they bear no significance to what is being said, and certain adjectives only when they will not impair the reader's ability to understand what is meant even though they are absent. For example, out of the words "Okay. Well, this was primarily due to, like, you know…" you may simply dictate "This was due to…" Other than dropping insubstantive words and adjectives, your dictation as a realtime court reporter must be verbatim, and you should use your verbatim summarization skills as sparingly as possible.

In all capacities of realtime voice writing work—whether for court reporting, CART, captioning, financial call reporting, et cetera—you will come across new words that are not possible to translate initially the first time you hear them. For example, if you encounter a unique word, acronym, or term that you know will not translate properly because it is not in Dragon's vocabulary, yet it is important that the word not be ignored or altogether dropped from your translations, a practice that will prove useful is to employ the **on-the-fly translations technique** (see the *On-the-Fly Translations* section of Chapter 3 for details). Such impromptu word occurrences may happen occasionally or regularly, depending on Dragon's familiarity with the subject matter. For example, in instances where the content you are reporting has to do with a corporate scandal involving an unusual name such as "Weizhen Tang," who happens to be a certain company's investment manager, and you know that name is not translatable since it doesn't exist in your vocabulary, you could use a general description in place of that person's name the first time you hear it, then use the on-the-fly translation technique to produce his actual name each time it occurs thereafter. A possible replacement choice may be "an investment manager." Or, if you are uncertain about this person's specific job title or role, you could use a more generic description like "company representative." The idea is to first convey proper meaning so that it is readably accurate, then render the actual name in your realtime output from that point forward.

Capturing the majority of what speakers are saying at faster-than-standard realtime dictation speeds is more valuable than producing inaccurate text and words that do not make sense, especially to those who rely on your realtime skills as a means for understanding educational content, televised information, and making important legal and financial decisions.

Dictation Patterns

It is impressive to know that Dragon demonstrates accuracy results at up to 99% for general users who dictate at average rates, but how does this type of performance translate to those who are expected to perform accurate realtime transcription for a living in the reporting industry when a voice writer's dictation must be translatable at rates beyond Dragon's reportable performance limits?

If the high accuracy findings based on general Dragon users' experiences can be objectively quantified by the word-per-minute rate used during voice model creation, then we could conclude that high accuracy success is attainable at any speed; that is, accuracy output should be determinable based on the proportionate extent to which the speaking rate used during voice model creation is consistent with the rate applied while dictating within that voice model to produce translations. For example, if a 150-word-per-minute speech rate is used during voice model creation and the average translation accuracy

Chapter 2: Dictation Techniques

result is 99% when you dictate at 150 words per minute in that same voice model, it would seem logical that if you wanted to obtain 99% accuracy at 200 words per minute, you could achieve that 99% result simply by creating the voice model at 200 words per minute. But when the dictation rate is increased and the foregoing principle has been adhered to, the general accuracy results do not prove directly proportionate. That is not to say that Dragon cannot recognize fast speech rates as accurately as it does regular speech rates. Dragon will successfully recognize a human's speech to the extent a human can produce consistent enough sound patterns in accordance with the same criteria Dragon was programmed to expect during voice model creation. The problem is that humans have imperfect speech patterns that become especially pronounced beyond the average rates at which they are accustomed to speaking, and rarely does a human say a word or phrase in the same way twice. When the word-per-minute consistency rule in voice model creation is not applied, however, there is a marked degree of accuracy degradation, so obeying this principle does positively affect speech recognition accuracy at faster rates. In that light, we must know that the solution to producing consistently accurate results at realtime speeds rests in the hands of a human's ability to dictate in a way that is possible for Dragon recognize. When Dragon has difficulty deciphering our fast speech, we must consider it our responsibility to reduce the computational work involved in Dragon's recognition tasks.

We already covered the basics of the speech-to-text conversion process (explained in the *How Speech Recognition Works* section of Chapter 1), so let's now examine some additional factors that weigh into the recognition equation.

As we dictate, Dragon is also adjusting to our volume, pitch, and rate changes. When our speech patterns consist of wide-ranging variables in these areas, we further compound Dragon's already complex analysis tasks by requiring it to measure and manage exceptional fluctuations as it filters through those frequencies to choose median levels of operation during adjustment periods.

Though the scenario of combining all these variables together at high speeds becomes more complex, the word-selection formula remains unchanged, a situation which can overwhelm Dragon's maximum efforts and result in recognition errors. It is no mystery, then, considering all the adjustments Dragon must already make, why we experience accuracy degradation at faster dictation rates when our enunciations of words and their boundaries are significantly less clear than at average rates.

So how can we force Dragon to perform above 96% accuracy at above-average rates of speech? Our understanding of how Dragon processes speech into text has enabled us to form a dictation theory of working concepts that will help us maintain standard realtime accuracy levels. It is important for us to remove as many variables from our speech as possible and stabilize our dictation patterns to minimize fluctuating patterns Dragon must analyze.

We can practice high-speed dictation by reading text out loud to develop the linguistic skills necessary for clearer speech before creating a voice model. This will help introduce us to better enunciation habits at faster dictation rates and learn how to maintain the same word-boundary distinctions customarily made only during average speech. We can also make our volume, pitch, and rate more consistent. By removing these variables, Dragon will have a more balanced range of sound input fluctuations to measure and adjust for during the recognition task.

Controlling volume, pitch, and rate produces better results as you move into higher dictation speeds. The illustrations below represent fluctuating patterns that Dragon has to filter through during speech-to-text conversions.

Illustration 1 depicts uncontrolled patterns of volume, pitch, and rate occurring in regular conversational speech. Illustration 2 depicts a carefully controlled dictation method which will minimize Dragon's overall computational workload involved when measuring these same variables during high-speed dictation, resulting in maintenance of maximum speech recognition performance.

Illustration 1

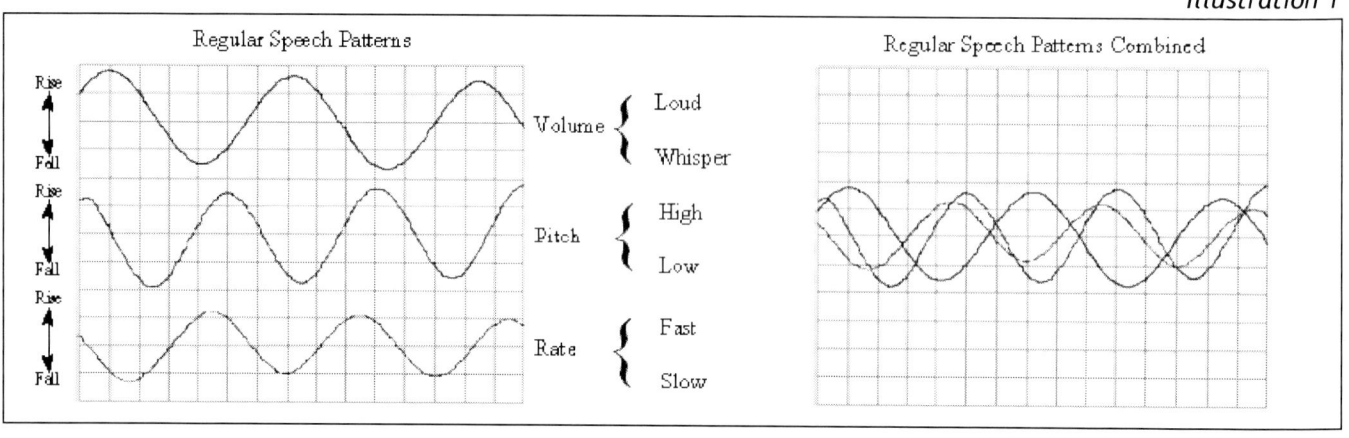

Illustration 2

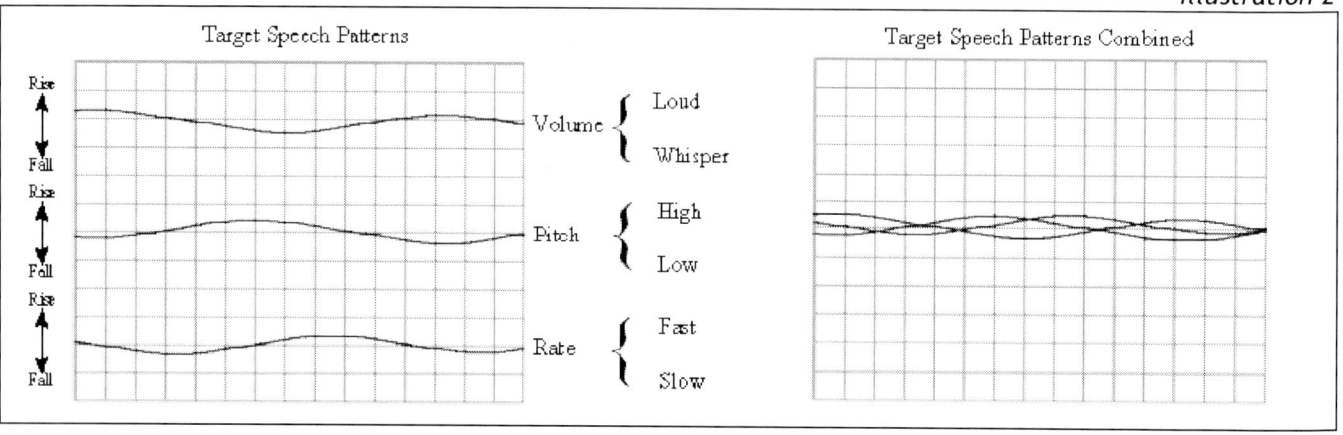

Chapter 3:
Voice Writing Theory

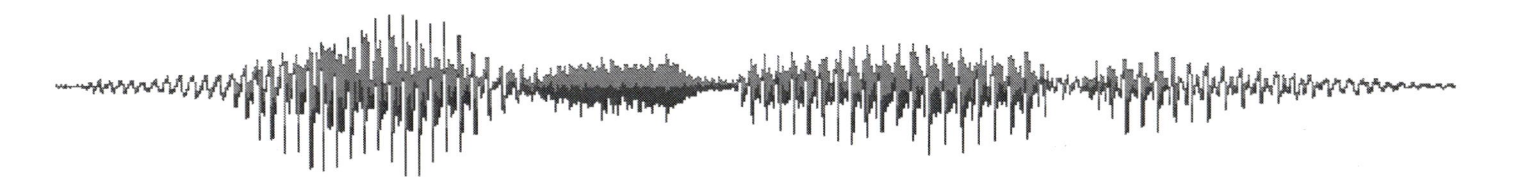

This chapter introduces the concept of **voice codes**, also periodically referred to as "code words," which are special utterances that form the basis of dictation shorthand methods for voice-to-text translations. The culmination of these voice codes and their conceptual usages together comprise a special dictation language known as a **voice writing theory**.

Lists of voice codes are provided within the tables of this chapter, along with their general descriptions and purposes. Whether you provide realtime services as a court reporter, CART provider, broadcast captioner, for financial calls, or for any other career specialty, you use this dictation method to keep track of speakers, insert punctuation and parentheticals, resolve basic word conflicts, use brief forms as shortcuts to output long phrases and blurbs without having to dictate every single word, and produce text in myriad other scenarios. Identify the voice codes that best suit your profession's requirements and customize them according to your preferences.

The idea of this dictation method is to combine a system of code words with the regular English language in order to quickly produce accurate content while simultaneously placing text in its proper format during realtime. So while voice codes are used only for the special purposes explained in this chapter, to produce all other words in your realtime transcripts, simply dictate regular English words, since the majority of the text you will produce as a realtime voice writer is plain English text.

In order for any word you dictate to be recognized and converted into text, those words must exist in your Dragon vocabulary. Since Dragon already contains thousands of common English words within its base vocabulary, you will not need to add most words you expect to dictate. The only words you will need to add are unique words, such as voice codes, uncommon words and acronyms, common names with uncommon spellings, highly-specialized technical terms, and special phrases.

Be aware that very careful consideration was given to forming this system of voice codes and that, although they work well for most based on their common success rates, accuracy results will vary among practitioners. If you do not have success with some of these voice codes, you may want to replace them with voice codes you create on your own. When designing them yourself, the first objective is to make sure their pronunciations do not sound similar to other words that already exist in Dragon's base vocabulary, or you will run into speech recognition conflicts between your code words and regular English words.

From the earliest stages of voice theory development, commonsense approaches were taken to come up with ways of creating unique sounding utterances, such as the idea to add an ending pronunciation of "mac" (for "macro") or "co" (for "command"). Usually, voice codes reflect in some way what they are represented to produce. You can apply these same concepts when creating your own voice codes. A simple way to create voice codes is to choose pronunciations that somehow correlate in sound or meaning to their output type. As an example, you can use "spee-1" for Speaker 1. This voice code is easy to remember because the word "speaker" contains the sound part "spee," and, of course, "1" represents the first speaker.

Other creation methods include the **lettering technique**, where letters of the alphabet are combined to represent whole words in phrases (e.g., "ITC" for "is that correct"), and the **doubling-up method**, where a word or word part is doubled up to effectively produce a distinctive sound based on the quality of the combination (e.g., "break break" for a parenthetical to indicate that a break is taken and "prise prise" to indicate surprise). You can even take an **ad-hoc combining** approach of piecing together word parts to represent word groups (e.g., "rezdegmed" for "reasonable degree of medical") and further extending the basic idea to produce a variety of other combinations (e.g., "rezdegmedcert" for "reasonable degree of medical certainty").

As far as spelling any new voice codes you create, just as a general rule and to keep things simple, always use a phonetic spelling and stick with lowercase letters, unless the voice code's pronunciation involves actually saying the names of the letters of the alphabet. As an example, the SRE will expect you to pronounce "ATTY1" as "A," "T," "T," "Y," "one"; whereas, "atty 1" would be pronounced *atty one*.

Please also note that although the one-syllable voice codes listed in this chapter have been demonstrated with success, stay away from creating your own one-syllable voice codes as much as possible since most of them will conflict with pronunciation sounds of regular English words during rapid dictation. That's not to say that one-syllable utterances can never be used; it is best to use two or more syllables when further developing your voice theory, because multisyllabic pronunciations are more distinctive than pronunciations containing only one syllable and are, therefore, less likely to conflict with regular English vocabulary words.

Lastly, for any voice codes that you intend to use with other applications (such as Microsoft Word or CAT software) in order to produce formatted output, you must know that additional work must be done in Dragon, other than simply entering them into Dragon's vocabulary, for Dragon to recognize them properly. This involves special preparation work involving the incorporation of such code words into documents representative of the type of voice writing work you will do; those documents should be run through Dragon's vocabulary-building utility for contextual analysis. (See the *Vocabulary Building* section of Chapter 6 for more information.)

Keeping all these considerations about voice theory development methods in mind, feel free to think outside the box and enjoy your freedom being creative!

Chapter 3: Voice Writing Theory

Punctuation

Tables 10 presents a suggested list of voice codes for punctuation. You may choose not to use these in your dictation, but doing so will make your realtime record more accurate by helping you to avoid punctuation-versus-word conflicts. For example: "period" conflicts with "."; "slash" conflicts with "/"; "dash" conflicts with "--"; "colon" conflicts with ":"; and so on. Alternatively, you can assign voice codes to the words instead of the punctuation marks if that is your preference.

An important benefit of using these voice codes is the ability to dictate faster by saying fewer syllables. For that reason, voice codes for the most commonly dictated punctuation marks contain only one syllable. The punctuation marks most frequently used in realtime reporting are: period (.), question mark (?), comma (,), and double dash (--).

These same punctuation voice codes can be used in any voice writing career specialty. Terminal punctuation in captioning, however, should produce a new line after each sentence, so just remember to make the appropriate adjustments in your CAT program or other application for that type of punctuation.

Table 10

Single Punctuation Marks

Description	Voice Code (or what you say)	Text, Formatting, and Use
,	just say "comma" (or *advanced users only* may say "kah")	Behaves as a comma.
.	peerk (or just say "period")	Behaves as a period.
?	kwee (or just say "question mark")	Behaves as a question mark.
!	excla (or just say "exclamation mark")	Behaves as an exclamation mark.
:	colo (or just say "colon")	Behaves as a colon.
;	semco (or just say "semicolon")	Behaves as a semicolon.
--	doesh (or just say "dash")	Behaves as a double dash.
-	hyka (or just say "hyphen")	Behaves as a hyphen.
/	slok (or just say "slash")	Behaves as a slash.

Table 10 (Continued)

Single Punctuation Marks

Description	Voice Code (or what you say)	Text, Formatting, and Use
open "	oco (or just say "open quote")	Behaves as an open quotation mark.
open " (capitalizes next letter)	ocap (or just say "open quote, cap")	Behaves as an open quotation mark and capitalizes the first letter of the next word.
close "	cloco (or just say "close quote")	Behaves as a closed quotation mark.
(	opar (or just say "open paren")	Behaves as an open parenthesis.
)	clopar (or just say "close paren")	Behaves as a closed parenthesis.

For shortcut methods of producing punctuation-mark combinations, the suggested list of voice codes below may be applied in your dictation. If you prefer not to dictate these, use regular pronunciations for each punctuation mark instead.

Table 11

Combination Punctuation Marks

Description	Voice Code (or what you say)	Text, Formatting, and Use
, " (capitalizes next letter)	kahcap (or just say "comma, open quote, cap")	Behaves as a comma with an open quotation mark and capitalizes the first letter of the next word.
,"	kahco (or just say "comma, close quote")	Behaves as a comma with a closed quotation mark.
."	peerco (or just say "period, close quote")	Behaves as a period with a closed quotation mark.
?"	kweeco (or just say "question mark, close quote")	Behaves as a question mark with a closed quotation mark.

Chapter 3: Voice Writing Theory

Speaker Identification

Tables 12, 13, 14, and 15 provide basic lists of speaker identifiers for you to begin using. Create and incorporate others as necessary.

If you are using Dragon in conjunction with other software, such as Microsoft Word or CAT software, the generic speaker voice codes you enter into Dragon's vocabulary will be replaced automatically with specific names when you update the speaker name fields within that specific application.

Please note that if you are using a CAT application, you may or may not need to use different voice codes to produce speaker names in colloquy versus byline formats, because your CAT software may automatically apply such formats according to the context in which these speaker names occur.

The voice codes presented in Table 12 on the next page are for speaker-identification methods to be used in court reporting. To keep track of speakers during a proceeding, you may want to draw a seating chart that portrays the positions of each speaker so that you do not have to rely on your memory alone to identify who is speaking. When taking a deposition at a law office, for example, your seating chart may be drawn as shown below.

Seating Chart for Speakers
(Court Reporting Sample Only)

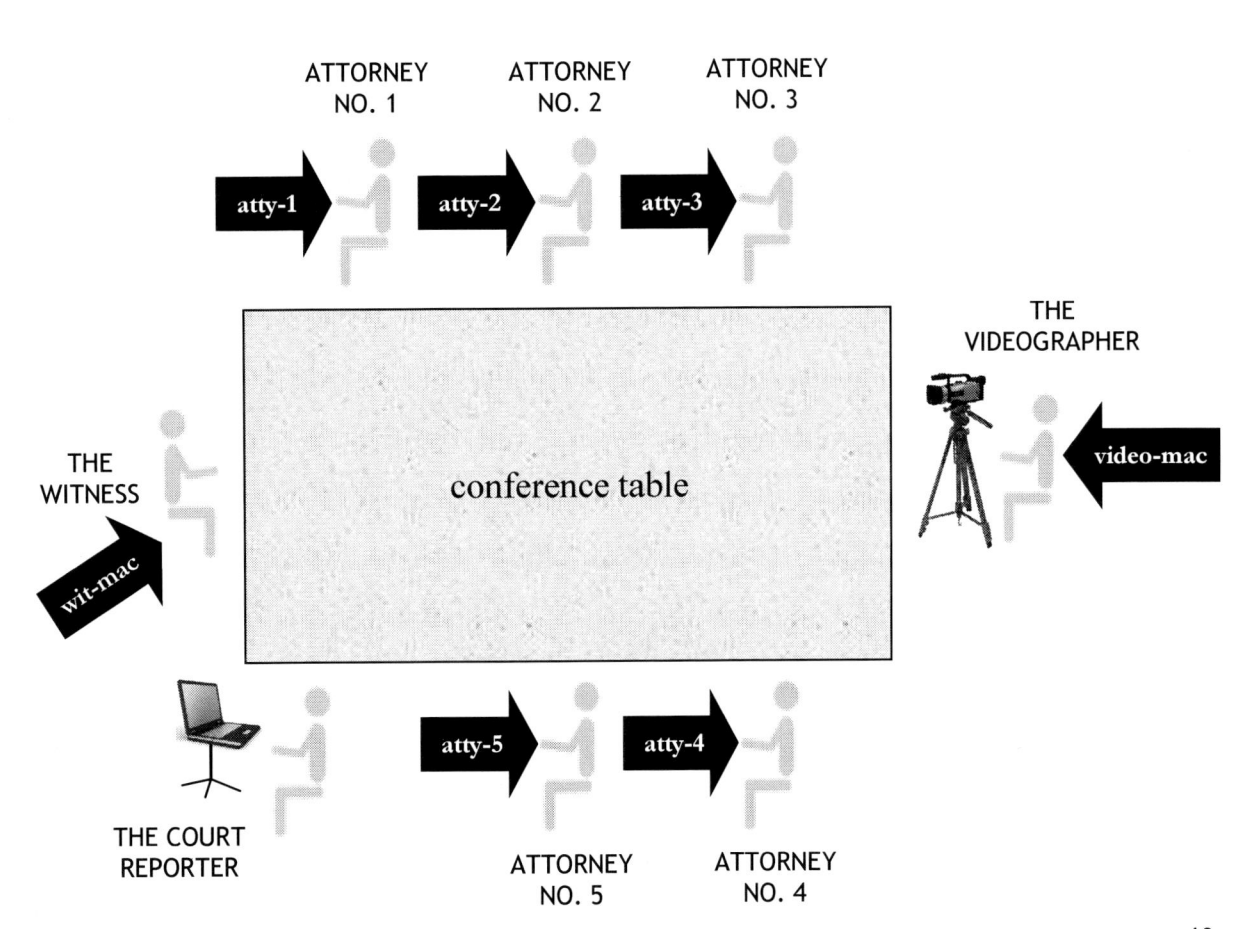

ATTORNEY NO. 1 — atty-1
ATTORNEY NO. 2 — atty-2
ATTORNEY NO. 3 — atty-3
THE VIDEOGRAPHER — video-mac
THE WITNESS — wit-mac
conference table
THE COURT REPORTER
ATTORNEY NO. 5 — atty-5
ATTORNEY NO. 4 — atty-4

Table 12

Speaker IDs for Court Reporting

Description	Voice Code	Text, Formatting, and Use
THE COURT: [1]	jud-mac [1]	Identifies the judge as **"THE COURT:"** in your specified format. Used when the judge begins speaking.
THE WITNESS:	wit-mac	Identifies the witness as **"THE WITNESS:"** in your specified format. Used when the witness begins speaking and it is not in response to an examiner's question.
THE JURY:	jur-mac	Identifies a juror as **"THE JURY:"** in your specified format. Used when a juror begins speaking.
THE BAILIFF:	baily-mac	Identifies the court's bailiff as **"THE BAILIFF:"** in your specified format. Used when the bailiff begins speaking.
THE CLERK:	clerk-mac	Identifies the court's clerk as **"THE CLERK:"** in your specified format. Used when the clerk begins speaking.
PLAINTIFF(S)COUNSEL: [1]	tiff-mac [1]	Identifies the attorney for the plaintiff either generically as **"PLAINTIFF(S) COUNSEL:"** or by his/her specific name (e.g., **"MR. SMITH:"**) in your specified format. Used when this attorney begins speaking.
DEFENSE COUNSEL: [1]	fense-mac [1]	Identifies the attorney for the defense either generically as **"DEFENSE COUNSEL:"** or by his/her specific name (e.g., **"MR. SMITH:"**) in your specified format. Used when this attorney begins speaking.
ATTORNEY 1: [1] through **ATTORNEY __:**	atty-1 [1,2] through atty-__	Identifies an attorney by his/her specific name (e.g., **"MR. SMITH:"**) in your specified format. Used when a specific attorney begins speaking.

Chapter 3: Voice Writing Theory

Table 12 (Continued)

Speaker IDs for Court Reporting

Description	Voice Code	Text, Formatting, and Use
SPEAKER 1: [1] through **SPEAKER __:**	spee-1 [1,3] through spee-__	Identifies any speaker by his/her specific name (e.g., "**MR. SMITH:**") in your specified format. Used when any other person—besides all other speaker types listed within this table—begins speaking. May also be used to identify names of attorneys.
THE VIDEOGRAPHER:	video-mac	Identifies the videographer as "**THE VIDEOGRAPHER:**" in your specified format. Used when the videographer begins speaking.
THE INTERPRETER:	terpret-mac	Identifies a language interpreter as "**THE INTERPRETER:**" in your specified format. Used when the interpreter speaks on behalf of him/herself as opposed to when he/she is speaking on behalf of a witness during language translations.
THE COURT REPORTER:	report-mac	Identifies you, the court reporter, as "**THE COURT REPORTER:**" in your specified format. Used when you begin speaking. This is inserted, for example, just prior to a parenthetical stating that you are performing a readback on the record.

1) If you do not have CAT software that automatically places speaker identifiers in colloquy or byline format according to context, you may need to have two separate sets of speaker voice codes for each format type: colloquy and byline. You can use the standard voice codes for colloquy, and then create additional voice codes by adding "by" to the beginning of the pronunciation of each one for a byline. For example, "jud-mac" and "by-jud-mac," "atty-1" and "by-atty-1," and so on. Your local court administration should have official guidelines governing formatting of colloquy, and each state and jurisdiction may have different formats.
2) Alternative voice codes for identifying attorneys are "torney-1" through "torney-__" and "spee-1" through "spee-__."
3) If Dragon has difficulty recognizing these voice codes, use "spok" or "spox" in place of "spee."

Speaker voice codes for CART services appear in Table 13 below. Be aware that this is only a basic list and that you should incorporate more as necessary. Voice codes of speaker identifiers for broadcast captioning (see Table 14 on the next page) may also be applicable in CART, for example, when providing realtime translations as a video picture is being viewed. Remember to use the creation methods explained at the beginning of this chapter when designing voice codes on your own.

If there are too many individual speakers to keep track of by memory, you may want to draw a seating chart for reference.

Table 13

Speaker IDs for CART [1]

Description	Voice Code	Text, Formatting, and Use
INSTRUCTOR:	struc-mac	Identifies the instructor as "**INSTRUCTOR:**" in your specified format. Used when the instructor begins speaking or performing actions.
PROFESSOR:	prof-mac [2]	Identifies the professor as "**PROFESSOR:**" in your specified format. Used when the professor begins speaking or performing actions.
STUDENT:	stud-mac [3]	Identifies a student as "**STUDENT:**" in your specified format. Used when a student begins speaking or performing actions.
PRESENTER:	prez-mac [4]	Identifies a speaker as "**PRESENTER:**" in your specified format. Used when the presenter of a meeting begins speaking or performing actions.
PASTOR: or **PRIEST:**	preach-mac	Identifies the pastor (as "**PASTOR:**") or the priest (as "**PRIEST:**") in your specified format. Used when the preacher begins speaking or performing actions.
SPEAKER 1: through **SPEAKER __:**	spee-1 through spee-__	Identifies a speaker by his/her specific name (e.g., "**MR. SMITH:**") in your specified format. Used when a specific person begins speaking or performing actions.

1) For a more comprehensive list of speaker identifiers for CART, refer to the CART website at www.cartwheel.cc where you will find a downloadable manual and other valuable information.
2) An alternative voice code is "pro-mac."
3) An alternative voice code is "stu-mac."
4) This voice code is spelled with the letter "Z" (instead of the letter "S" in the original word) to help you remember that its pronunciation sounds more like *prez* than *press*, but spelling it either way would be acceptable. As with all voice codes, choose the spelling method that best serves your memory.

Chapter 3: Voice Writing Theory

Table 14 presents a basic list of voice codes to use when identifying speakers in broadcast captioning. Create additional speaker codes as required.

Table 14

Speaker IDs for Broadcast Captioning [1]

Description	Voice Code	Text, Formatting, and Use
>>: (speaker change)	spee-mac	Indicates a speaker change as ">>:". Used when a new speaker begins talking or performing actions.
>>>: (new subject)	new-mac	Indicates a subject change as ">>>:". Used when a speaker begins talking about a new subject.
>>SPEAKER 1: through >>SPEAKER __:	spee-1 through spee-__	Identifies a speaker by his/her specific name (e.g., ">>JOHN SMITH:"). Used when a specific person begins speaking or performing actions.
>>NARRATOR:	nar-mac	Identifies the narrator as ">>NARRATOR:". Used when the narrator begins speaking or performing actions.
>>VOICE:	voice-mac	Identifies a generic voice as ">>VOICE:". Used when an unidentified person begins speaking or performing actions.
>>MAN:	he-mac	Identifies a generic man as ">>MAN:". Used when an unidentified male begins speaking or performing actions.
>>WOMAN:	she-mac	Identifies a generic woman as ">>WOMAN:". Used when an unidentified female begins speaking or performing actions.
>>CHILD:	chy-mac [2,3]	Identifies a generic child as ">>CHILD:". Used when an unidentified child begins speaking or performing actions.
>>EVERYONE:	all-mac	Identifies everyone as ">>EVERYONE:". Used to indicate the words and actions of everyone as a whole.

1) For more information about identifying speakers and the captioning process, go to www.fcc.gov and search that website's database using the keywords "closed captioning."
2) Alternative voice codes are "chil-mac" and "child-mac."
3) This voice code is spelled with the letter "Y" (instead of the letter "I" in the original word) to help you remember that its pronunciation sounds more like *chy* than *chee*, but spelling it either way would be acceptable as long as you can remember to pronounce the first syllable as *chy* and not *chee*, as the sound *chee* in this pronunciation would conflict with the pronunciation of the voice code "she-mac". As with all voice codes, choose the spelling method that best serves your memory.

In financial call reporting, realtime is performed by listening to audio; you will not be able to see any of the speakers as they are talking. Learning how to properly identify them in a timely manner will only come through practice as you gain experience listening to financial calls.

The order of speakers is usually as follows: Operator > Investor Relations Person > CEO, CFO, and/or any other company speakers > CEO and/or CFO along with a number of analysts during a Q&A session. During initial introductory monologues and presentations, speakers will usually state who they are turning the call over to before transitioning to a new speaker. To gain familiarity with how speakers take turns speaking during financial calls, listen to audio recordings found on the Internet by doing Google searches using keywords such as "audio of earnings calls," "audio of financial calls," et cetera.

Speaker voice codes for financial call reporting are provided below (Table 15). For identifying speakers during question-and-answer sessions, refer to Table 18 in the *Q&A Markers* section of this chapter.

Table 15

Speaker IDs for Financial Call Reporting

Description	Voice Code	Text, Formatting, and Use
Operator	op-mac	Identifies the operator as **"Operator"** in your specified format. Used when the operator begins speaking.
Investor Relations Person	vest-mac	Identifies the investor relations person by his/her specific name (e.g., **"John Smith, Investor Relations"**) in your specified format. Used when the investor relations person begins speaking.
CEO	lead-mac	Identifies the CEO by his/her specific name (e.g., **"Jack Simmons, CEO"**) in your specified format. Used when the CEO begins speaking.
CFO	money-mac	Identifies the CFO by his/her specific name (e.g., **"Michelle Thompson, CFO"**) in your specified format. Used when the CFO begins speaking.
Company Management Speakers	comp-1 through comp-__	Identifies a speaker by his/her specific name (e.g., **"Susan Nelson,** *(title)***"**) in your specified format. Used when a specific company management person begins speaking.
Generic or Unnamed Company Speaker	comp-spee	Inserts the generic title of **"Company Speaker"** in your specified format. Used when the exact name of the company speaker cannot be identified.

Chapter 3: Voice Writing Theory

Examination Lines

This section addresses how to indicate examinations of witnesses in transcripts of legal proceedings and relates strictly to the court reporting profession.

Table 16

Examination Lines for Court Reporting

Description	Voice Code	Text, Formatting, and Use
EXAMINATION or **DIRECT EXAMINATION** or **CROSS-EXAMINATION**	exi [1] or drexi or crexi	Produces an **examination line** in your specified format *(e.g., centers "EXAMINATION," "DIRECT EXAMINATION," or "CROSS-EXAMINATION" on a line by itself)* to indicate an examination is about to begin. This is inserted just before the examiner's name and/or byline in a transcript. Used when it is his/her first turn examining the witness during a proceeding.
REEXAMINATION or **REDIRECT EXAMINATION** or **RECROSS-EXAMINATION**	rexi or redrexi or recrexi	Produces a **reexamination line** in your specified format *(e.g., centers "REEXAMINATION," "REDIRECT EXAMINATION," or "RECROSS-EXAMINATION" on a line by itself)* to indicate a reexamination is about to begin. This is inserted just before the examiner's name and/or byline in a transcript. Used when he/she returns to examine the witness a second time during a proceeding.
FURTHER REEXAMINATION or **FURTHER REDIRECT EXAMINATION** or **FURTHER RECROSS-EXAMINATION**	furexi or furdrexi or furcrexi	Produces a **reexamination continuance line** in your specified format *(e.g., centers "FURTHER REEXAMINATION," "FURTHER REDIRECT EXAMINATION," or "FURTHER RECROSS-EXAMINATION" on a line by itself)* to indicate a continuing reexamination is about to begin. This is inserted just before the examiner's name and/or byline in a transcript. Used when he/she returns to further examine the witness during a proceeding a third time or any time subsequently thereafter.

1) If you have difficulty getting Dragon to recognize "exi" with consistency, use "exi-exi" instead. Doubling up on the pronunciation for this voice code gives it more sound distinction.

Q&A Markers

Q&A markers are inserted in transcripts during question-and-answer sessions. Many different voice code concepts can be applied, and below are some ideas from which to borrow. If you do not have successful recognition with any of these or have trouble remembering them, experiment with other ideas to create your own voice codes.

Two-Syllable Voice Codes:

    quesco (for question command)
    respo (for response)

    Q-mar (for question marker)
    A-mar (for answer marker)

    Q-par (for question paragraph style)
    A-par (for answer paragraph style)

One-Syllable Voice Codes:

    quex (for question + hard sound of X)
    kax (for hard sound of K + answer + hard sound of X)

    keek (rarely-used English word which begins and ends with hard K sound)
    kak (for hard sound of K + answer + hard sound of K)

Although the tables in this section provide Q&A voice codes strictly for use in court reporting and financial call reporting, they may be applied during question-and-answer segments in CART and captioning as well.

Table 17

Q&A Markers for Court Reporting

Description	Voice Code	Text, Formatting, and Use
Q	quesco [1]	Produces a **question marker** in your specified format *(e.g., inserts a period, a new line, capital letter Q, tab, and capitalizes first letter of next word)*, to indicate a question is being asked of a witness during an examination.
A	respo [1]	Produces an **answer marker** in your specified format, *(e.g., inserts a question mark, a new line, capital letter A, tab, and capitalizes first letter of next word)*, to indicate an answer is being given by a witness during an examination.

1) See also the alternative voice codes mentioned above this table.

Chapter 3: Voice Writing Theory

The Q&A voice codes listed in Table 18 may be used for generically indicating questions and answers and/or identifying specific speaker names during question-and-answer sessions of financial calls.

Depending on the formatting guidelines used by the financial call transcription services company for which you're working, you may be expected to do one of the following:

- Translate only generic Q&A markers
- Insert only specific speaker names
- Insert Q&A markers which incorporate specific speaker names

To conform to your job's requirements, choose from among the voice code possibilities offered in both the *Speaker Identification* section of this chapter (see Table 15) and the table below.

Table 18

Q&A Markers for Financial Call Reporting

Description	Voice Code	Text, Formatting, and Use
Q (unidentified questioner)	**quesco**	Produces a **question marker** in your specified format. Used to generically indicate a question is being asked by an unidentified analyst.
A (unidentified responder)	**respo**	Produces an **answer marker** in your specified format. Used to generically indicate an answer is being given by either a CEO, CFO, or other unidentified company management representative.
Q (by identified analysts)	**spee-1** through **spee-__**	Produces an **analyst-identified question marker** in your specified format. Used to indicate the exact name of the analyst asking a question.
A (by CEO)	**rezlead** [1] (for <u>res</u>ponse + <u>lead</u> for CEO)	Produces a **CEO-identified answer marker** in your specified format. Used to indicate the exact name of CEO giving an answer.
A (by CFO)	**rezmoney** [1] (for <u>res</u>ponse + <u>money</u> for CFO)	Produces a **CFO-identified answer marker** in your specified format. Used to indicate the exact name of CFO giving an answer.

1) Concept for remembering these voice codes: The sound "rez" comes from the word "response," but it is spelled with a "Z" instead of an "S" to better represent its pronunciation; *rez* is more distinctive sounding than *res*.

Parentheticals

Tables 19, 20, and 21 provide basic lists of parentheticals for the main career specialties covered in this book. Create and incorporate more of your own as necessary.

In court reporting, the main purpose of parentheticals is to objectively report facts and nonverbal communications of what cannot be seen by a person who reads the transcript, without including details that may mislead the reader by an interpretation of those facts. A court reporter is restricted from using parentheticals to describe the demeanor of others.

Table 19

Parentheticals for Court Reporting

Description	Voice Code	Text, Formatting, and Use
(OFF THE RECORD)	frec-frec	Indicates an **off-the-record** period during a proceeding and appears in your specified format. It may appear, for example, as (OFF THE RECORD) on a centered line by itself.
(BREAK TAKEN)	break-break [1]	Indicates a **break** period during a proceeding and appears in your specified format; e.g., centers "(BREAK TAKEN)" on a line by itself.
(INTERRUPTION)	ruption [2]	Indicates an **interruption** during a proceeding and appears in your specified format; e.g., centers "(INTERRUPTION)" on a line by itself.
(Enters.)	enter-mac	Appears after a speaker's name in a transcript to indicate when that person **enters the room** during a proceeding. This is not used prior to the time a proceeding begins; it's only used when the person has exited the proceeding after it has already begun, then returns for participation.
(Exits.)	exit-mac	Appears after a speaker's name in a transcript to indicate when that person **exits the room** during a proceeding.

1) Alternative voice codes are "breaky-breaky" and "break-mac."
2) Alternative voice codes are "ruption-ruption" and "ruption-mac."

Chapter 3: Voice Writing Theory

Table 19 (Continued)

Parentheticals for Court Reporting

Description	Voice Code	Text, Formatting, and Use
(Nods head up and down.)	nod-nod [3]	Indicates a **head nod** as an affirmative, nonverbal response given. Used when a person nods his/her head up and down instead of saying, "Yes."
(Shakes head side to side.)	shake-shake [4]	Indicates a **head shake** as a negative, nonverbal response given. Used when a person shakes his/her head side to side instead of saying, "No."
(Complies.)	compco	Indicates a nonverbal response of **complying with instructions** during a proceeding and appears in your specified format. Used, for example, when a person signs a document upon instruction to do so, yet does not verbalize that fact and this action is the only response given.
(No response.)	norpo	Indicates **no response given** either verbally or nonverbally.
(Reviews file.) and (Reviews files.)	revie-file and revie-files	Indicates the nonverbal response of **reviewing a single file** or **multiple files** and appears in your specified format. Used when no one speaks any words to verbalize this fact and this action is either the only response given or noting this fact simply makes the transcript more understandable.
(Reviews note.) and (Reviews notes.)	revie-note and revie-notes	Indicates a nonverbal response of **reviewing a single note** or **multiple notes** and appears in your specified format. Used when no one speaks any words to verbalize this fact and this action is either the only response given or noting this fact simply makes the transcript more understandable.

4) An alternative voice code is "nods-mac." Note that "nod-mac" is not a recommended voice code option, as this pronunciation may tend to pose recognition conflict with "9-mac," which is the voice code used to produce the number 9 (see the *Producing Single Digits* table under the *Numbers* section of this chapter).

Table 19 (Continued)

Parentheticals for Court Reporting

Description	Voice Code	Text, Formatting, and Use
(Reviews photograph.) and (Reviews photographs.)	revie-photo and revie-photos	Indicates a nonverbal response of either **reviewing a single photograph** or **multiple photographs** and appears in your specified format. Used when no one speaks any words to verbalize this fact and this action is either the only response given or noting this fact simply makes the transcript more understandable.
(Reviews document.) and (Reviews documents.)	revie-doc and revie-docs	Indicates a nonverbal response of either **reviewing a single document** or **multiple documents** and appears in your specified format. Used when no one speaks any words to verbalize this fact and this action is either the only response given or noting this fact simply makes the transcript more understandable.
(Indicating.) and (indicating)	cat-cat [5] and cat-mac [5]	Inserts the word "**indicating**" as a parenthetical in a transcript; for example, when a person either points to an object or uses hands in some way to convey a size or measurement of something. Used when no one speaks any words to verbalize this action taking place and it is either the only response given or noting this fact will simply make the transcript more understandable. Say *cat-cat* for this to appear as a standalone sentence, as follows: (Indicating.) Say *cat-mac* for this to appear inside a sentence with surrounding words, as follows: (indicating)

5) Voice code alternatives for "(Indicating.)" and "(indicating)" are "cate-cate" and "indi-mac." Note that "cate-mac" is not presented as a voice code option here, as it is likely to pose recognition conflict with "K-mac," the voice code used to produce the letter K (see the *Spelling Technique* section of this chapter).

Chapter 3: Voice Writing Theory

Table 19 (Continued)

Parentheticals for Court Reporting

Description	*Voice Code*	*Text, Formatting, and Use*
(Demonstrating.) and **(demonstrating)**	**strat-strat** [6] and **strat-mac** [6]	Inserts the word "**demonstrating**" as a parenthetical in a transcript. An example is when a person uses any part of his body to illustrate movements by signaling, gesturing, or any type of physical motioning. Used when no one speaks any words to verbalize such action taking place and it is either the only response given or noting this fact will simply make the transcript more understandable. Different capitalization and punctuation is used depending on context given. Say *strat-strat* for this to appear as a standalone sentence, as follows: (Demonstrating.) Say *strat-mac* for this to appear inside a sentence with surrounding words, as follows: (demonstrating)
(Reading.) and **(as read):**	**read-read** and **read-mac**	Indicates when someone is **reading aloud** from a document instead of composing his own thoughts when speaking. Used when material being read cannot be placed within quotes in a transcript because it is not being read verbatim; therefore, including this parenthetical makes the transcript more understandable. Different words, capitalization, and punctuation are used depending on context given. Say *read-read* for this to appear as a standalone sentence, as follows: (Reading.) Say *read-mac* for this to appear as the ending of a sentence, at the point when material is about to be read aloud: (as read):

6) Voice code alternatives for "(Demonstrating.)" and "(demonstrating) are "strate-strate" and "strate-mac."

Table 19 (Continued)

Parentheticals for Court Reporting

Description	Voice Code	Text, Formatting, and Use
(Drawing.) and **(drawing)**	draw-draw and draw-mac	Indicates when someone is **drawing** a picture or diagram as part of his response in a transcript. Used when no one speaks any words to verbalize this fact and this action is either the only response given or noting this fact simply makes the transcript more understandable. Different capitalization and punctuation is used depending on context given. Say *draw-draw* for this to appear as a standalone sentence, as follows: (Drawing.) Say *draw-mac* for this to appear inside a sentence with surrounding words, as follows: (drawing)
(inaudible)	sound-mac	Inserts the word "**inaudible**" as a parenthetical in a transcript. Used sparingly to indicate that a speaker's words were imperceptible and could not be transcribed due to audibility factors.
(indiscernible)	disc-mac	Inserts the word "**indiscernible**" as a parenthetical in a transcript. Used sparingly to indicate that a speaker's words were imperceptible and could not be transcribed due to unclear speech.

Chapter 3: Voice Writing Theory

In CART and broadcast captioning, parentheticals are generally used to report how actions and words are conveyed and to give clarity to non-hearing readers relying on written content for comprehension and clarity of communications taking place in live environments or video pictures. Unlike court reporting, interpretations of demeanor are allowed. This includes emotive gestures and moods which may be necessary to help the reader understand verbal communications when they are not to be interpreted literally. Parentheticals for nonverbal communications are primarily used in captioning rather than CART; viewers in CART environments are usually present in the live environment, so there is already an awareness of actions taking place.

Table 20 provides only a basic list of parentheticals you may want to apply in CART and captioning. Use your best judgment in determining whether you need to add more parentheticals to this list, and employ concepts similar to the ones shown here when creating voice codes for them. Other examples of parentheticals you might find useful include indications of words being expressed with a sly or arrogant tone, in a stuttering manner, et cetera; additional sound effects may include thunder, a thump, a boom or explosion, et cetera.

Table 20

Parentheticals for CART and Broadcast Captioning

Description	Voice Code	Text, Formatting, and Use
[ENTERS]	enter-mac	Appears after a speaker's name to indicate when that person **enters**.
[EXITS]	exit-mac	Appears after a speaker's name to indicate when that person **exits**.
[EXCITED]	cite-cite	Indicates a **specific person is excited**. Appears after a specific speaker's name.
[SURPRISED]	prise-prise	Indicates a **specific person is in a state of surprise**. Appears after a specific speaker's name.
[JOKING]	joke-joke	Indicates a **specific person is joking**. Appears after a specific speaker's name.
[NONCHALANTLY]	care-care	Indicates a **specific person is being nonchalant**. Appears after a specific speaker's name.
[CONFUSED]	fuse-fuse	Indicates a **specific person is in a state of confusion**. Appears after a specific speaker's name.
[LOUD SOUND]	sturb-mac	Indicates a **loud sound or noise disturbance**.
[GUN SHOTS]	bang-bang	Indicates **gun shots** being heard.

Table 20 (Continued)

Parentheticals for CART and Broadcast Captioning

Description	Voice Code	Text, Formatting, and Use
[♪MUSIC PLAYING♪]	muse-mac	Used to indicate **music playing**. Appears when music begins to play; appears just before the words of a song (if this music is accompanied by a song).
[SINGS] and [SINGING]	sing-sing and sing-mac	"Sings" is used to indicate a **specific person singing** and appears after a specific speaker's name. "Singing" is used to indicate **general singing** and appears on a line by itself.
[APPLAUSE] and [APPLAUDING]	clap-clap and clap-mac	"Applause" is used to indicate **general applause** and appears on a line by itself. "Applauding" is used to indicate a **specific person applauding** and appears after a specific speaker's name.
[CHEERS] and [CHEERING]	cheer-cheer and cheer-mac	"Cheers" is used to indicate **general cheers** and appears on a line by itself. "Cheering" is used to indicate a **specific person cheering** and appears after a specific speaker's name.
[LAUGHTER] and [LAUGHING]	laugh-laugh and laugh-mac	"Laughter" is used to indicate **general laughter** and appears on a line by itself. "Laughing" is used to indicate a **specific person laughing** and appears after a specific speaker's name.
[SHOUTS] and [SHOUTING]	shout-shout and shout-mac	"Shouts" is used to indicate **general shouts** and appears on a line by itself. "Shouting" is used to indicate a **specific person shouting** and appears after a specific speaker's name.

Chapter 3: Voice Writing Theory

Table 20 (Continued)

Parentheticals for CART and Broadcast Captioning

Description	Voice Code	Text, Formatting, and Use
[SCREAMS] and [SCREAMING]	scream-scream and scream-mac	"Screams" is used to indicate **general screams** and appears on a line by itself. "Screaming" is used to indicate a **specific person screaming** and appears after a specific speaker's name.
[CRIES] and [CRYING]	cry-cry and cry-mac	"Cries" is used to indicate **general cries** and appears on a line by itself. "Crying" is used to indicate a **specific person crying** and appears after a specific speaker's name.
[SCARED] or [FRIGHTENED]	scare-scare or fright-fright	Used to indicate a **specific person is scared or frightened**. Appears after a specific speaker's name.
[ANGRY] or [FURIOUS]	gry-gry (pronounced *gree gree*) or furi-furi	Used to indicate a **specific person is angry or furious**. Appears after a specific speaker's name.
[SILENCE] and [SILENT]	sil-mac and si-si (pronounced *sie sie*)	"Silence" is used to indicate **general silence** and appears on a line by itself. "Silent" is used to indicate a **specific person is being silent** and appears after a specific speaker's name.

Note the creation methods used in designing the parenthetical voice codes listed in Table 20. Meaningful words and/or word parts were either doubled up or a "mac" ending was attached to the end of a meaningful word or word part to create distinctive sounding utterances. Build your own voice codes from these concepts when further extending this list of parentheticals.

Table 21 shows parentheticals that may be used during voice writing of financial calls. The main purpose of these parentheticals is to communicate facts about information that either should not be included or cannot be effectively interpreted based on the audibility of what is being heard.

Depending on the formatting guidelines of the company for which you're providing financial call transcription services, these exact parentheticals may or may not apply. You may be required to use alternative parenthetical forms that are somehow similar to the ones listed below or parentheticals that are altogether different.

Table 21

Parentheticals for Financial Call Reporting

Description	Voice Code	Text, Formatting, and Use
[Operator Instructions]	struc-struc	Used in place of the actual words spoken by an operator regarding instructions to callers about how to operate the system they are listening to.
[Question Inaudible]	sound-quest	Used to indicate that a speaker's question could not be heard during a Q&A segment.
[Audio Gap]	audio-mac	Indicates a portion of audio that was dropped from the call.
[Inaudible]	sound-mac	Indicates that a speaker's words were imperceptible and could not be transcribed due to unclear audio. This is to be used sparingly.
[Indiscernible]	disc-mac	Indicates that a speaker's words were imperceptible and could not be transcribed due to unclear speech. This is to be used sparingly.

To view samples of financial call transcripts and the variety of formats used, simply do a Google search using key words such as "financial call transcripts," "earnings call transcripts," et cetera.

Numbers

To produce most numbers, you can usually dictate them the same way you speak in normal conversation. For example, the word "and" in "five hundred and thirty-six" will be treated as part of a number and produce the number string as digits (536). If you say "dollars" while dictating numbers, the dollar sign ($) will be inserted and produce numbers as currency. Telephone numbers, social security numbers, times, and numbers within dates—such as the *month + day + year*, *month + day*, and *month + year*—will appear in their proper formats as well.

Dragon can be set to display numbers as digits or spelled out, depending upon the context in which they occur, or be set to display numbers in one fixed format. Dragon user options allow you to choose how you want your numbers to be formatted. Some CAT programs have these same capabilities as well, allowing you to benefit from an additional layer of AI that improves the way numbers are produced in special contexts, such as spelling out numbers when they begin sentences, automatically inserting hyphens (-) or slashes (/) between numbers in dates, and other special formatting circumstances that are not provided by Dragon.

As a general rule—except for sports scores, percentages, ratios, and other measurements—we spell out numbers under 10 and use digits for all other numbers. So we need to choose the appropriate settings within our Dragon user and/or CAT program to consistently produce that type of format. However, there are times when digits must be produced for numbers under 10 and the AI components within Dragon and/or our CAT software will not be able to guess the correct output for every number formatting scenario. This applies to instances such as names of exhibits, businesses, and forms containing a single digit. In those situations, you can use the voice codes listed in Table 22 to override the spelled-out versions of numbers and force-produce them as digits.

Table 22

Producing Single Digits (for Numbers Under 10)

Number	Voice Code
0	0-mac
1	1-mac
2	2-mac
3	3-mac
4	4-mac
5	5-mac
6	6-mac
7	7-mac
8	8-mac
9	9-mac

When relying on the number-formatting capabilities of Dragon alone, you are catering to rules set forth by only one set of AI rules. But if you are using Dragon in conjunction with a CAT program that has its own set of AI rules to work with, that additional layer of AI will change the mechanics involved in how numbers are produced, and depending on your CAT software, possibly even pose conflicts with certain number formats. In combining the functionality of both AI components, the form of number output from Dragon must work in harmony with the CAT program's formatting capabilities to further enhance accuracy of the final realtime result.

If a CAT program has automatic-number-formatting attributes built within its AI system, this may, in effect, actually interfere with the final number-formatting process that would ordinarily take place if you were to use Dragon alone. For example, if you said "May thirty-first" using Dragon alone, Dragon would produce "May 31st." But if you said "May thirty-first" when using Dragon in conjunction with a CAT program that has its own set of AI rules for number formatting, your result could instead appear as either "May 31" or "May 301st." Number formatting results will vary depending upon the CAT program you are using. If you do experience such peculiar problems with your numbers, voice codes are not required to resolve those problems; they can easily be corrected through your CAT program's utilities so that numbers will be produced in their proper format in the final realtime result. It is best to rely on the CAT program's number-formatting capabilities in those instances to avoid AI conversion conflicts. If your CAT program provides a dictionary module for making number-conversion entries, refer to your CAT program's manual for help in determining how to make dictionary entries to avoid number-conversion inconsistencies. If your CAT program does not have its own system of AI, you do not have to worry about these particular types of number-conversion inconsistencies.

Another limitation you should be aware of regarding the AI system within Dragon and/or a CAT program is the inability to make correct determinations between what should be a number and what should be a word when there is contextual ambiguity involved. For example, suppose you intend to produce "five five-story buildings"; if you said *five five-story buildings*, you would likely end up with "55 story buildings." To resolve this problem, you would need to insert a space between each "five" in your dictation to end up with "five five-story buildings." If your CAT program has a dictionary module, you could add a voice code such as "sert-sert" (see Table 23) to Dragon's vocabulary and make an entry in that dictionary which corresponds to "sert sert" so that your CAT program will know to insert a space in your dictation. You would then be able say *five sert-sert five story buildings* and get your intended result.

(Note: Dragon already has the ability to insert a space by using a voice command such as "space bar," but voice commands may not be recognizable if you are operating a CAT program in conjunction with Dragon. Most CAT programs turn off Dragon's command functions in order to prevent them from interfering with continuous realtime text production during your dictation.)

Table 23

Inserting a Space Between Numbers

Description	Voice Code	Text, Formatting, and Use
insert space command	sert-sert	Forces a space. Useful for separating numbers during conversions.

Spelling Technique

The purpose of the spelling technique is to allow you to produce individual letters of the alphabet during realtime. Common examples involve capital letters occurring as part of a company name (e.g., "Big G Automotive") or exhibit name (e.g., "Exhibit A").

Dragon allows you to produce individual letters by either dictating the phonetic or international communications alphabet or saying each letter as you normally would when reciting the alphabet. However, those methods are unreliable for realtime voice writing. The problem with using the phonetic or international communications alphabet as a spelling method is that it's a mouthful to say a string of utterances such as *Victor Oscar India Charlie Echo*—for example, just to spell the word "voice"—not to mention the additional level of difficulty imposed by having to dictate those words at high speeds. As far as dictating the actual letters of the alphabet (A, B, C, etc.), you would run into conflicts with homophones in many cases (such as "a" versus "A," "be" versus "bee" versus "B," "see" versus "sea" versus "C," etc.). It is also important to be aware that pronunciations of certain letters conflict with each other as well. Letter groups prone to conflict are in the table below.

Table 24

Letters Prone to Conflict

B, D, and P • B and V • C and Z
D and T • F and S • G and J • M and N

The most reliable method for spelling individual letters in realtime voice writing is to use the spelling technique presented in Tables 25a and 25b (see the next page).

When needing to produce single-digit and -letter constructions, combine your ideas of using the spelling technique along with the method of producing single digits. For example, you would say *1-mac A-mac* for "1A," *2-mac B-mac* for "2B," *3-mac C-mac* for "3C," and so on.

Depending on the capabilities offered to you through a particular CAT program, you can produce letters in either uppercase, lowercase, or mixed casing and/or have them appear with spaces in between, glued, stitched together, et cetera. Refer to your CAT program's manual for specific instructions on how to produce your letters in a variety of cases and formats.

Do not use the spelling technique as a means of spelling words. Refer to the *On-the-Fly Translations* section of this chapter to learn how to output words during realtime when they do not exist in Dragon's vocabulary.

Voice Writing Method • **FIFTH EDITION**
Dragon NaturallySpeaking® 12

Table 25a

A through M	
Letter	Voice Code
A	A-mac
B	B-mac
C	C-mac
D	Don-mac
E	E-mac [1]
F	F-mac
G	G-mac
H	H-mac
I	I-mac
J	Jack-mac
K	K-mac
L	L-mac
M	Mike-mac

Table 25b

N through Z	
Letter	Voice Code
N	N-mac
O	O-mac
P	Paul-mac
Q	Q-mac
R	R-mac
S	S-mac
T	T-mac
U	U-mac
V	Vick-mac
W	Will-mac [2]
X	X-mac
Y	Y-mac
Z	Zack-mac

1) Should you be required to use the speaker voice code "he-mac" for captioning, this will conflict with "E-mac." In that case, you would use "Ed-mac" as the voice code for the letter "E" instead.
2) Although "W" does not conflict in sound with any other letter of the alphabet, it is easier to use the two-syllable utterance of "Will-mac" than "W-mac" (pronounced "double-U-mac") which contains four syllables.

Conflict Resolution

Dragon can usually distinguish the difference between most words that have identical or similar pronunciations based on the contexts in which they occur. Proper word selections are derived from algorithms based on analyses of bigrams (two-word combinations), trigrams (three-word combinations), and so on, to output the most highly-probable words based on their occurrences in relationship to other words and how they fit in to those *n*-grams. For example, the proper form of a word within a homophone category such as "they're/there/their" will appear correctly in the context of other words in phrases such as "they're here" (a bigram), "there were two" (a trigram), and "it's at their house" (a quadgram). In the phrase "there were two," the word "two" will appear correctly, because the words "to" or "too" within this particular sequence is a less likely grammatical choice.

There are instances, however, when Dragon will not be able to predict proper word selections. This usually happens when there isn't enough context. For example, suppose you say *we have two*, followed by a pause. Dragon is more likely to recognize that phrase as "we have to" instead of "we have two." Both "to" and "two" are common words, but "to" is more frequently used throughout the English language than "two," so "to" is the most probable choice. For Dragon to understand that you meant "two," you would need to either extend the phrase a little further to give Dragon additional context (e.g., "we have two children") or use a voice code.

Also, as previously mentioned, Dragon has more trouble recognizing smaller, one-syllable words than it does larger, multisyllable words. When small words are spoken together too quickly and/or not enunciated clearly, words such as "and" and "in" can sound similar enough to pose just as many conflicts as homophones. Incorporating small words and homophones into phrases as vocabulary entries may help you resolve such recognition problems. (Refer to the *Phrases* section in this chapter for recommendations on which types of phrases to create.)

Be aware that certain conflicts are not resolvable through Dragon alone. For additional conflict-resolution capabilities, you would need to have CAT software with advanced-level AI programming sophisticated enough to override Dragon's translation mistakes.

Homophones

There are contexts where voice codes must be used for proper recognition results to occur with certain homophones. For example, the words "two" and "four" will usually be recognized as "to" and "four" if they are not surrounded by other words spoken together within a phrase. The word "too" just about never appears when it occurs at the end of a sentence. And when the words "to" and "for" are dictated along with numbers, they may be recognized as the digits "2" and "4."

Here are some common homophone conflict examples:

> Q   How many people were with you in the vehicle at the time of the accident?
> A   **Four.**
> (You could end up with "for".)
>
> Q   How many children do you have?
> A   **Two.**
> (You could end up with "to".)
>
> Q   Were your children with you **too**?
> A   Yes.
> (You could end up with "to".)
>
> Q   How fast were you traveling?
> A   About 40 **to** 50 miles per hour.
> (You could end up with "4250".)
>
> Q   How long were you going at that speed?
> A   **For** 15 or 20 minutes.
> (You could end up with "415".)

Instead of using regular pronunciations for the homophones shown in the previous example, use the voice codes below when those homophones occur in such contexts.

Table 26

Most Common Homophones
Resolution for Ambiguous Contexts

Word	Voice Code	To avoid conflict
to	tah-tah	Use this when needing to insert the word "to" in the midst of dictating numbers at a fast rate.
two	twah-twah	Use this when the context given is too ambiguous for grammatical rules alone to apply in properly interpreting of the word "two."
too	twee-twee	Use this when the context given is too ambiguous for grammatical rules alone to apply in properly interpreting the word "too."
for	fah-fah	Use this when needing to insert the word "for" in the midst of dictating numbers at a fast rate.
four	frah-frah	Use this when the context given is too ambiguous for grammatical rules alone to apply in properly interpreting the word "four."

Remember that it is not necessary to use the above voice codes for Dragon to make the correct homophone selections if the contexts in which they occur are clear, as follows:

> The general manager received a total of **four** complaints, **two** of which had **to** do with company representatives talking **too** much **to** each other when they should have been handling services **for** customers.

Less common conflicts arise among homophones such as "their," "there," and "they're." But in the event Dragon does not accurately recognize them, even when they are spoken together with other words within phrases that form proper contexts, preventing these errors from occurring again in the future is usually a simple matter of making corrections using the correction tool (covered in Chapter 6). When the context given to Dragon is unclear, Table 27 gives you a way to force-produce those words without having to rely on context.

Table 27

Other Homophone Examples
Resolution for Ambiguous Contexts

Word	Voice Code	To avoid conflict
there	there-mac	Use these voice codes to ensure that these words will be recognized by Dragon regardless of proper context being given.
their	their-po	
they're	they're-co	

Chapter 3: Voice Writing Theory

The creation approaches taken for the preceding voice codes involved the use of a meaningful word part ("po" to represent the possessive word form "their") and adding the familiar endings of "mac" and "co." The doubling-up method cannot be used as voice codes for any of these homophones, as Dragon would use its intelligence in contextual analyses to interpret those doubled-up sounds as "they're there."

Additional conflict examples include words such as "been" and "bin," "would" and "wood," and "one" and "won." Of course, you would not have to assign a voice code to the words within these pairs which are most common (e.g., "been," "would," and "one"), because Dragon will probably already correctly recognize them. Use voice codes for the less common words (e.g., "bin," "wood," and "won") when the contexts within surrounding words are unclear, but remember to be careful not to create more conflicts for yourself. You wouldn't be able to use a voice code like "won-won" for "won" because the result may likely appear as "11." Neither would you be able to use "won-mac" to produce "won" because it would interfere with Dragon's recognition of "1-mac" which is used to force-produce a digit; your leftover voice code choices would be to use either the ending sound "co" for "command" or to create your own unique ending sound, using a meaningful word part such as "con" to represent that it is a "conflict."

Use this same technique of voice code application with other sound-alike words as necessary.

Words with Soft Pronunciations

For a word with a pronunciation that is not distinctive enough for Dragon to consistently recognize, use a different pronunciation with a stronger sound. One example is the affirmative response "uh-huh." If this word occurs frequently enough for you to justify using a voice code, you may choose an utterance such as "yay-yay," and it should appear correctly every time.

Following is a chart to help you produce some common written expressions whose regular pronunciations are too soft for Dragon to distinguish.

Table 28

Pronunciation Distinctions for Soft-Sounding Expressions

Expression	Voice Code	Meaning
uh-huh	yay-yay	An affirmative response.
uh-uh	nay-nay	A negative response.
huh	wuh-wuh	Usually an inquisitive response in the form of a question. May also express surprise or indifference.
ha-ha	wah-wah	An expression usually intimating laughter or amusement.
uh-oh	wuh-woh	Usually expresses alarm about awareness of something or imparts a type of foreboding or dismay.

Undesirable Words

It can be very embarrassing when incorrect words appear in a realtime record, especially when those words are offensive. If you are to produce a verbatim record as a court reporter, you will probably encounter the use of foul language in your career, but it is advisable for you to remove all such words from Dragon's vocabulary (see the *Deleting an Entry* section in Chapter 5). You would not want any regular words to be misrecognized as swear words. If you do want Dragon to have the ability to recognize offensive words, you should create voice codes that are so extraordinarily unique they cannot be misrecognized for any other words even one percent of the time during your dictation. It is not recommended that you use a "mac" or "co" ending, because the majority of voice codes are already designed that way and you do not want to risk having any possible sound conflicts whatsoever. The best way to produce these words only when intended is to use the on-the-fly translation technique (see the *On-the-Fly Translations* section of this chapter) to prevent foul words from appearing by mistake.

Small Words

When conflicts arise between small words such as "a" and "the," "were" and "where," "in" and "and," et cetera, people are very surprised at the high degree of errors Dragon makes, even when they are being careful to clearly enunciate those words. We must realize that our top-down processing method for understanding language clouds our opinion about SR's effectiveness in recognizing our speech correctly. We must try to think from a bottom-up processing perspective to unlock the mysteries surrounding small-word conflicts when finding a way to resolve them.

When we performed Experiments 1 through 5 in Chapter 2, we learned that inflective accents help Dragon with detecting differences in words that would ordinarily sound similar during high-speed dictation. If Dragon still does not recognize your small words correctly while you are making clear sound distinctions, it is probably due to grammatical judgment factors. You should always dictate small words together with other words in phrase groupings. Rules used to select words based on grammar override the rules used to select words based upon sound alone when a grammatical modeling algorithm does not comport with contextual probabilities. This can be very problematic when you are trying to make an accurate realtime record with a speaker who does not obey rules of English grammar.

Here is a good example of when Dragon will misrecognize a clearly enunciated small word, because it does not recognize the context as being correct:

Table 29

Small-Word Misrecognition
Caused by Improper Grammar Usage

If you said...	*Instead, you may get...*
She gave me **a** orange.	She gave me **the** orange. or She gave me **an** orange.

Chapter 3: Voice Writing Theory

Even though Dragon "hears" your clear pronunciations of words, attempts to accurately recognize them will fail if Dragon determines those words to be contextually improbable. This happens because Dragon uses a collection of criteria related to three sets of data: 1.) sound probability match, 2.) word probability match, and 3.) grammatical probability match. If any one of these three data points is in disagreement toward a probable match, it does not matter how perfect your speech is; the result will be a misrecognition. As an example, let's use the sentence from the previous table, "She gave me *a* orange." When you say the word "a," Dragon extracts a batch of all probable word choices from within its word bank that either identically or closely match your pronunciation of "a." This batch may include "a," "an," "and," "the," "of," et cetera. Even though "a" may be the top priority word match, Dragon will select a different word if "a" is not a good contextual fit for the sentence. That is why, in this context, Dragon would probably select the word "an" or "the."

Dragon's biggest hurdle with selecting words, after correctly identifying proper matches based upon sound, has to do with its method of selecting words based upon context. Recalling the fact that people do not always use proper grammar in ordinary speech, let's address how to overcome conflicts that arise due to improper grammar usage. One way is to build the grammatical model with text samples of improper language syntax (refer to the *Vocabulary Building* section of Chapter 6). Doing this will enhance Dragon's ability to recognize words occurring in improper contexts. But unless you can provide the grammatical model with every instance of improper language usage, Dragon will still have difficulty recognizing them.

The best way to resolve conflicts arising from improper grammar usage is to use voice code pronunciations (see Table 30 below) instead of the actual pronunciations of those words in your dictation, but you must give at least a slight pause both before and after dictating the voice code. The longer the pause, the higher are the probabilities for Dragon to not combine them with other words in a phrase. This will override the grammatical model's influence of word selections and force-produce your desired result.

Table 30

Common Small Words for Improper Grammar Use and/or Ambiguous Contexts

Word	Voice Code	Meaning
a	a-co	Use this to insert the word "a" in contexts of improper grammar usage. (Example: "I got a apple" instead of "I got an apple")
an	an-con	Use this to insert the word "an" when the context given is not ample enough to rule out selection of similar-sounding words "and" or "in." (Example: "push an effort" instead of "push and effort" or "push in effort")
be	be-co	Use this to insert the word "be" in contexts of improper grammar. (Example: She be silly.)

Note that the voice code for "an" is spelled "an con" ("con" for "conflict") and not "an co" or "an mac," because those pronunciations would conflict with the common English phrase "and go" and the spelling-technique voice code "N-mac." Similarly, "a-mac" and "be-mac" could not be used to produce the small words "a" and "be," because they would pose conflicts with "A-mac" and "B-mac." Neither could the doubling-up method be used to create voice codes for these small words, because it would create context conflicts; "and-and," "an-an," and "in-in" could be recognized as the word combinations "and an," "and in," and "in an."

Another problem you will encounter is the additional production of words due to over-enunciation. A common example is when you say "and," but you get the words "and the" or "and a" instead, resulting from a typical effort to make the ending *d* sound in "and" distinctive. Even if you correct Dragon a hundred times to recognize your over-enunciated pronunciation of "and," you may still have trouble producing that word by itself; if you over-enunciate "in," you could end up with "in a" or "in the." You can lessen over-enunciation errors by using spoken-form spellings for those small words. (Written and spoken forms are addressed in the *Written- and Spoken-Form Spelling Rules* section of Chapter 5.) For example, vocabulary entries would be made for the written form of "and" accompanied by the spoken form of "anduh," "in" accompanied by "inuh," "on" accompanied by "onuh," and so on; you must be careful when recording pronunciations for them that you do not place so much emphasis on the ending portion of the spoken form ("uh") that you make it into a distinctive second syllable when over-enunciating the ending consonant sounds of those small words.

Similar-Sounding Words

For similar-sounding words containing letters prone to conflict (see Table 24 in the *Spelling Technique* section of this chapter), you may not be able to permanently prevent misrecognitions from occurring between them. For example, "pill" and "bill" are always prone to conflict, because the pronunciations of "P" and "B" are too similar. For words containing letters prone to conflict but that otherwise sound identical, you can create voice codes to make their sounds more distinguishable.

In keeping with the idea of incorporating "mac" or "co" with a word or word part when creating a voice code, you can apply that technique to the words "bill" and "pill," as shown in Table 31. Realize that you must assign to each word a different distinguishing sound ("mac" for one and "co" for the other) so the voice codes themselves do not pose conflict. Exercise your own judgment as to when voice codes should be used for such words, basing your decision on factors concerning the speed of your dictation, your ability to properly enunciate at those speeds, and the frequency of conflicting-word occurrences during realtime sessions.

Table 31

Similar-Sounding Words
Alternative Method for Creating Sound Distinctions

Word	Voice Code
pill	pill-mac
bill	bill-co

Chapter 3: Voice Writing Theory

Words versus Acronyms

Another type of conflict that may arise is the pronunciation of an acronym that conflicts with the pronunciation of a common word, such as "cat," for example. If you said "CAT scan," that would not pose a translation problem since that is a common enough term for Dragon to properly identify.

But suppose you will be providing realtime services on a large screen in front of an audience of reporters during a seminar about computer-aided transcription (CAT). You know that the acronym "CAT" is going to be regularly encountered and that it will not arise in connection with the word "scan." Therefore, you know Dragon will probably recognize you to mean "cat," as in the animal, not "CAT" as in computer-aided transcription. So what do you do?

You could enter phrases (see the *Phrases* section in this chapter) for the acronym "CAT" to be dictated in conjunction with particular words such as "program" or "software" and make the spoken-form spellings and vocal recordings match the way those phrases are actually pronounced (i.e., "cat program" and "cat software"). But if you were to encounter "CAT" in conjunction with the word "product," Dragon would likely produce the regular English word "cat" instead, unless there exists a stronger relationship between "CAT" and "product" within its grammatical model.

To get "CAT" to translate properly according to the mechanics of SR and commonly applied practices for increasing word-recognition capabilities, you would have to vocabulary-build with documents containing the acronym "CAT" as it will occur in various contexts. This process involves grammatical-model customization. (See the *Vocabulary Building* section of Chapter 6.) Those documents, ideally, would have to contain all instances of words that could possibly occur alongside and around the acronym "CAT" for Dragon to recognize it based on context. Proper recognition results should occur in most cases.

The techniques of adding phrases and vocabulary building, however, are not fool-proof solutions. Just suppose that "CAT" had to be produced all by itself, and not in connection with any other words, for example, "CAT" being the only word in a sentence in response to a question. "CAT" would likely never translate properly in that instance. Or suppose you dictated the sentence, "What does CAT stand for?" You may get either "CAT" or "cat." Who knows! It depends on whether Dragon has strong enough contextual information linking a relationship between the words "does" and "CAT" and/or "CAT" and "stand."

Further suppose, after properly doing all your vocabulary-building work, that you are in court the following day and the testimony in the matter has to do with a dispute regarding someone's pet "cat." You would most likely get the word "cat," but now there is a potential risk of Dragon producing "CAT." The result of all your hard work will actually end up creating a conflict between the word "cat" and the acronym "CAT" because they both share the same pronunciation. So do not fool yourself into thinking that adapting the grammatical model is the best way of getting around errors just because that is the traditional approach to take. There is a much simpler way.

To ensure "cat" and "CAT" are free from conflict, you would simply use the pronunciation *cat* for the word "cat" and recite the capital letters *C, A, T* when making a vocal recording for your pronunciation of the acronym "CAT." (See Table 32.)

Not only does this save you the time of having to manually input a ton of phrases into Dragon's vocabulary and reinforce the acronym's recognition by analyzing documents; you have removed a pronunciation conflict between a word and an acronym and do not have to worry about typing an additional spelling for a spoken form, because Dragon will accept a vocal pronunciation based on the written-form spelling. (For more information about written and spoken forms, see the *Written- and Spoken-Form Spelling Rules* section of Chapter 5.)

Table 32

Acronym vs. Word
for Acronyms and Words with Identical Pronunciations

Word	Pronounced	Comments
cat	cat	This is already a base vocabulary word with an assigned pronunciation. To produce the word "cat," as in an animal, just say *cat*.
CAT	CAT	This is also a base vocabulary word with an assigned pronunciation, but its pronunciation is the same as "cat," as in an animal. You will want to create a new vocabulary entry for "CAT" that is pronounced by reciting letters *C, A, T* to avoid pronunciation conflicts.

All Other Words

All other types of errors encountered are resolved by using either the correction tool (see the *Correcting Misrecognitions* section in Chapter 6) to improve Dragon's ability to recognize the differences between pronunciations of words or through vocabulary building (see the *Vocabulary Building* section in Chapter 6) to enhance Dragon's ability to recognize the differences of words based on context.

For persistent errors, you can train words individually by locating them in Dragon's vocabulary and recording vocal pronunciations directly. (See the *Working with the Vocabulary* section of Chapter 5.)

You can also delete any words from the vocabulary that you know you will never have a use for, should they pose conflicts with more necessary words. (See the *Deleting an Entry* subsection under *Working with the Vocabulary* section of Chapter 5.)

Phrases

Adding phrases to Dragon's vocabulary is an important step in attuning your voice model to more concisely represent nuances in your pronunciation of words that commonly occur together during high-speed dictation. It also enhances Dragon's word-selection performance when locating best-match guesses. The task of recognition becomes easier to accomplish because you are giving Dragon shortcut instructions for locating a single-phrase entry, as opposed to multiple individual words.

For example, when you dictate "do you know," Dragon must scan for a match of three different words within its vocabulary base before finally reaching a conclusion. But if you entered the words in the vocabulary together as a phrase, there is a higher likelihood that those words will be recognized together because they exist as a unit. Distinguishing word sets by combining them in groups prioritizes them for selection within the vocabulary. And just as Dragon more accurately recognizes words with multiple syllables, it will likewise more accurately recognize phrases containing multiple words.

Identifying common phrases is a lengthy process that is not necessary to complete all at once. Limit the addition of phrases to those word groups that Dragon commonly misrecognizes when you say them together. You may want to add these phrases to your vocabulary only as you encounter them through actual dictation work. Or you could compile a base set of phrases by skimming through materials representative of the type of material you expect to dictate in your realtime work and manually identify those you want to add to Dragon's vocabulary. (Also refer to the *Adding Phrases* section in Chapter 6 to learn how the *N-Gram Phrase Extractor* will automatically identify common phrases found in documents.)

Since Dragon has more trouble recognizing smaller words than it does larger, more complex words, find as many phrases as you can containing small-word combinations so that when you record your speech, Dragon will learn how you say those small words when you run them together. Remember that it isn't necessary to add every common phrase you identify; only add the phrases you think Dragon will have trouble recognizing when you dictate at realtime speeds.

Be sure when recording your speech for phrases that you say them the same as you would during mid-dictation to give Dragon an idea of how it should expect you to say them again in the future.

Following are some ideas of the types of phrases you could enter into Dragon's vocabulary to increase your recognition accuracy:

Table 33

Examples of Common Phrases		
based on the	the impact of	and that is
over the last	been able to	in terms of
you can see	as well as	whether or not
do you have	how do you	insofar as

You can create additional phrase entries by combining prepositions (such as "of" and "to"), articles (such as "a" and "the"), verbs (such as "are" and "am"), conjunctions (such as "but" and "so"), pronouns (such as "you" and "they"), and so on. You can even throw in contractions as part of those phrases.

You may also want to add entries to your vocabulary for phrases that you tend to slur a lot or for which you use slangful pronunciations. For example, if you commonly pronounce "did you" and "would you" as "didja" and "woojew," Dragon may misrecognize them as "digit" or "which you." To increase Dragon's ability to recognize those phrases based on the way you actually pronounce them, you would include spoken-form spellings (e.g., "didja" and "woojew") for those phrases when entering them in your vocabulary. The table below gives examples of such phrases.

Table 34

Examples of Slurred Phrases

Phrase (Written Form) [1]	Slurred Pronunciation (Spoken Form) [1]
did you	didja
would you	woojew
let me ask	lemyask
I don't know	idunno
do you know	dooyunno
have you ever	havyeffer

1) For an explanation about written and spoken forms, see the *Written- and Spoken-Form Spelling Rules* section of Chapter 5.

Other phrase entries you may want to add are for commonly punctuated word groups. This will increase the speed of your realtime translations by removing the need to constantly dictate punctuation marks for phrases like "yes, sir" and "no, sir," because punctuation is automatically built in. If you have CAT software, you may not need to enter pre-punctuated phrases into Dragon's vocabulary, as your CAT program may be able to automatically handle all the punctuation for you; refer to your CAT software manual for details.

To input a pre-punctuated phrase entry in Dragon's vocabulary, you would type the actual spelling of the phrase just as it should appear (including punctuation) in the *Written form* field, type just the words of the phrase (excluding punctuation) in the *Spoken form* field, and record a vocal pronunciation of just the words. You should know that making a pre-punctuated phrase entry (e.g., "yes, sir") in Dragon's vocabulary only increases the likelihood that Dragon will produce it in your dictation because of its priority within the vocabulary list, but there is no guarantee that the phrase ("yes, sir") will be chosen over the individual words within that phrase ("yes" and "sir") a hundred percent of the time. The individual words, excluding the punctuation, ("yes sir") will occasionally be produced.

If you have a CAT program with a dictionary module, the most effective way to ensure that punctuation will always be part of a phrase in the final realtime output is to make an adjustment for it through your CAT dictionary. This way, whether Dragon recognizes the actual pre-punctuated phrase or recognizes each individual word independently, the result will always include punctuation unless you pause too long between words while dictating those phrases. Pausing between words often causes Dragon to break up your words into separate batches; whole phrases must be processed together as one batch for your CAT dictionary entries which correspond to those phrases to effectively render output. For example, if you enter the pre-punctuated phrase "if, however," but you pause between the words "if" and "however," the commas (,) will not appear because Dragon will produce the text the same way you spoke it ("if" and "however"), in two separate batches. Thus, the CAT program ends up reading each word individually and not the entire phrase as one unit. Ordinarily, all words within phrases must be dictated into Dragon without pausing in order for them to be produced and processed as a single batch. Some CAT programs offer time-delay settings to allow multiple batches to be received before words are processed, so these types of translation problems may not be a problem for you if your CAT program has this capability.

Table 35 contains a sample list of commonly punctuated phrases that you can program Dragon and/or your CAT program to produce without having to actually dictate the punctuation. These are just some general ideas for you to begin working with. Add more pre-punctuated phrase entries later as you encounter them in your dictation.

Table 35

Examples of Pre-Punctuated Phrases	
and/or	if, in fact,
yes, sir *(or ma'am)*	if, however,
no, sir *(or ma'am)*	if, for any reason,
oh, okay	good morning, sir *(or ma'am)*
oh, well	good afternoon, sir *(or ma'am)*
oh, yes	any way, shape, or form
oh, no	year-to-date basis
yes, I do	risks, uncertainties, and assumptions
no, I don't [1]	please continue, sir *(or ma'am)*
yes, it is	you may proceed, sir *(or ma'am)*
no, it isn't [1]	sir *(or ma'am)*, you may continue
as an example,	sir *(or ma'am)*, you may proceed

1) If your CAT program has a dictionary and AI capabilities, you can set up these entries as conflicts to produce the proper results if Dragon makes incorrect choices between "no, I don't" versus "know I don't" and "no, it isn't" versus "know it isn't."

You will also want to add proper names as phrase entries. This is an important part of the preparation work that must be done before you enter a realtime session. For example, if the subject matter you will be reporting involves two individuals with the same last name spelled in different ways ("Tammy Green" and "Bob Greene"), you would enter each full name as a phrase to increase the likelihood that the proper spelling of each name will appear when first and last names are dictated together. Note, however, that if you have CAT software with a built-in dictionary module, you may only need to make such phrase entries in your CAT dictionary.

Brief Forms

The majority of words you dictate do not require different pronunciations because Dragon recognizes plain English. But at speeds of 180 to 200 words per minute and higher, maintaining clear enunciation and keeping up the pace can be challenging when having to simultaneously identify speakers, insert punctuation, and dictate multisyllable voice codes to resolve common conflicts among certain words.

An effective technique for increasing the word translation rate without diminishing the quality of your speech is to dictate brief forms. Using brief forms makes it easier for you keep up the pace in your dictation at higher speeds without having to sacrifice accuracy. Brief forms can either be shortcut versions of common word groups represented by a condensed number of syllables, conflict-free versions of common word groups represented by alternate pronunciations, or common word groups represented by alternate pronunciations as a means of differentiating between phrases which include punctuation versus phrases which do not include punctuation.

Because the subject matter in realtime reporting is so vast, all of the possible brief forms for common word groups are not presented here. Only the concept of how to design them is addressed, accompanied by select examples and explanations, to allow you the freedom to determine through your own experience which groups of words justify application of brief forms in the subjects you will report.

After you have identified the groups of words for which you would like to create brief forms, you can use two simple methods to design voice codes on your own: 1.) the lettering technique (using capital letters to represent whole words) and 2.) ad-hoc combining (combining together selective word parts from within phrases).

When recording pronunciations of capital letters that form part of a voice code, pronounce them the way you say the alphabet (*A*, *B*, *C*, etc.). Whether you choose to designate capital letters or word parts to your system of brief-form voice codes, make sure that the pronunciations of them do not sound similar to the pronunciations of regular English words or recognition conflicts will arise between them.

The examples in Table 36 on the next page identify brief forms that can be used for either producing pre-punctuated phrases or for resolving conflicts between word groups Dragon has trouble recognizing at high speeds.

Chapter 3: Voice Writing Theory

Table 36

Brief Forms Lettering Technique	
Phrase	**Voice Code**
; is that correct [1]	ITC
, you know, [1]	YK
thank you	TK

1) If your CAT program has a dictionary and AI capabilities, you can make entries for these phrases in your CAT dictionary so your CAT program will render different punctuation results according to different contexts in which these phrases may occur. For example, "ITC" could result in "Is that correct," if it begins a sentence ", is that correct," in the middle of a sentence, or "; is that correct?" at the end of a sentence. Likewise, "YK" could result in either "You know," ", you know," or ", you know."

Now let's review the first group of sample brief forms. "ITC" represents each letter in a common phrase ("is that correct"). If you want to produce a common phrase without punctuation, you would ordinarily just dictate the regular words ("is that correct"). But if you want to produce a phrase with punctuation, you could dictate a brief form ("ITC") as a means of differentiation. "YK" is convenient to use during high-speed dictation because people say "you know" so often during ordinary conversation. Constantly dictating punctuation before and after those words in the middle of sentences drains your reporting stamina. You'll also notice when dictating at high speeds that the words "thank you" often sound similar to the words "think you." If Dragon produces the words "think you" instead of "thank you," you would have the ability to say "TK" instead to achieve a correct translation.

Table 37 provides other examples of brief forms that may be created by combining word parts.

Table 37

Brief Forms Ad-Hoc Combining Approach	
Phrase	**Voice Code**
ladies and gentlemen	laizgent
if, for any reason,	ifrez
any way, shape, or form	wayshorm
in your opinion and in my opinion	inyop and inmop
beyond a reasonable degree of	bayrez
reasonable degree of medical	rezdegmed

77

Voice Writing Method • FIFTH EDITION
Dragon NaturallySpeaking® 12

Table 37 (Continued)

Brief Forms
Ad-Hoc Combining Approach

Phrase	Voice Code
to the best of your knowledge and to the best of my knowledge	boyledge and bomledge
do you actually recall	doyactal
let me ask you	laskou
for the record	forec
state your name	stairm
state your name for the record	stairmrec
let the record reflect	lerft
would you please	woplez
no further questions	noferquest

Let's examine the design of the voice codes for the briefs in the above table by extracting their parts for analysis:

laisgent	=	ladies (**laiz**) and gentlemen (**gent**)
ifrez	=	if (**if**), for any reason (**rez**)
wayshorm	=	any way (**way**), shape, or form (**shorm**)
inyop	=	in (**in**) your (**y**) opinion (**op**)
inmop	=	in (**in**) my (**m**) opinion (**op**)
bayrez	=	beyond (**bay**) a reasonable (**rez**) degree of
rezdegmed	=	reasonable (**rez**) degree (**deg**) of medical (**med**)
boyledge	=	to the best (**b**) of (**o**) your (**y**) knowledge (**ledge**)
bomledge	=	to the best (**b**) of (**o**) my (**m**) knowledge (**ledge**)
doyactal	=	do (**do**) you (**y**) actually (**act**) recall (**al**)
laskou	=	let (**l**) me ask (**ask**) you (**ou**)
forec	=	for (**fo**) the record (**rec**)
stairm	=	state (**stai**) your (**r**) name (**m**)
stairmrec	=	state (**stai**) your (**r**) name (**m**) for the record (**rec**)
lerft	=	let (**le**) the record (**r**) reflect (**ft**)
woplez	=	would (**wo**) you please (**plez**)
noferquest	=	no (**no**) further (**fer**) questions (**quest**)

This system of voice theory development will lead to a consistent pattern in your thinking to help you remember your brief forms quickly during realtime. Once you have identified which word parts you would like to represent whole words, you can extend your base phrases and blurbs to include more words which combine their parts together in different ways.

Just as *med* may be incorporated into a variety of briefs containing the word "medical," *rez*, *cert*, and *prob* may be used within briefs to indicate "reason" or "reasonable," "certainty" or "certification," and "probable" and "probability." For example, "reasonable degree of medical certainty" can become *rezdegmedcert* and "reasonable degree of medical probability" can become *rezdegmedprob*. Extending this a bit further, "beyond a reasonable degree of medical certainty" could be dictated *bayrezmedcert*. These ideas can be applied to just about any common phrase you can think of. It's easy to remember how to spell brief forms when word-part spellings are consistent. But when the number of words in a group get too large, you may not always be able to keep certain word parts if trying to save on syllabic density. Use your best judgment to determine whether it is smarter to keep all syllables for consistency's sake or take a shorter dictation route by removing certain syllables altogether.

In conclusion, it is recommended that you limit your designation of brief forms to only those you will benefit from most often in your dictation. You do not want to create unnecessary conflicts for yourself—in an effort to make your job easier—only to end up with the numerous briefs you've created actually hindering your accuracy. Heavy usage of too many briefs may cause Dragon to misrecognize them as regular English words instead, due to the weight of the grammatical model's influence in placing regular English words in higher-selection priority due to their more frequent occurrences in common language contexts.

On-the-Fly Translations

The only words that are possible for Dragon to recognize from your dictation are words that exist in its vocabulary. If you dictate a word that isn't in its vocabulary, Dragon will end up recognizing it as one or more of the words from its existing vocabulary that most closely matches your word's pronunciation. As you already know, Dragon's base vocabulary contains thousands of the most common words in the English language, so the majority of the words you speak already exist in Dragon's vocabulary, and if you have a properly trained voice model, most of them will usually be recognized.

But what happens when you encounter a word that you are uncertain about whether or not to dictate because you're not sure if it is a part of Dragon's vocabulary? You can dictate a general, descriptive word or term in place of that specific word or term. This way, you do not have to worry about the risk of producing an error in your translation, and it will make your realtime record more readable in the process.

Here is an example of one such situation:

Example of a Newly-Encountered Word (Scenario 1)

    Q Who was the bill from?
    A **Cingular Wireless.**

Example of What You Could Say Instead (Scenario 1)

Q Who was the bill from?
A **A cell phone company.**

Now, that's a great way to make a transcript more readable, but suppose you want to be more specific, as in the following example:

Example of a Newly-Encountered Word (Scenario 2)

Q What is the name of the doctor who performed your exam?
A **Dr. Zugawuga.**

You obviously will not be able to dictate an unusual proper name like "Zugawuga" and have it translate correctly unless you have first entered it into Dragon's vocabulary. And you do not want to use a generic term (e.g., "a doctor") in lieu of the actual doctor's name, because the question is specific as to what the doctor's name is.

When you encounter a word that does not exist in Dragon's vocabulary and you need to have a method of translating it anyway, you can insert it on the fly if you have CAT software. To force-produce any word or group of words that do not exist in Dragon's vocabulary while you are voice writing in realtime, you would use generic translation voice codes that will generate, at whim, whatever text you desire. This is referred to as the **on-the-fly translations technique**, where you dictate code words (*tran-1*, *tran-2*, *tran-3*, etc.) that your CAT program converts to your custom words. Performing an on-the-fly translation usually involves typing your custom text into a translation field and dictating the corresponding voice code. The custom text is then output onto your screen and also sent in a realtime feed. You CAT program's translation utility is usually kept visible on your screen so you can easily keep track of your custom text entries as you dictate. If you have CAT software, refer to your user's manual to learn how to operate this feature. If you do not have CAT software, you will not be able to perform on-the-fly translations.

To produce" Dr. Zugawuga," for example, you would type it into one of your CAT program's generic translation fields (e.g., field 1) and translate it as text by simply saying *tran-1*. Thereafter, each time that name is encountered during your realtime session, you would continue to dictate *tran-1* to produce "Dr. Zugawuga."

Example of What You Could Say Instead (Scenario 2)

Q What is the name of the doctor who performed your exam?
A **tran-1.**

The on-the-fly translations technique is also handy for force-producing words that pose persistent conflicts with other words and for being able to discriminate between selective spellings of identical-sounding words (e.g. "green," "Green," and "Greene"). If your CAT program allows you to create specialized job dictionaries and/or provides the capability to perform different types of globaling functions, you may want to also consider using those features for translation needs.

Table 38 lists the voice codes you can use for on-the-fly translations. Note that the number of allowable translations may vary with different CAT programs.

Table 38

On-the-Fly Translations

Generic Translation Field	Voice Code	Text, Formatting, and Use
(field 1)	tran-1	During dictation, these voice codes are converted to whatever custom text you specify.
(field 2)	tran-2	
(field 3)	tran-3	
(field 4)	tran-4	
(field 5)	tran-5	
(field 6)	tran-6	
(field 7)	tran-7	
(field 8)	tran-8	
(field 9)	tran-9	

Delete Command

If you made a mistake in your dictation and would like to delete the last text that was produced, Dragon will allow you to say a command like *scratch that*. But if you are using a CAT program, this function may not be available because most CAT programs disable many Dragon commands. This is to ensure that speech input intended to produce text is never misrecognized for speech input to produce commands. It is helpful to have Dragon's standard delete command disabled; otherwise, if you said *scratch that* and wanted to produce those actual words, Dragon may end up deleting the last dictated text even when that wasn't your intention.

If you have the option of using a special delete command in your CAT software, you can program the delete function to respond to the voice code "del-mac." If you find it regularly beneficial to delete more than one word at a time, you should take this delete command a step further by using a series of deletion voice codes (i.e., *del-1, del-2, del-3*, etc.). It is easiest to use a numbering system for identifying the number of words to delete when having to quickly remember your voice codes in the midst of dictation. (See Table 39.)

A delete command is very useful for removing text that was produced as the result of a dictation error. Even when Dragon misrecognizes dictation that is perfectly clear—many times due to the word not existing within Dragon's vocabulary—using a delete command will remove the error and allow you the opportunity to insert either the correct text or other descriptive word(s) (such as the cell phone example) as an on-the-fly translation. In fact, using a delete command will help you produce an on-the-fly translation faster when it is dictated simultaneously as you prepare the translation.

Table 39

Deleting Previously Translated Text by Voice

Command	Voice Code
delete last word	del-mac or del-1
delete last two words	del-2
delete last three words	del-3
delete last four words	del-4
delete last five words	del-5

Part 3

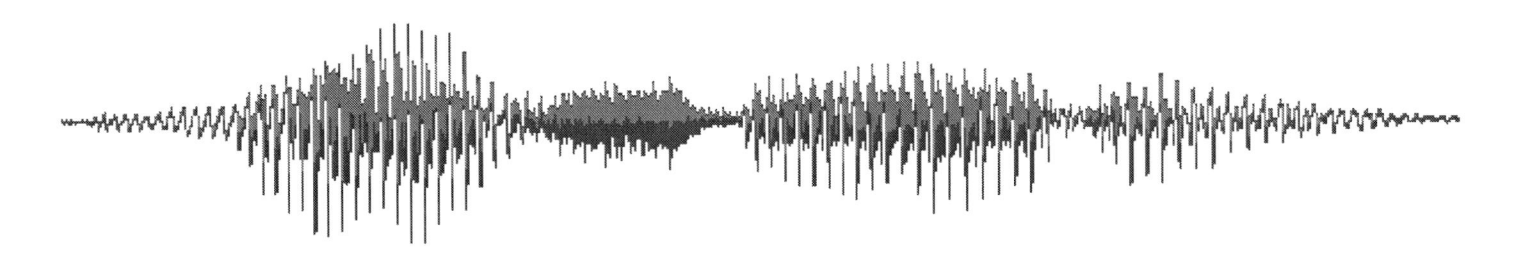

Getting Set Up and Doing Realtime

Chapter 4:
Getting Started

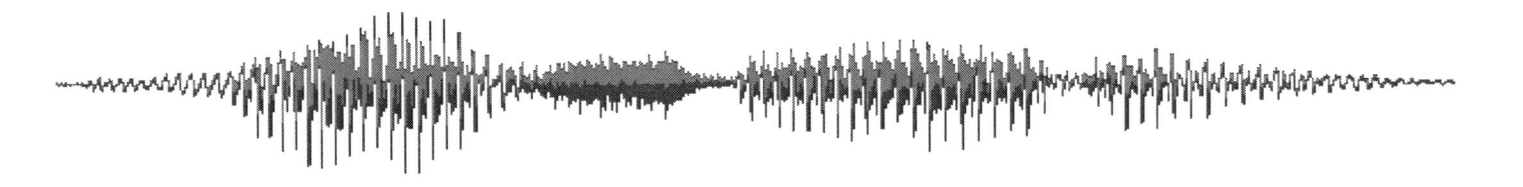

This chapter takes you through the steps of setting up your equipment and Dragon NaturallySpeaking® software for realtime voice writing. Dragon setup work involves creating a user and setting user options. Once complete, you will do a practice dictation session, correct any speech recognition errors encountered, and learn how to save, close, and re-open Dragon files.

You must have Dragon NaturallySpeaking® software installed before proceeding further.

Computer and Equipment Setup

Follow the instructions below to set up your computer and other peripheral devices, then be sure to perform the steps in the next two subsections to properly prepare your computer for uninterrupted realtime performance.

Step 1

Plug the cord of your surge-protection power strip into an electrical outlet.

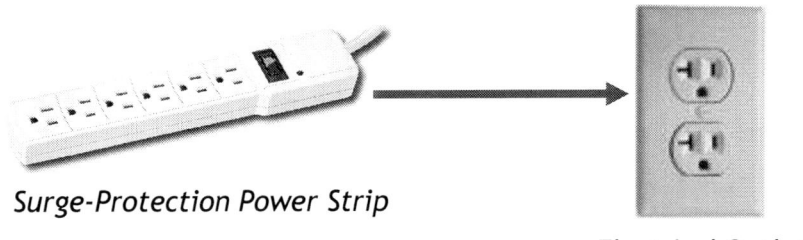

Surge-Protection Power Strip

Electrical Outlet

Step 2

Connect one end of your computer's power cord to your computer and plug the other end into a surge-protection power strip.

Computer   *Power Cord*   *Surge-Protection Power Strip*

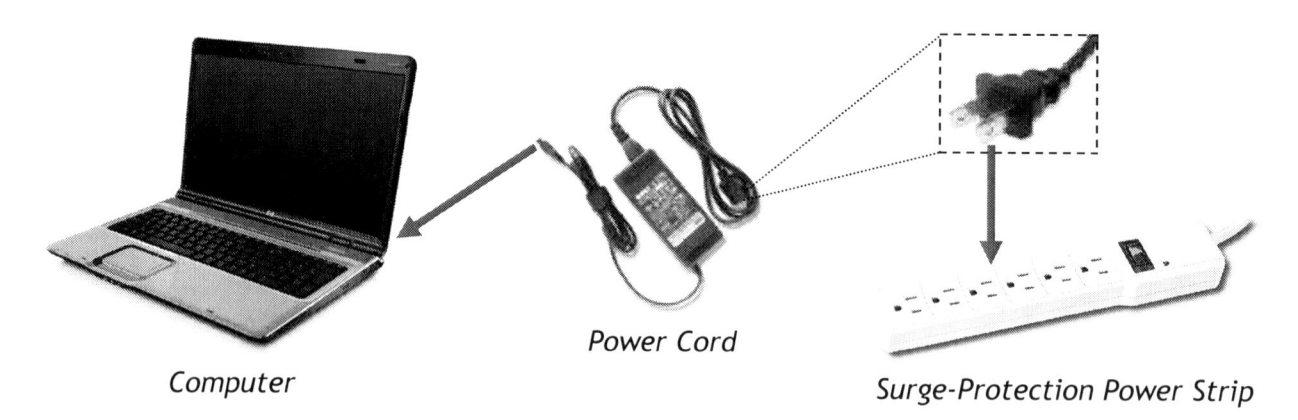

(Note: It is best not to connect your computer directly to an electrical outlet. When your computer receives electricity through a surge-protection power strip, damage to your computer will be prevented in the event power outages and electrical shortages occur.)

Step 3

Turn your computer on and wait for the boot-up process to be fully complete before moving on to the next step.

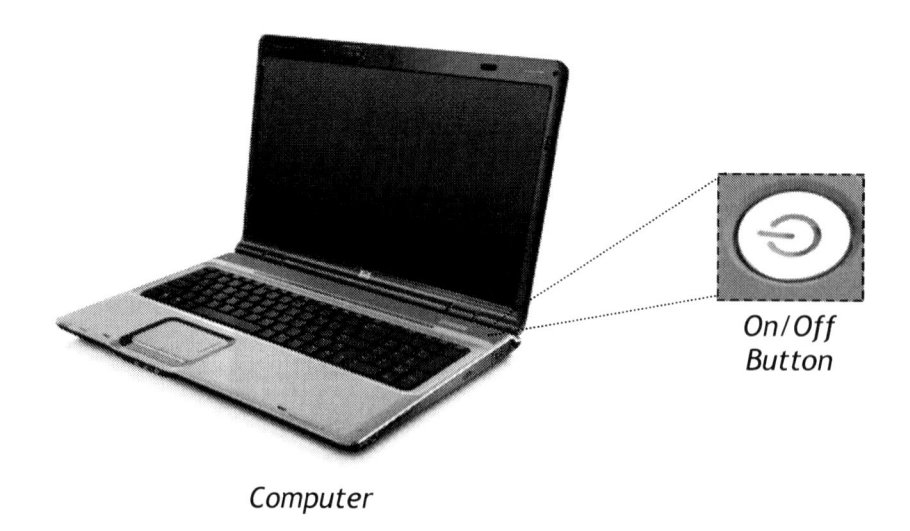

On/Off Button

Computer

Chapter 4: Getting Started

Step 4

Insert the microphone plug of either your open-mic headset or speech silencer—whichever device you will be using—into the hole next to the picture of the microphone on the USB sound card.

Open-Mic Headset

Microphone Plug

Speech Silencer

USB Sound Card (back)

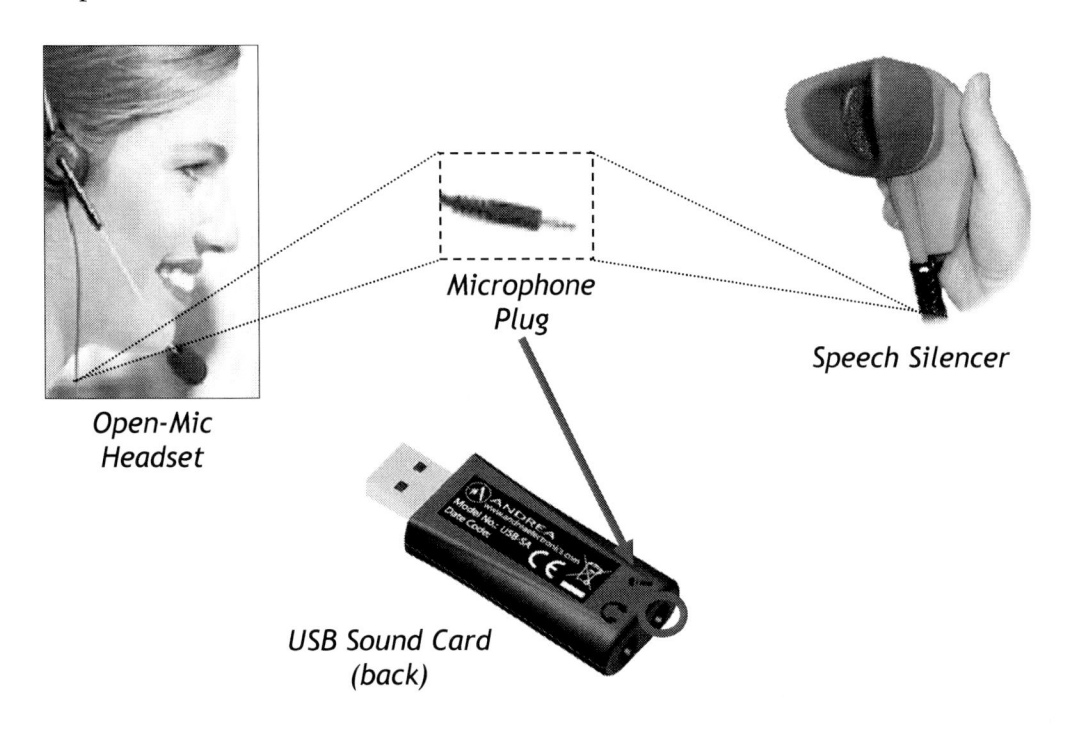

Step 5

Plug the USB sound card into your computer through a dedicated USB port.

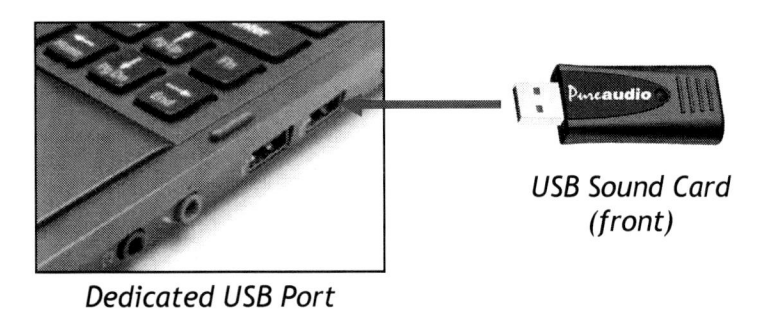

Dedicated USB Port

USB Sound Card (front)

(Note: Designate <u>one USB port</u> for your USB sound card. Your computer tends to route itself to the port most frequently used by this device. This means in the future you should <u>always plug your USB sound card into the same port</u> you're plugging it into now.)

87

Voice Writing Method • FIFTH EDITION
Dragon NaturallySpeaking® 12

USB Connectivity Settings

To ensure that the computer does not periodically disable USB connections in an effort to conserve resources during intensive speech-recognition processing tasks, perform the steps below to set your USB hubs' power management settings for continuous operation.

(Note: These steps may vary with different operating systems. For additional help in locating Windows' screens related to USB connections, refer to your Windows Help menu.)

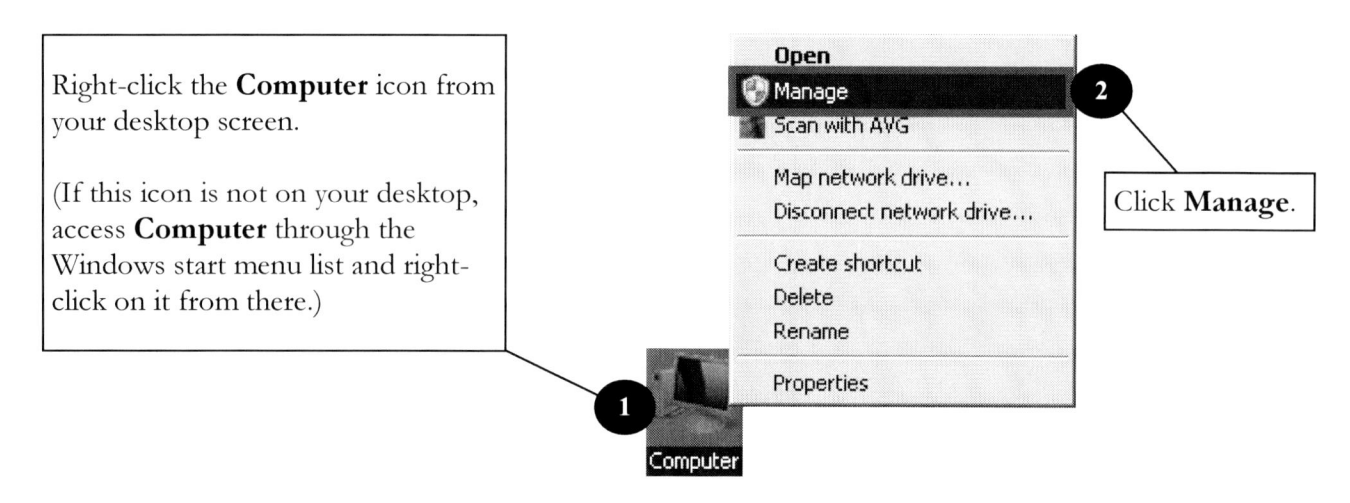

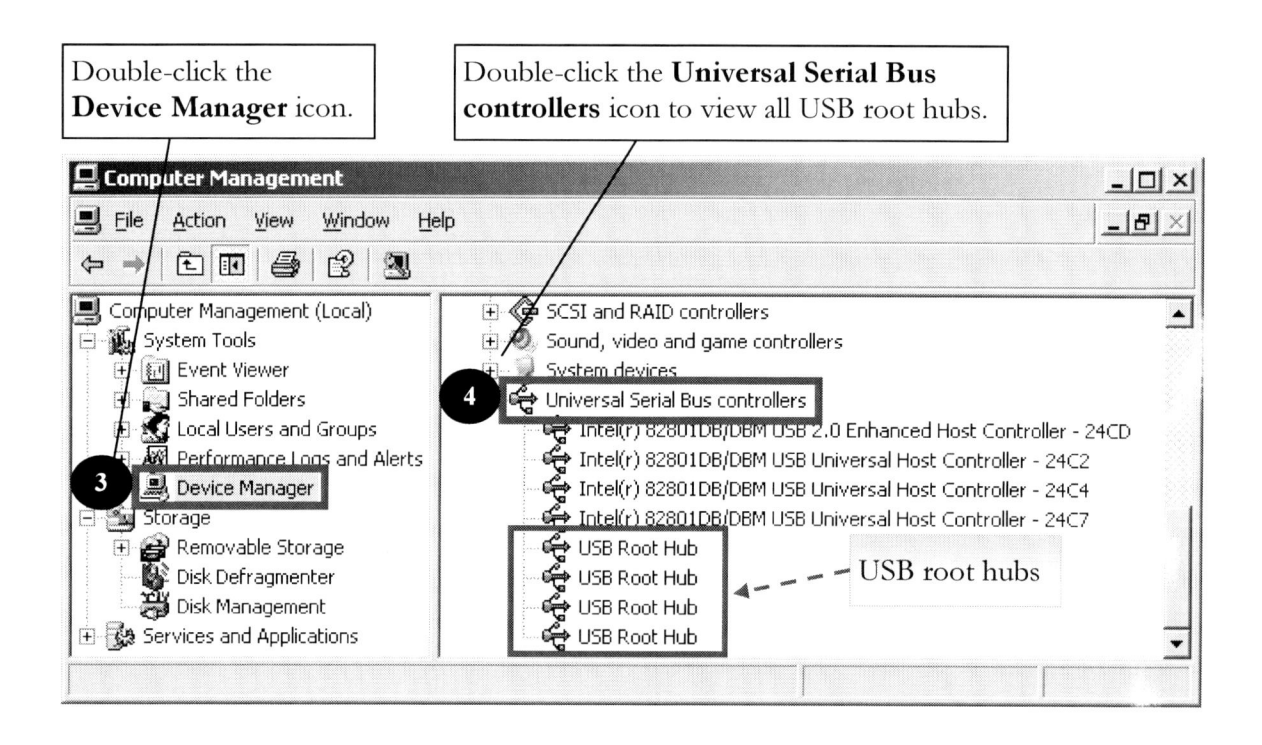

Right-click the **Computer** icon from your desktop screen.

(If this icon is not on your desktop, access **Computer** through the Windows start menu list and right-click on it from there.)

Click **Manage**.

Double-click the **Device Manager** icon.

Double-click the **Universal Serial Bus controllers** icon to view all USB root hubs.

USB root hubs

88

Chapter 4: Getting Started

Double-click on the first **USB Root Hub** icon. — 5 — USB Root Hub / USB Root Hub / USB Root Hub / USB Root Hub

6 — Go to the **Power Management** tab.

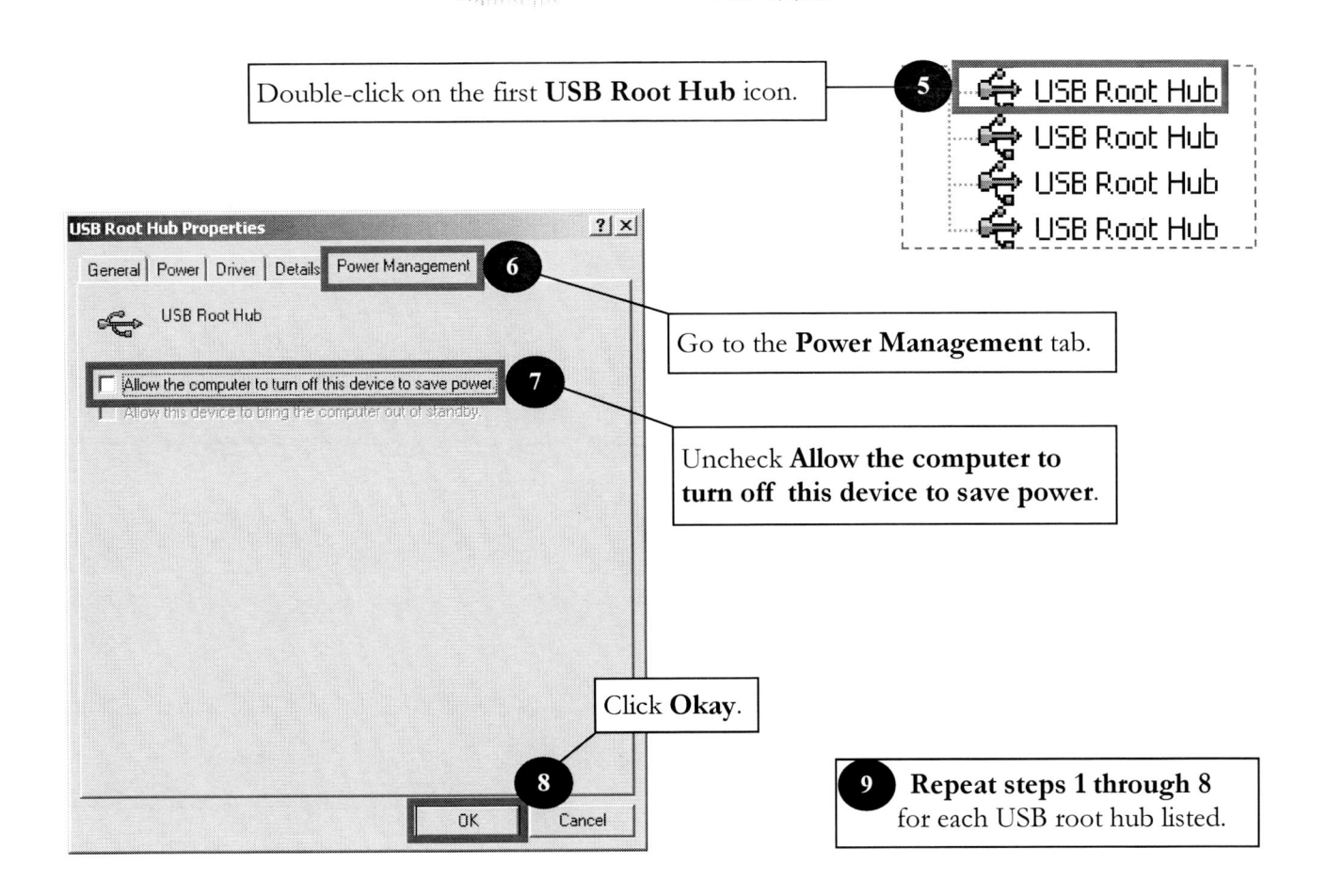

7 — Uncheck **Allow the computer to turn off this device to save power**.

8 — Click **Okay**.

9 — **Repeat steps 1 through 8** for each USB root hub listed.

Turning Off Automatic Updates and Disabling Anti-Virus Software

Prior to using your computer for realtime, you should turn off Windows Automatic Updates and disable anti-virus software. Removing these tasks will free up memory for optimum speech recognition performance, prevent software updates from being installed, and eliminate the inconvenience of automatic virus scans occurring while you are in the middle of dictation sessions. Anti-virus software may even interfere with the proper operation of Dragon by blocking certain program functions. You may turn Automatic Updates back on and re-enable your anti-virus software when you intend to use your computer for any general computing tasks unrelated to speech recognition.

Steps for turning off and disabling these applications will vary depending on the version of Windows that is installed on your computer and the type of anti-virus software you are using. For assistance in locating Windows Automatic Updates, refer to your Windows Help menu. An icon for your anti-virus software is usually displayed in the System Tray of your computer's desktop screen (located next to the time display). To disable it, right-click on the program icon, then click Close or Exit.

Other functions you should disable on your computer are screen savers, sleep mode settings, and power options that automatically shut down processes to conserve on resources.

Dragon NaturallySpeaking® Setup

You are about to be guided through the steps of setting up a user in Dragon to create a voice model through a process called **general training**, where you will read stories aloud so Dragon can learn the way you speak.

Dragon contains a variety of stories for you to choose from, but you do not have to read all of them. The recommended time for completing the general training process is a minimum of five minutes, but no more than 15. Performing more than 15 minutes' worth of initial training will not significantly increase your recognition accuracy. The remaining voice model improvement work will be accomplished by correcting misrecognitions and vocabulary building, topics which will be covered later.

When working in a voice model, always use the same dictation microphone (open-mic headset or speech silencer) and USB sound card combination that was originally used to create it. If you decide to use different or newer dictation input devices in the future, you must create a new voice model with your new combination of components to obtain the highest accuracy results. Design qualities among different devices vary, and this configuration must be kept consistent at all times so your accuracy results will be consistent.

As you perform general voice training, speak clearly and evenly, but do not run words together by tying in endings to beginnings through slurred speech, and do not chop them up either by pausing unnaturally between each. Dragon is not just analyzing how you say words and word parts, but also how you say them in connection with other words and word parts.

Use a normal pace in the beginning. Dragon may get hung up on words as you read the first story, but it will more readily accept your speech input at faster speeds as you progress. If you are prompted to repeat a word more than once, say the word once, then pause for Dragon to accept it before reading further. Should you inadvertently say a word the wrong way (e.g., "the" when you should have said "that") and Dragon accepts your speech input without prompting you to repeat the word, the erroneous pronunciation will become a permanent part of your voice model and corrupt the system's judgment during future speech-to-text conversions of other words which contain the same phonemes of the word you mispronounced. So if this type of mistake occurs while performing general training, it is important to click the **Redo** button and re-speak that section.

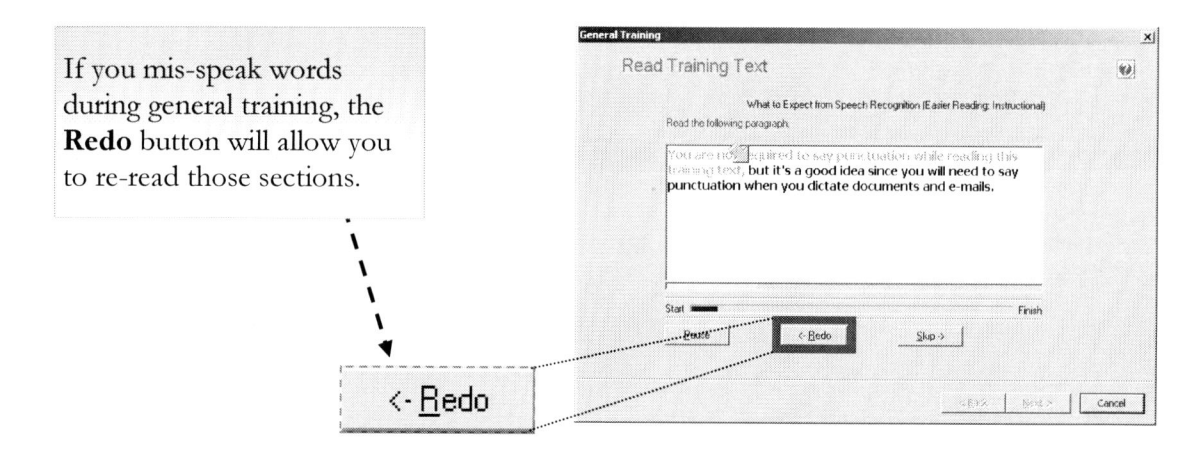

If you mis-speak words during general training, the **Redo** button will allow you to re-read those sections.

Chapter 4: Getting Started

Dragon incrementally accepts your speech input at higher speeds, continuously adapting your voice model as it acquires more acoustic data. Since you will be dictating at realtime speeds—between 180 to 200 words per minute—when providing realtime services, it would be best for you to perform your general training within this same range in order for Dragon to know what rate of speech to expect from you in the future.

To determine the realtime dictation range for any given general voice training story without having to guess at word-per-minute reading speeds, it is recommended that you open a Dragon voice training story in SpeedMaster™ and practice dictating it before you actually begin performing general training in Dragon.

For SpeedMaster™ installation instructions, see the *How to Install* subsection under the *SpeedMaster™ Software* section of Chapter 7. All of Dragon's basic general training stories may be found in the *SpeedMaster* folder on your computer's hard drive (*C:\Program Files\Speedmaster*). You can set the text display to 180 and make incremental increases up through 200 words per minute as you move from one story to the next.

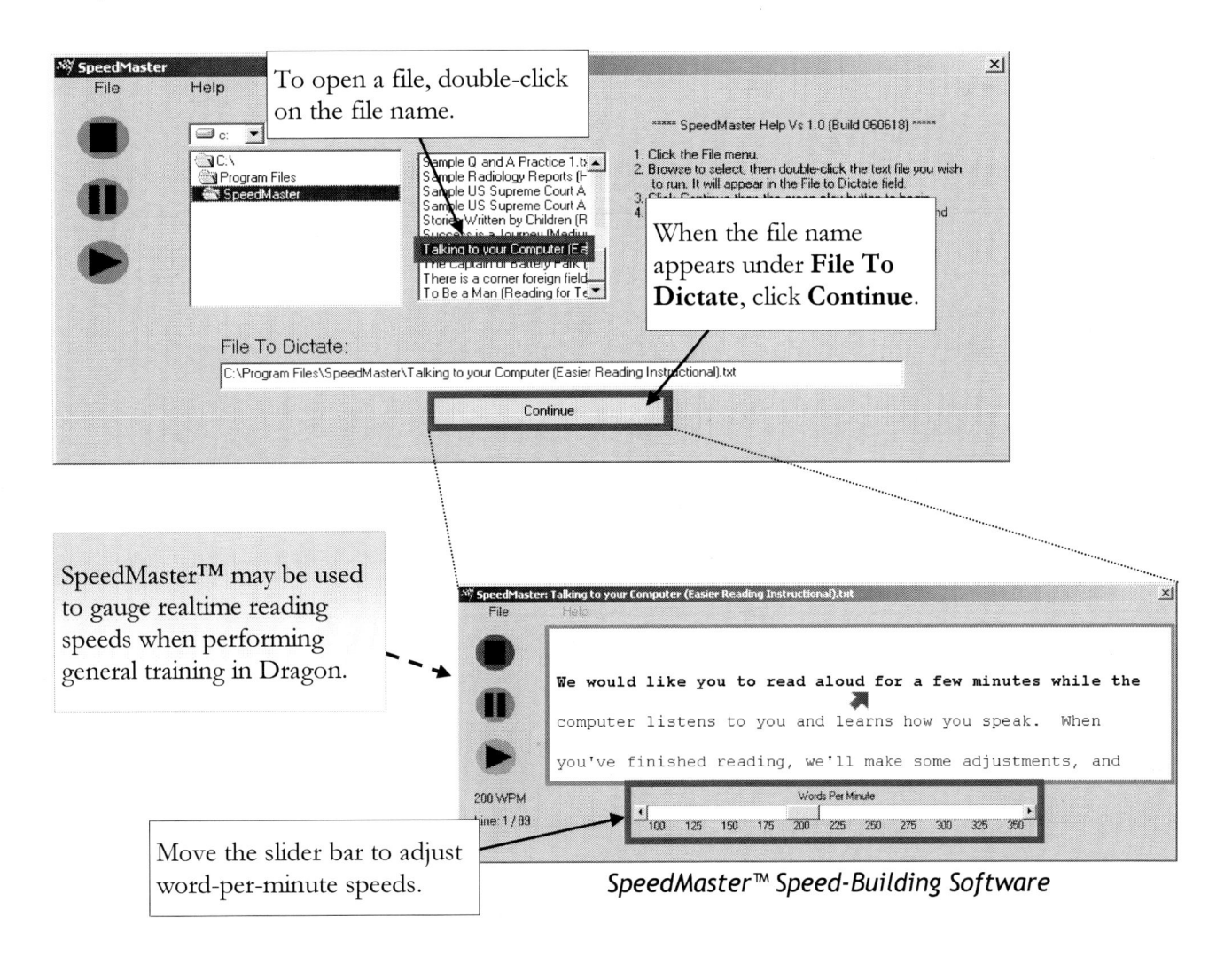

To open a file, double-click on the file name.

When the file name appears under **File To Dictate**, click **Continue**.

SpeedMaster™ may be used to gauge realtime reading speeds when performing general training in Dragon.

Move the slider bar to adjust word-per-minute speeds.

SpeedMaster™ Speed-Building Software

Once you have completed general voice training, set your user options and back up your user files. Store your backup onto your computer's hard drive for easy retrieval in case of file corruption, but also be sure to export your user files to an external storage device—such as a backup hard drive, flash drive, or CD—in the event of critical damage to your computer. Preserving a backup or exported copy of your user files will save you the effort of having to reproduce all your prior work from scratch. For instructions, go to the *Backing Up/Restoring User Files* and *Exporting/Importing User Files* subsections under the *User File Management* section of Chapter 8.

Summary of Steps to Complete When Creating a User in Dragon NaturallySpeaking®

Step 1

Speaking at realtime speed levels (180 to 200 words per minute), perform approximately fifteen minutes' worth of general voice training, or simply read three stories. If you mis-speak and Dragon accepts your speech input for a portion of text that you incorrectly pronounced, click the *Redo* button and re-speak that section. Once you have successfully completed general voice training, move on to Step 2.

Step 2

Follow the steps in the *Setting User Options* section of this chapter, then go to Chapter 8 to back up and export your user files.

One final note: After you have completed general training and begun dictating and performing corrections to misrecognitions, it is not recommended that you perform further general training under common circumstances. However, if your voice changes due to a common cold or upper-respiratory infection in the future, you may find temporary benefit from performing additional general training—just for adjusting your voice model to those differences at the time—being sure that you back up your user files before doing additional training so you can restore your original user files later when your vocal qualities return to normal.

Now let's get started!

Chapter 4: Getting Started

Creating a User

First, make sure your dictation input devices are properly connected and that the USB sound card is plugged into one dedicated USB port of your computer (remember to use the same USB port every time). Then go through the Start menu and select **Dragon NaturallySpeaking 12.0** from the Programs list, or double-click the **Dragon NaturallySpeaking 12.0** icon on your desktop screen to enter Dragon.

Open **Dragon**.

If you are entering Dragon for the first time, you will automatically be guided through the process of creating a user. Steps for setting up your user appear in the next section.

If this is not your first time working in Dragon and you've previously created a user, go through the following menus to create a new one: **Profile** menu > **New User Profile** > **New** button.

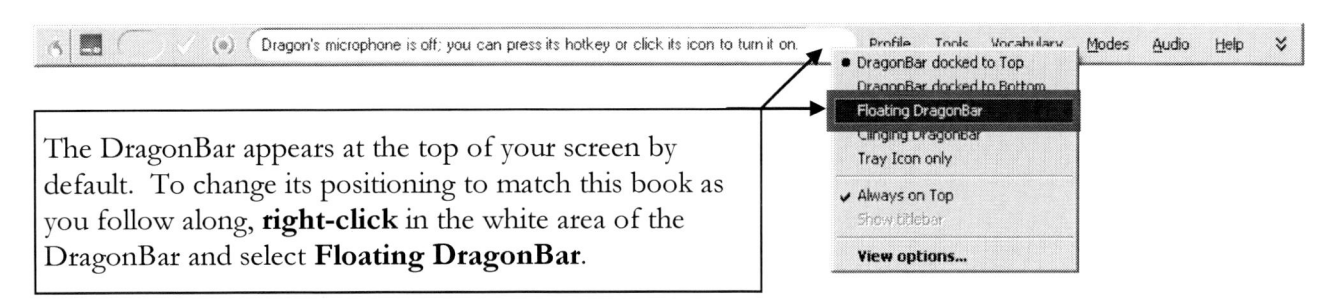

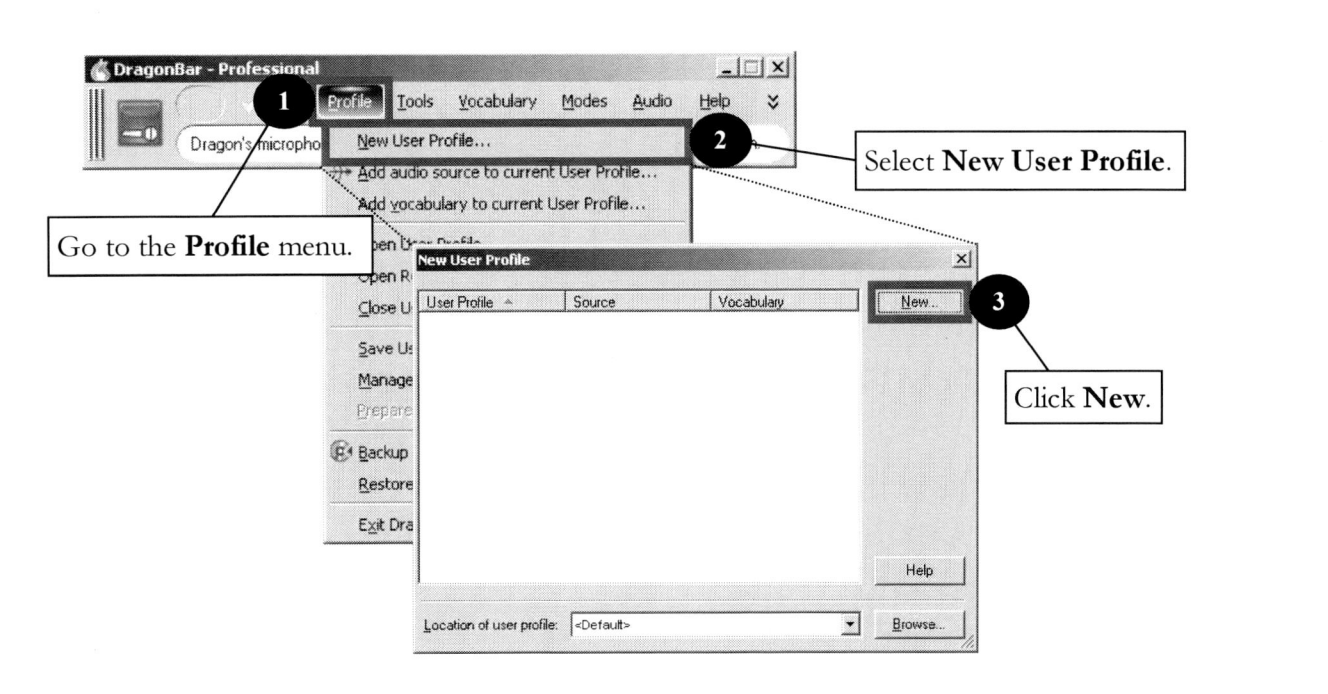

The DragonBar appears at the top of your screen by default. To change its positioning to match this book as you follow along, **right-click** in the white area of the DragonBar and select **Floating DragonBar**.

1. Go to the **Profile** menu.
2. Select **New User Profile**.
3. Click **New**.

93

Voice Writing Method • FIFTH EDITION
Dragon NaturallySpeaking® 12

Begin Creating a User Profile

Click **Next**.

Naming Your User Profile

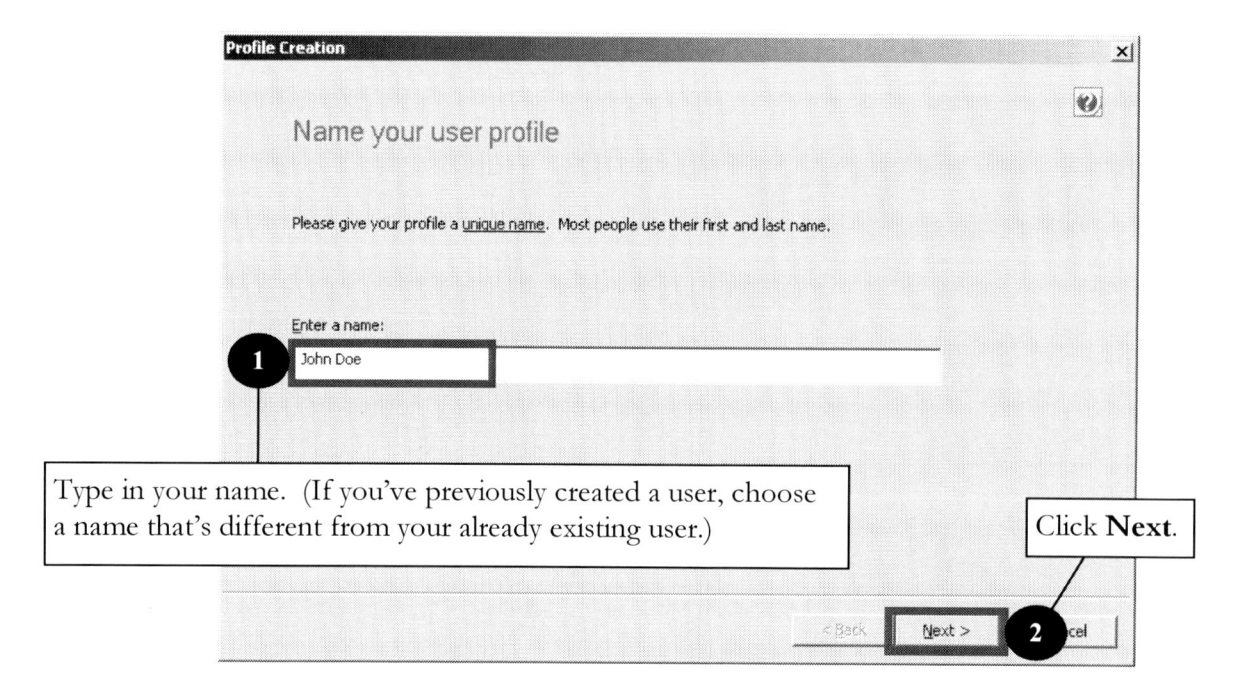

1. Type in your name. (If you've previously created a user, choose a name that's different from your already existing user.)

2. Click **Next**.

Chapter 4: Getting Started

Age Group Selection

(Screenshot: Profile Creation dialog — "Select your age group" with Age dropdown showing 22-54 selected among 13 or under, 14-21, 22-54, 55+, Prefer not to say.)

1. Choose your age group.
2. Click **Next**.

Selecting English Vocabulary Type

The option in the screen below determines the spellings of vocabulary words and language model (a.k.a. grammatical model) that will be applied to the user you are creating. For example, the formal written style of certain words in American English is different than in British English (e.g., "realize" versus "realise"), and so is context in many cases (e.g., "baseball" versus "fly ball").

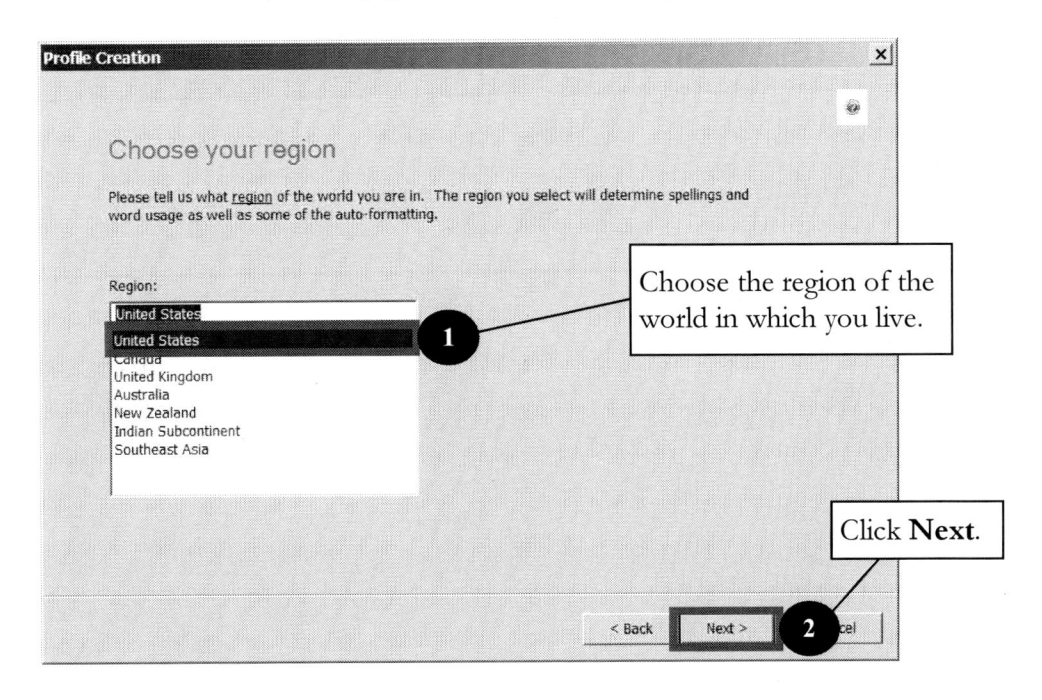

(Screenshot: Profile Creation dialog — "Choose your region" with Region dropdown showing United States selected among United States, Canada, United Kingdom, Australia, New Zealand, Indian Subcontinent, Southeast Asia.)

1. Choose the region of the world in which you live.
2. Click **Next**.

95

Choosing an Accent

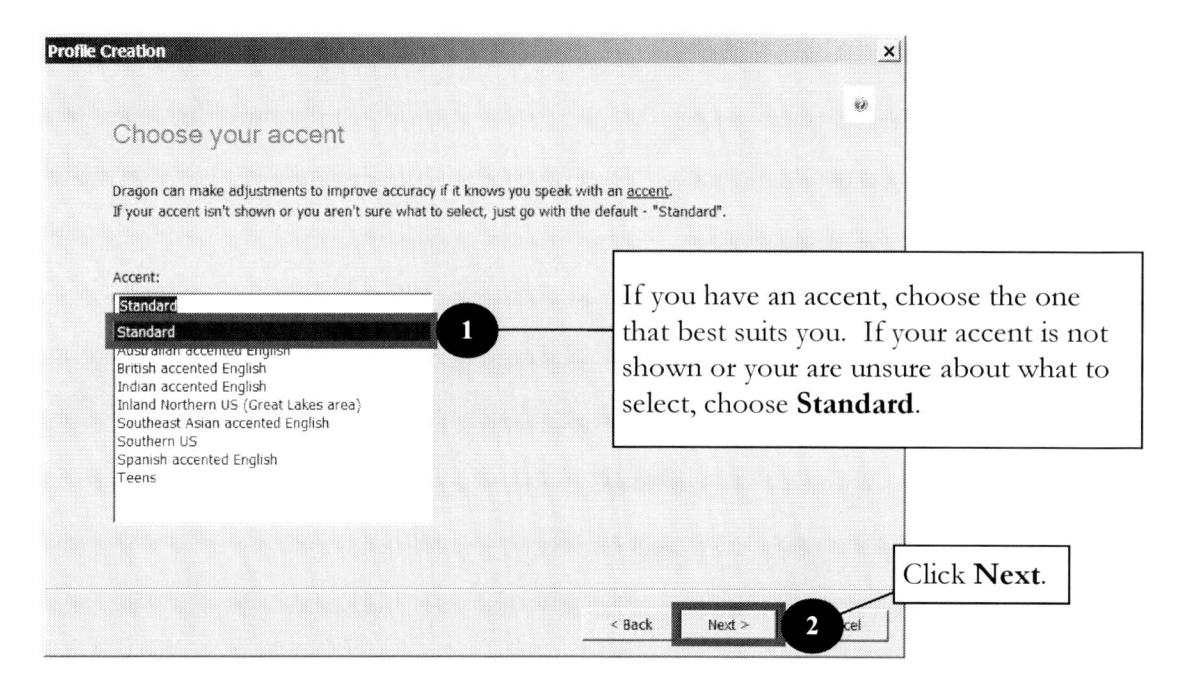

*If you have an accent, choose the one that best suits you. If your accent is not shown or your are unsure about what to select, choose **Standard**.*

*Click **Next**.*

Dictation Input Selection

(Note: Premium edition users will notice occasional differences in the availability of options, with fewer options present in this and other screens. Bear in mind that all NaturallySpeaking® images in this book were taken from the Professional edition, which, having additional features, show additional options. Regardless of whether you are a Premium or Professional edition user, the selections to make are still the same either way. So do not worry if your screens do not exactly match the images in this book.)

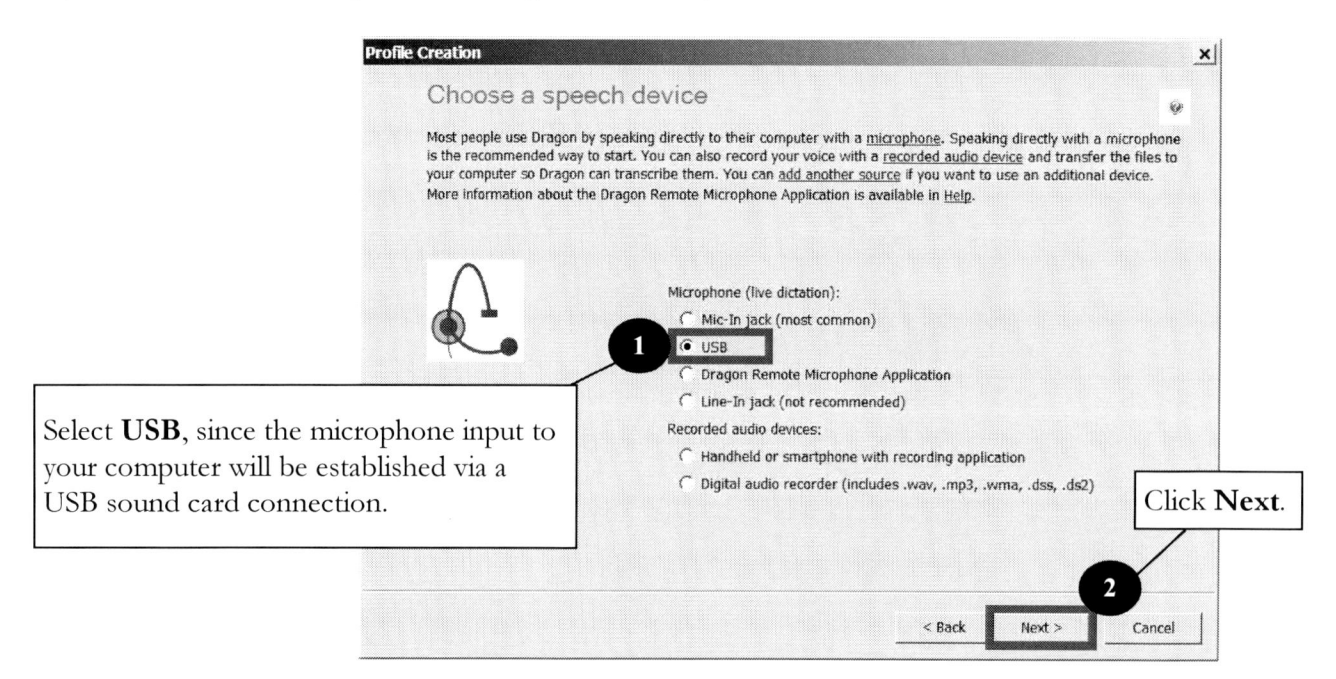

Select **USB**, since the microphone input to your computer will be established via a USB sound card connection.

Click **Next**.

Chapter 4: Getting Started

Finalizing Profile Creation

2 If your computer has less than 8GB RAM, select **BestMatch IV***.
If your computer has 8GB RAM or more, leave this at **BestMatch V***.

(*If you have an accent, be sure to choose the appropriate accent category when making this selection.)

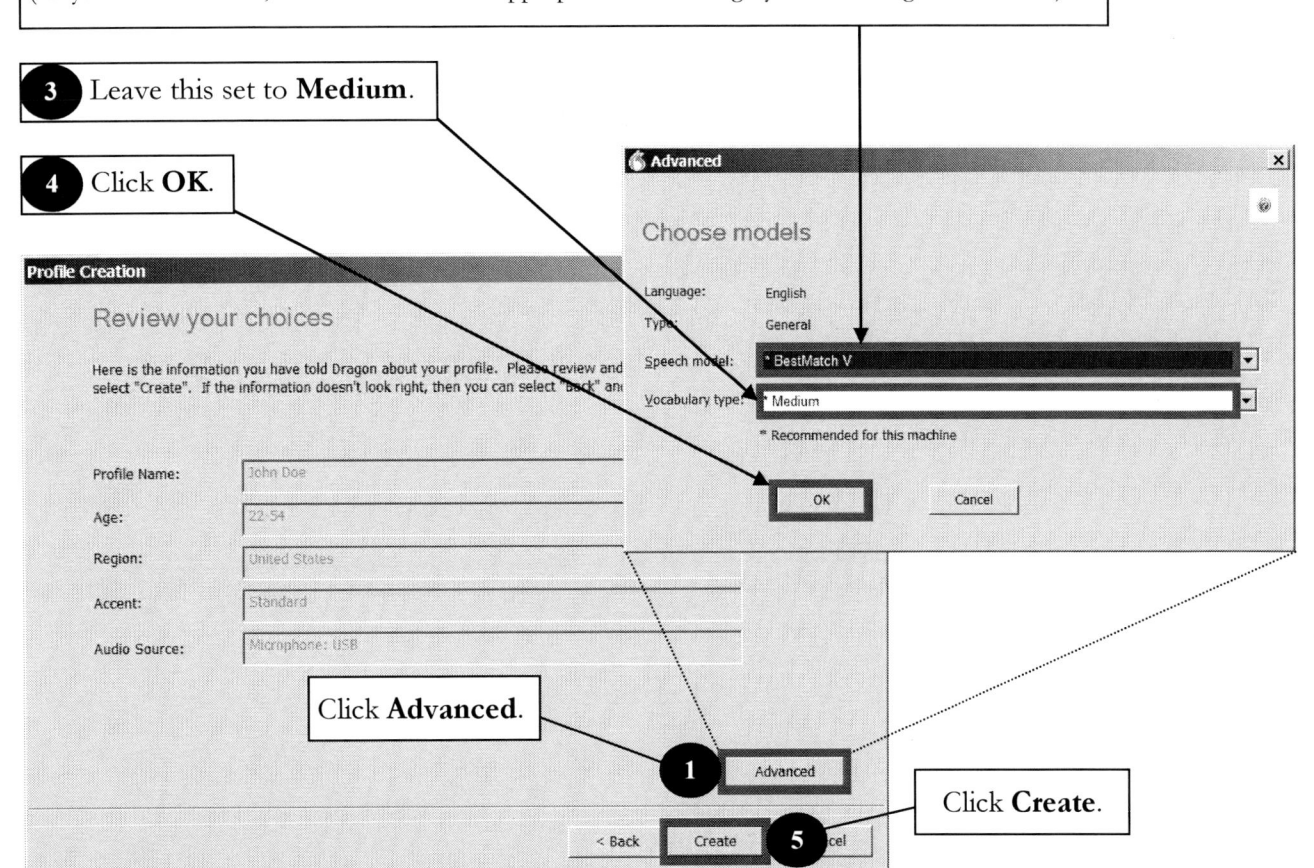

3 Leave this set to **Medium**.

4 Click **OK**.

Click **Advanced**.

1

Click **Create**.

5

Note About Speech Model Selection

BestMatch V is the newest speech model, introduced by Dragon 12, which offers more complex algorithms than in previous versions of Dragon. BestMatch V is automatically selected when your computer has a minimum of 4GB RAM, although a minimum of 8GB RAM is actually what is needed (with 16GB being recommended) for voice writers wanting to enjoy the accuracy boost this speech model offers.

Note About Vocabulary Types

Medium is the best vocabulary type choice for most voice writers, since using the *Large* vocabulary results in more speech recognition conflicts and requires more RAM. *Empty Dictation* is an option unique to the Professional edition and is for advanced users who intend to build their entire vocabulary from scratch.

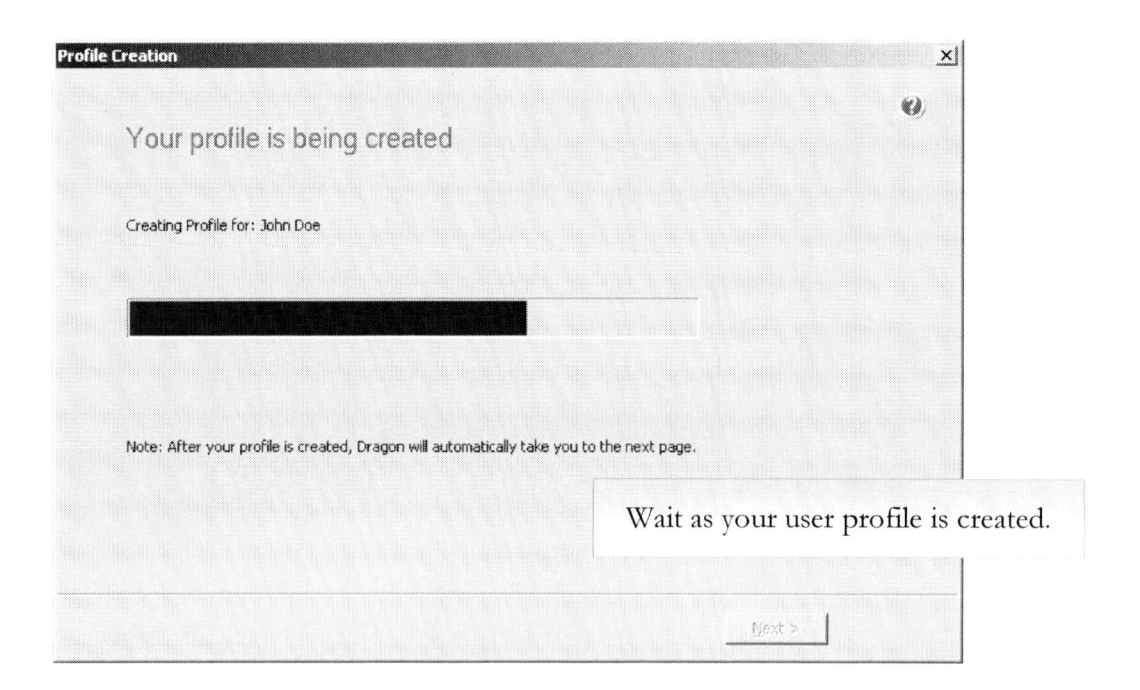

Voice Writing Method • FIFTH EDITION
Dragon NaturallySpeaking® 12

Wait as your user profile is created.

Positioning Your Mic-Input Device

If you are using an open-mic headset, position it at a slight angle to the side (not directly in front) of your mouth as shown below, and remember to keep it in that same position during dictation. If you are using a speech silencer, be sure when positioning it around your mouth that it forms a secure seal so the internal microphone will not get exposed to sounds in the outer environment during your dictation.

Open-Mic Headset Positioning    *Speech Silencer Positioning*

Chapter 4: Getting Started

If you have the Sylencer® model by Talk Technologies Inc. that contains a SmartMic™, follow the recommended settings for male and female voices shown in the diagram below using the tuning adjustment tool provided with your speech silencer purchase. These are only beginning setting recommendations. You can always change this setting if you find an adjustment is needed in the future. Be sure to read this product's user's manual for complete instructions.

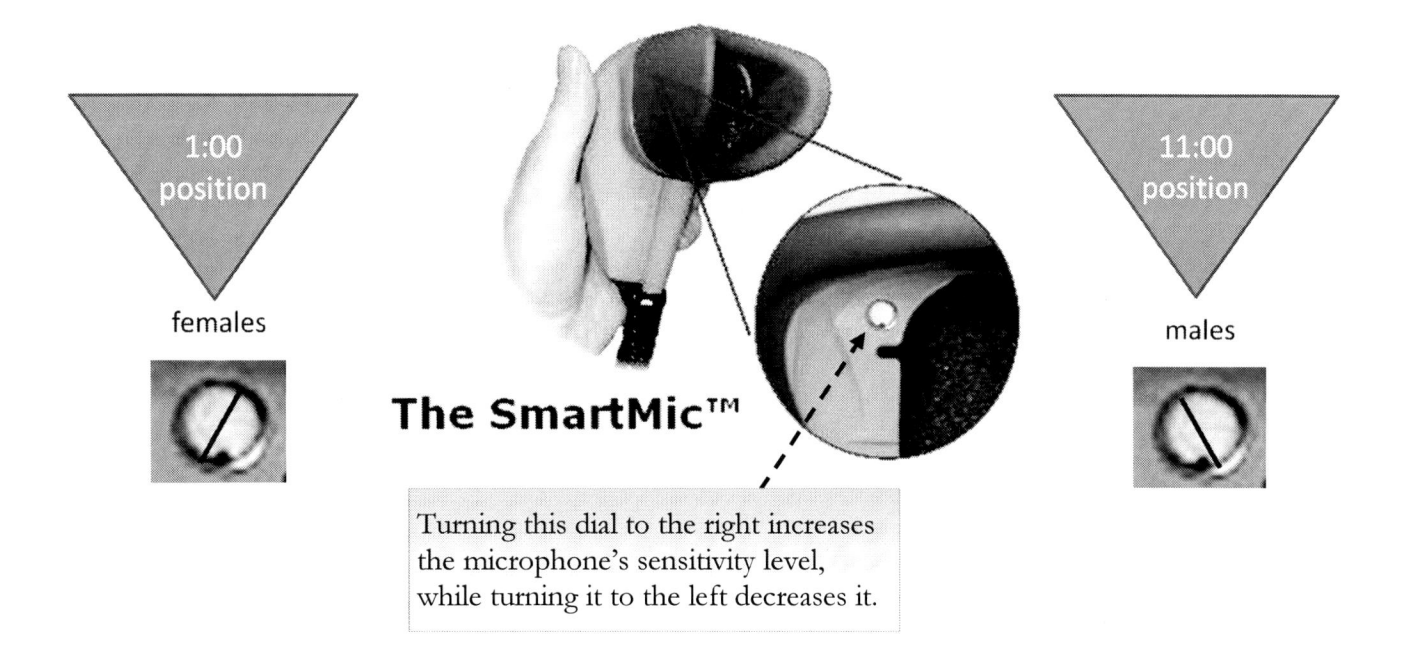

1:00 position — females

11:00 position — males

The SmartMic™

Turning this dial to the right increases the microphone's sensitivity level, while turning it to the left decreases it.

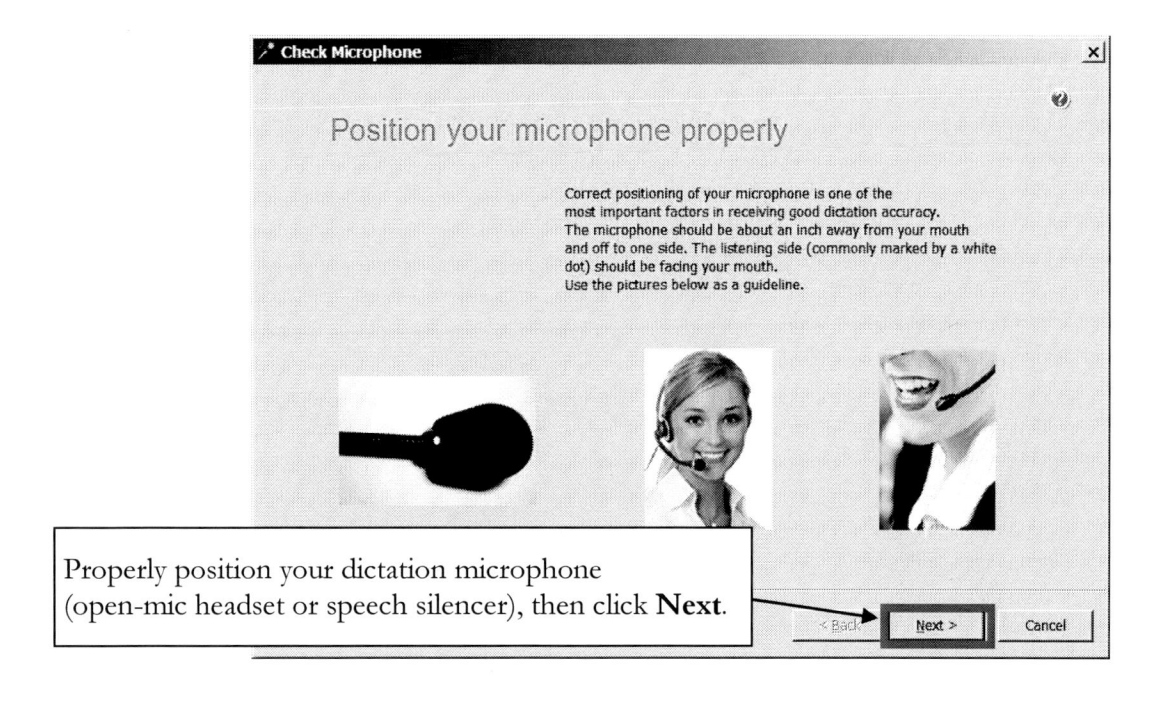

Check Microphone

Position your microphone properly

Correct positioning of your microphone is one of the most important factors in receiving good dictation accuracy. The microphone should be about an inch away from your mouth and off to one side. The listening side (commonly marked by a white dot) should be facing your mouth. Use the pictures below as a guideline.

Properly position your dictation microphone (open-mic headset or speech silencer), then click **Next**.

99

Voice Writing Method • FIFTH EDITION
Dragon NaturallySpeaking® 12

Adjusting Your Volume Level

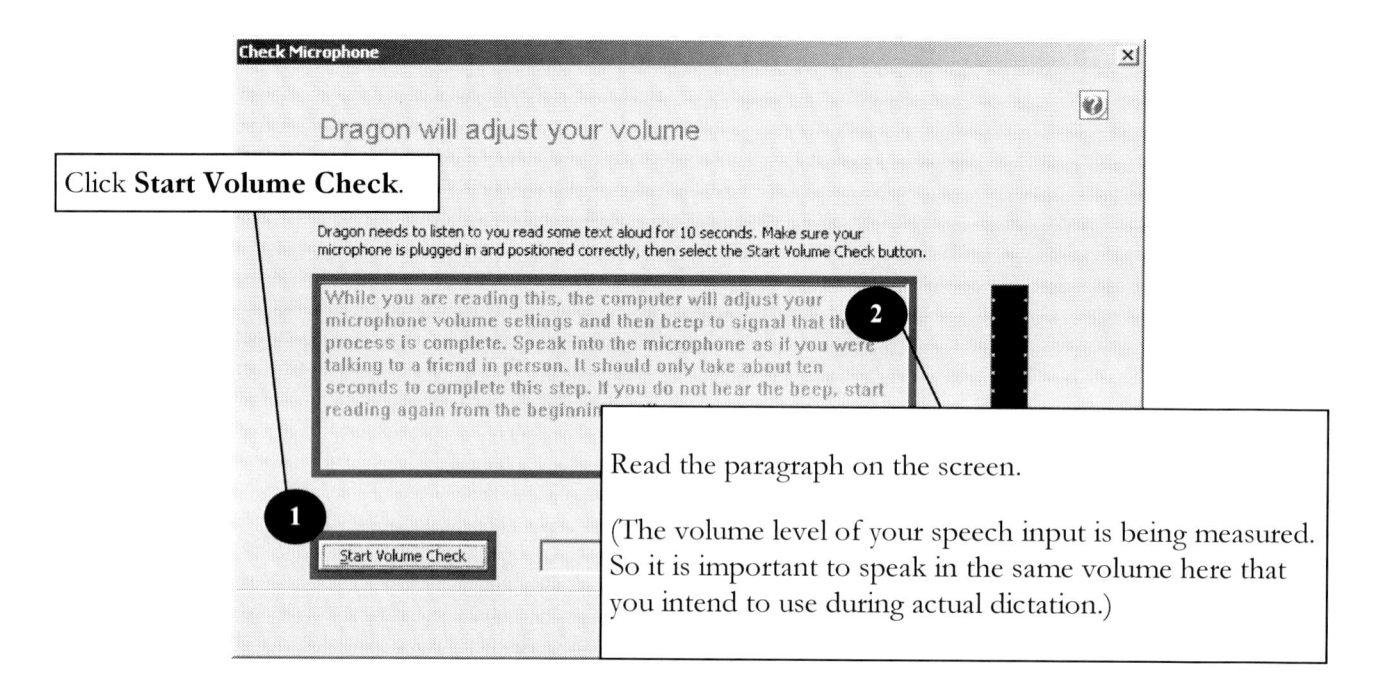

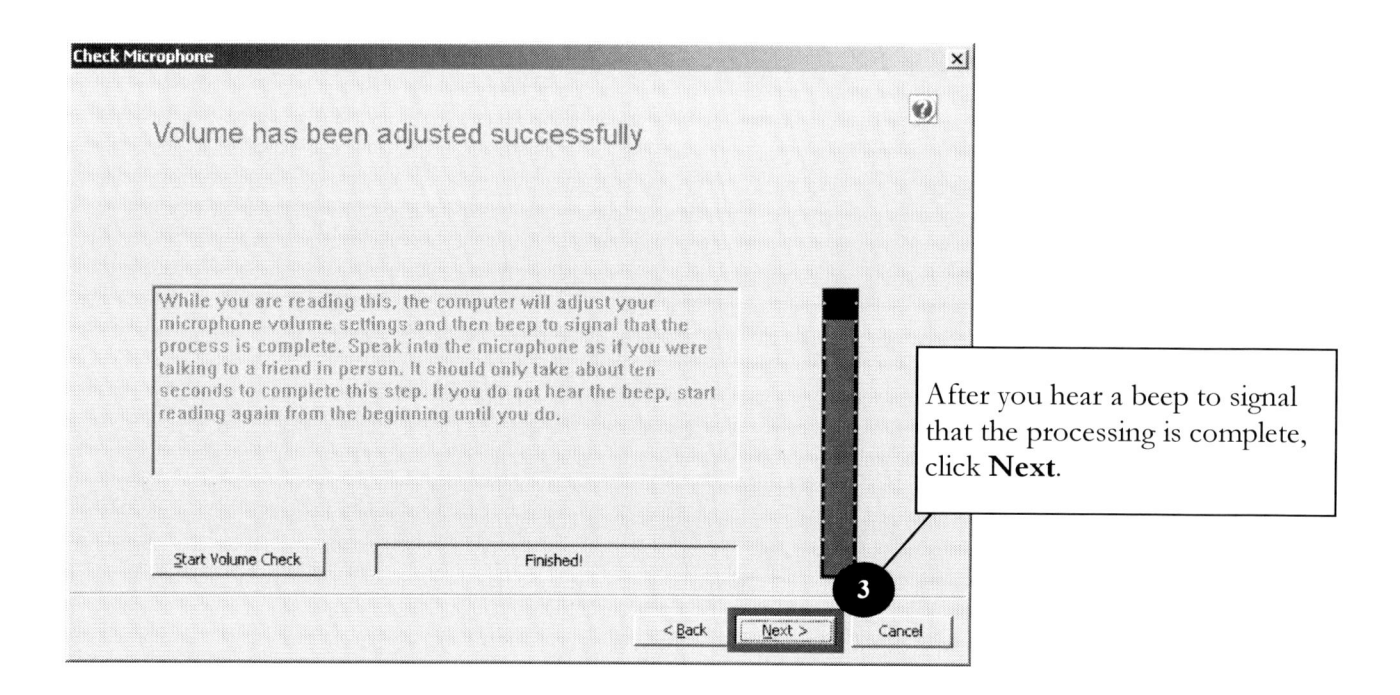

Click **Start Volume Check**.

1

Read the paragraph on the screen.

(The volume level of your speech input is being measured. So it is important to speak in the same volume here that you intend to use during actual dictation.)

2

After you hear a beep to signal that the processing is complete, click **Next**.

3

100

Chapter 4: Getting Started

Measuring Your Speech-to-Noise Signal Ratio

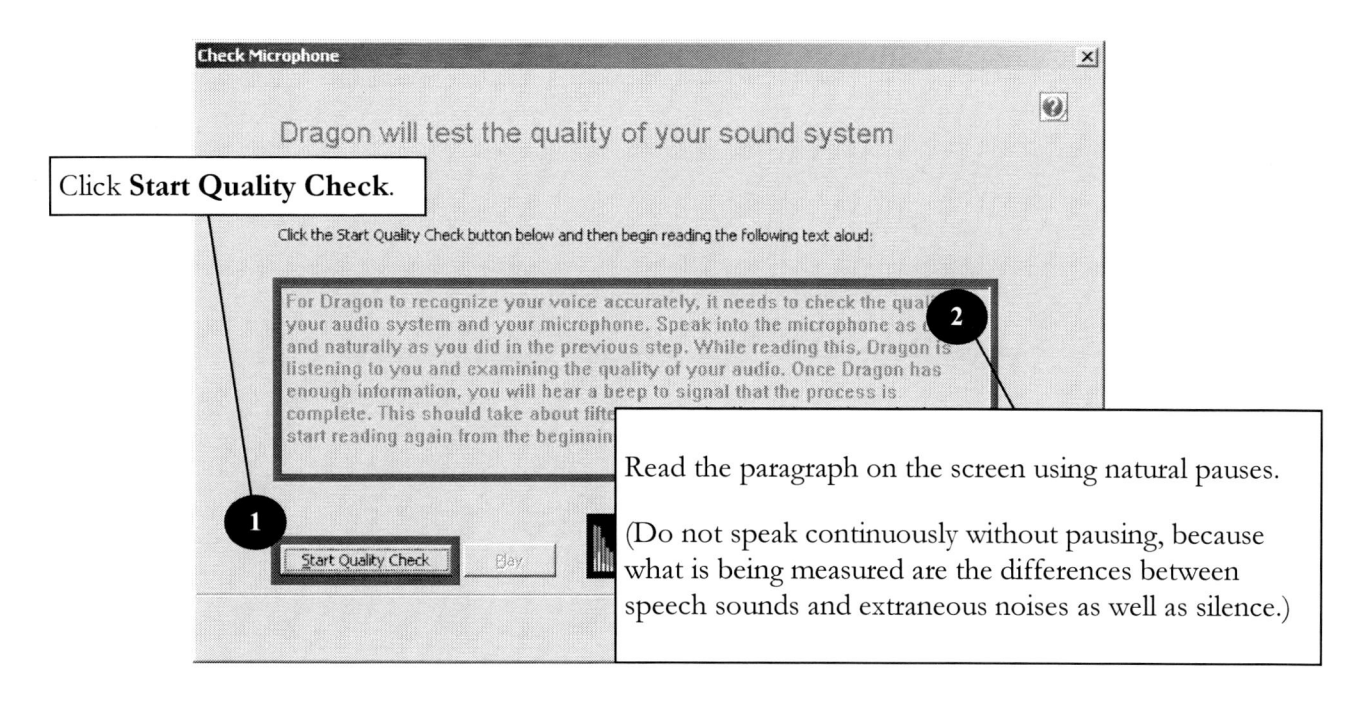

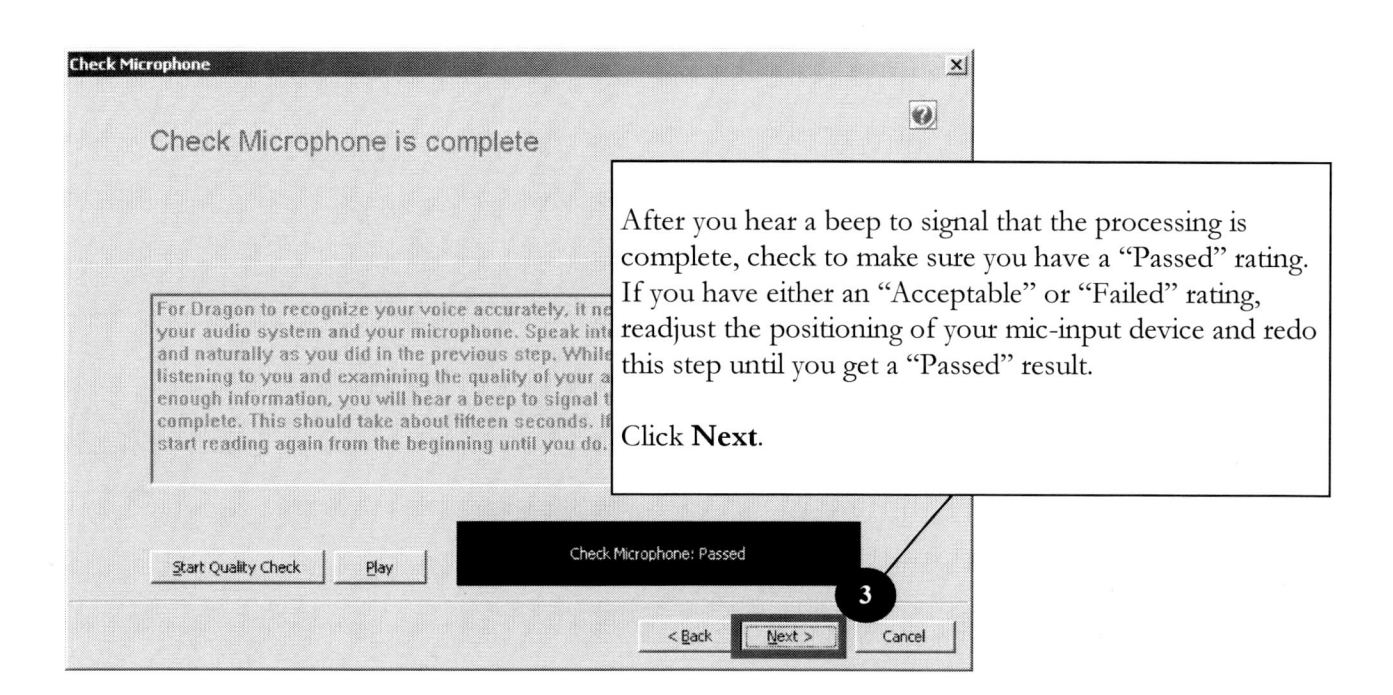

Click **Start Quality Check**.

1

2 Read the paragraph on the screen using natural pauses.

(Do not speak continuously without pausing, because what is being measured are the differences between speech sounds and extraneous noises as well as silence.)

3 After you hear a beep to signal that the processing is complete, check to make sure you have a "Passed" rating. If you have either an "Acceptable" or "Failed" rating, readjust the positioning of your mic-input device and redo this step until you get a "Passed" result.

Click **Next**.

Voice Writing Method • FIFTH EDITION
Dragon NaturallySpeaking® 12

General Training Options

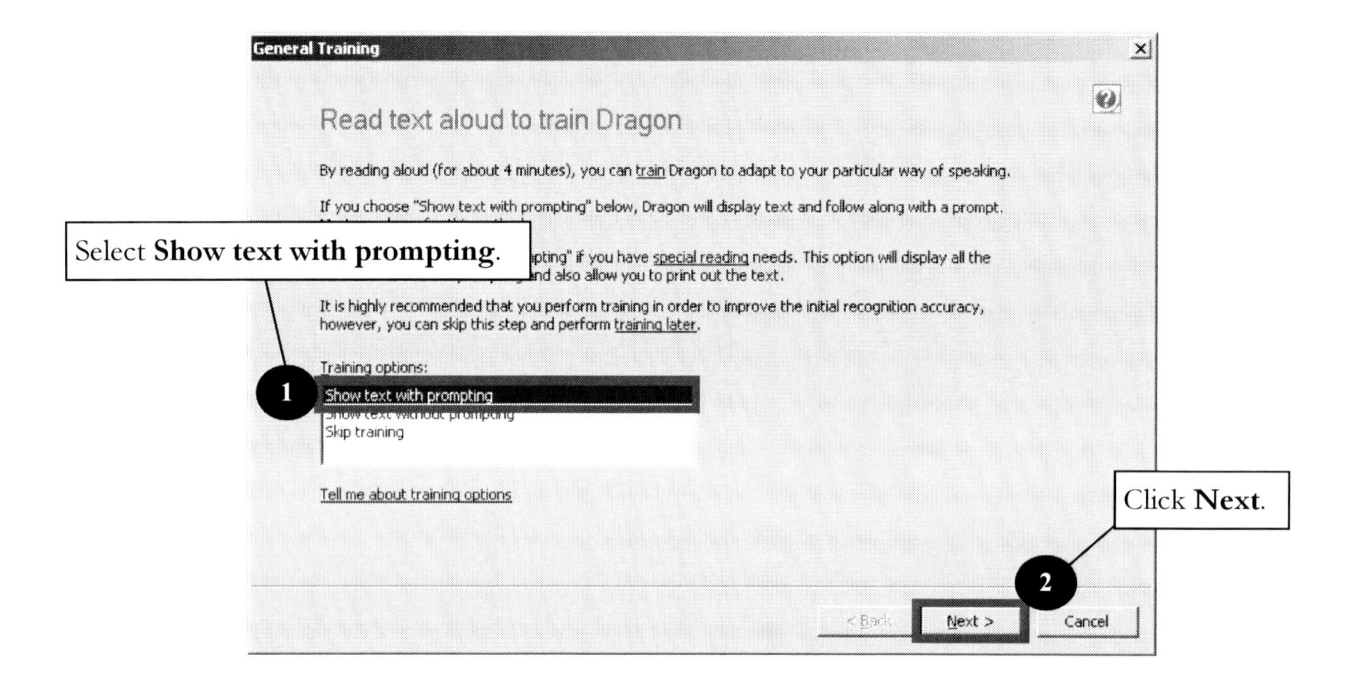

Select **Show text with prompting**.

1

Click **Next**.

2

Reading Introductory Text

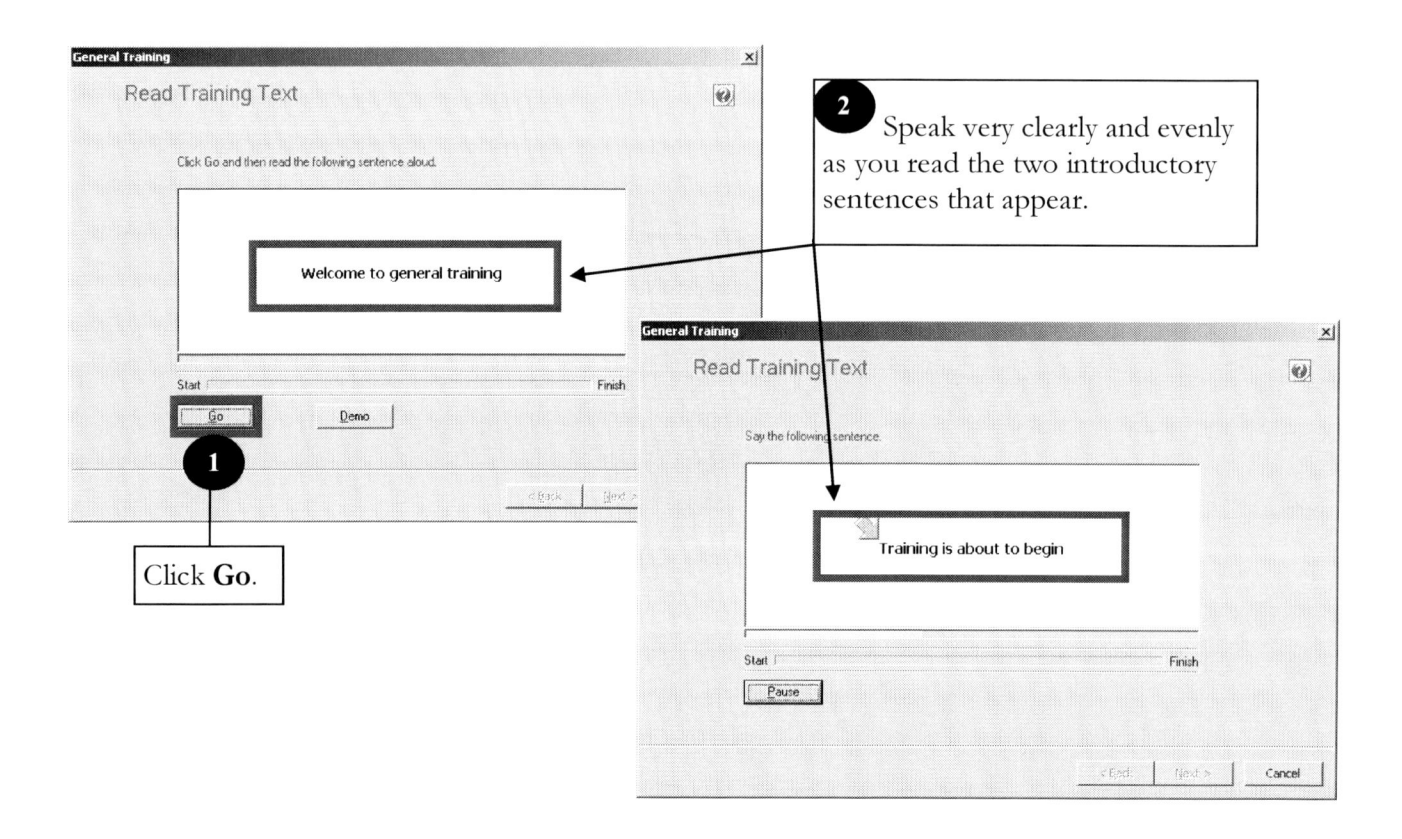

Click **Go**.

1

2 Speak very clearly and evenly as you read the two introductory sentences that appear.

102

Chapter 4: Getting Started

Selecting a Voice Training Story

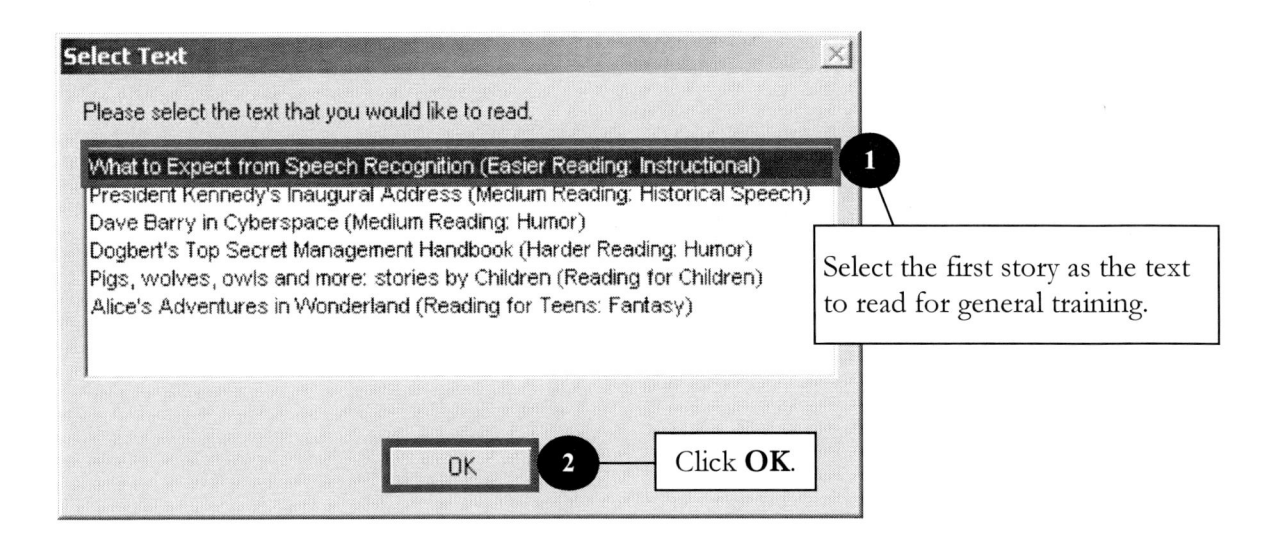

1. Select the first story as the text to read for general training.
2. Click **OK**.

Performing Voice Training

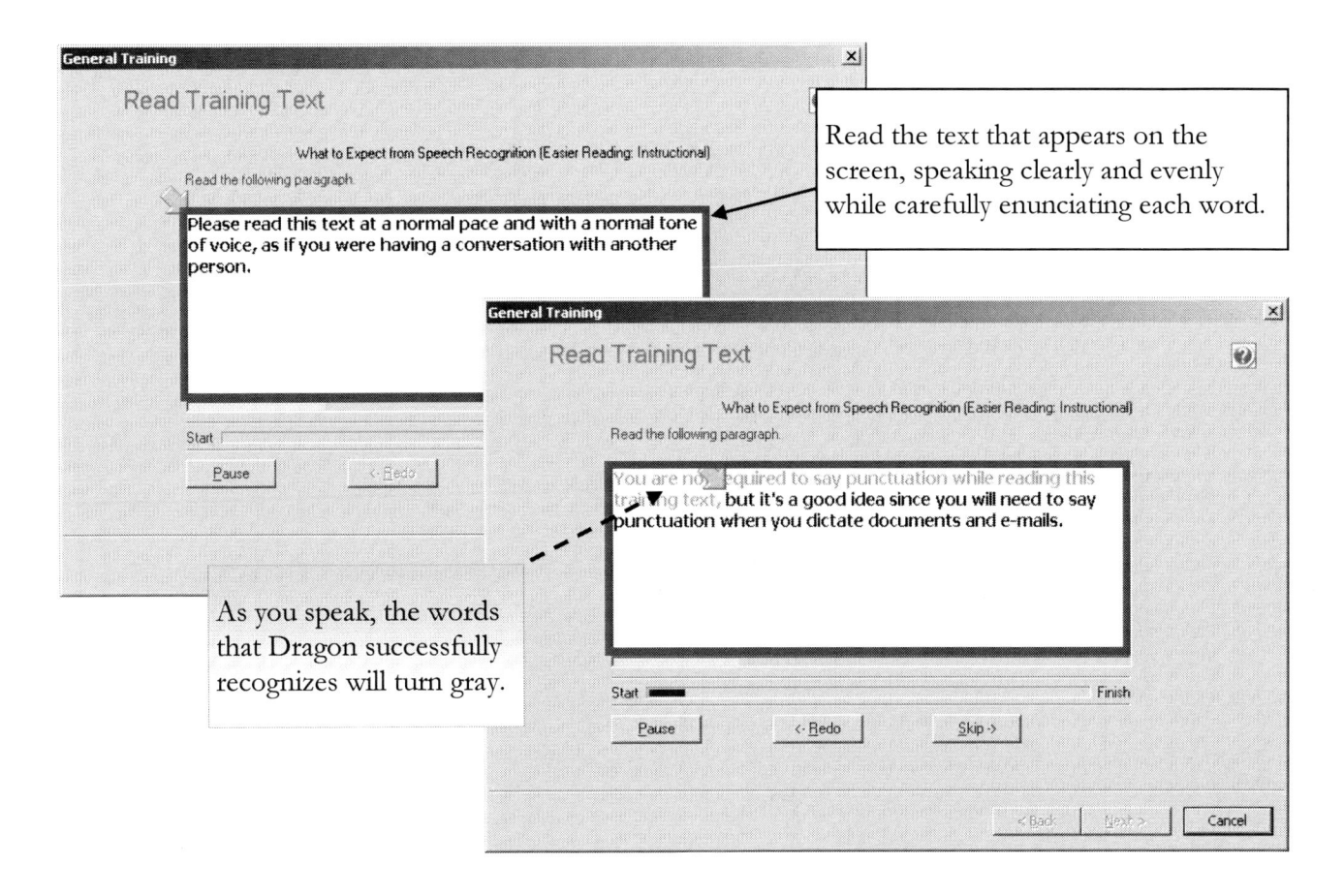

Read the text that appears on the screen, speaking clearly and evenly while carefully enunciating each word.

As you speak, the words that Dragon successfully recognizes will turn gray.

Voice Writing Method • FIFTH EDITION
Dragon NaturallySpeaking® 12

Adapting and Saving User Files

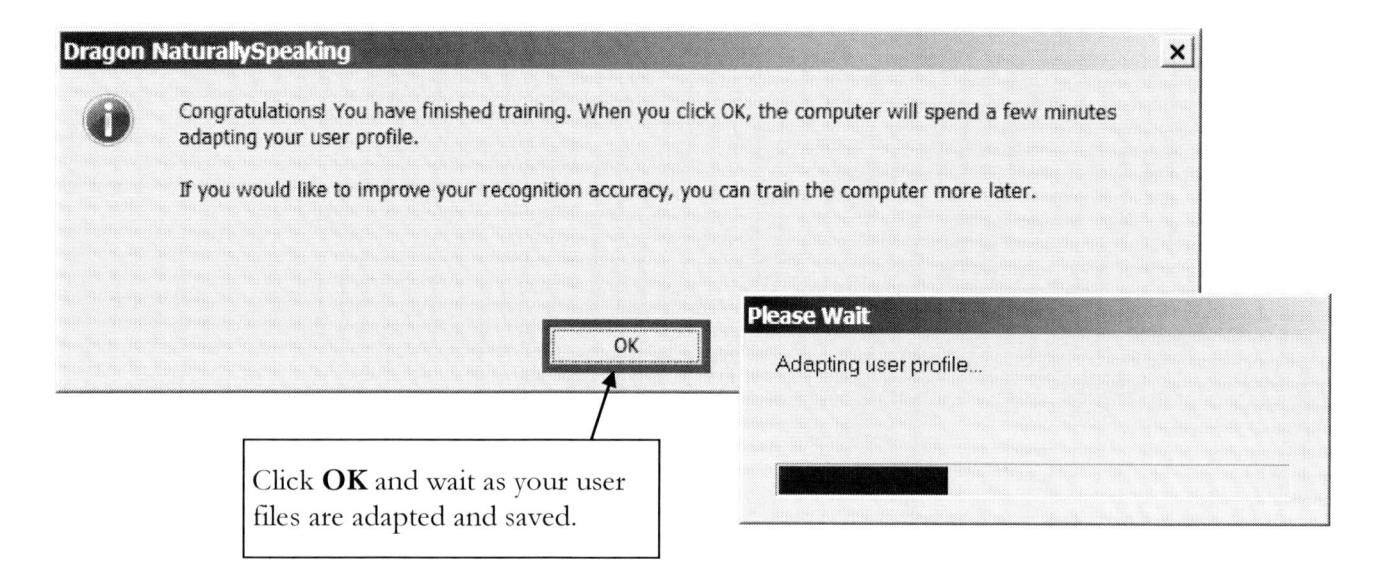

Click **OK** and wait as your user files are adapted and saved.

Option to Adapt to Writing Style of Documents

Dragon gives you the option to add new words its finds within documents on your computer and also to customize its grammatical model according to the context in which those words are used, but you will be preparing special documents for Dragon to analyze by a separate process explained in a later chapter.

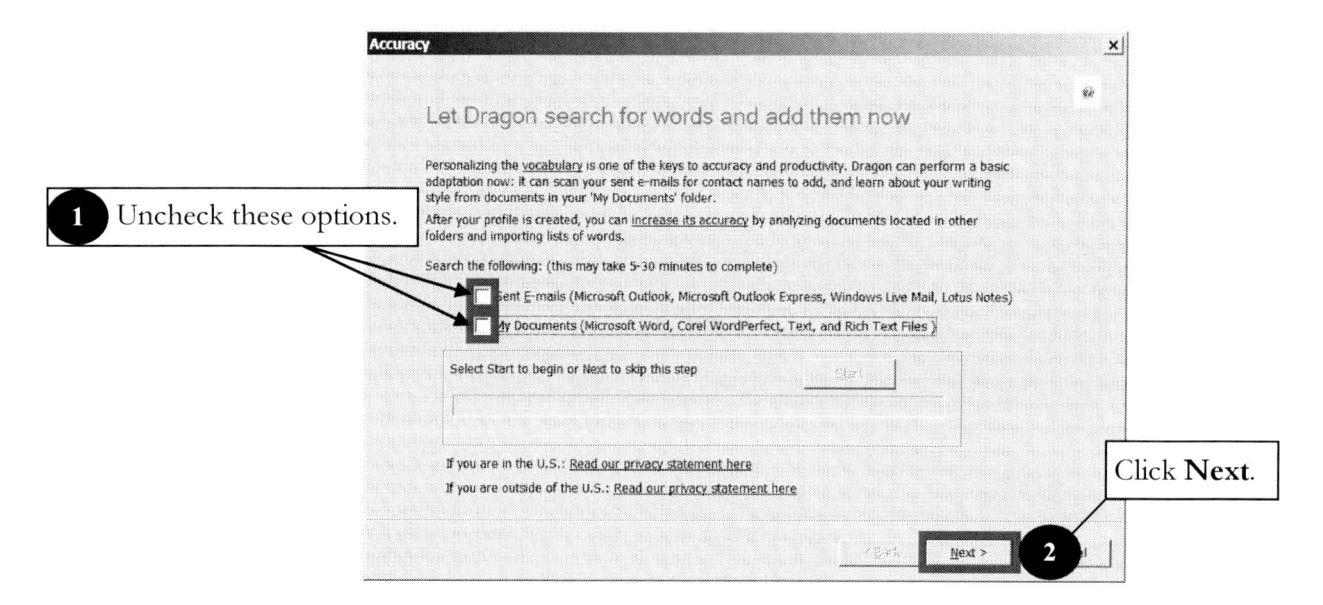

1 Uncheck these options.

2 Click **Next**.

104

Chapter 4: Getting Started

Accuracy Tuning

While this option to automatically improve accuracy is beneficial, it should be disabled. Generally, a voice writer's work computer must be clear from performing any prescheduled automatic processes, as you always want to keep your computer readily accessible for regular work tasks.

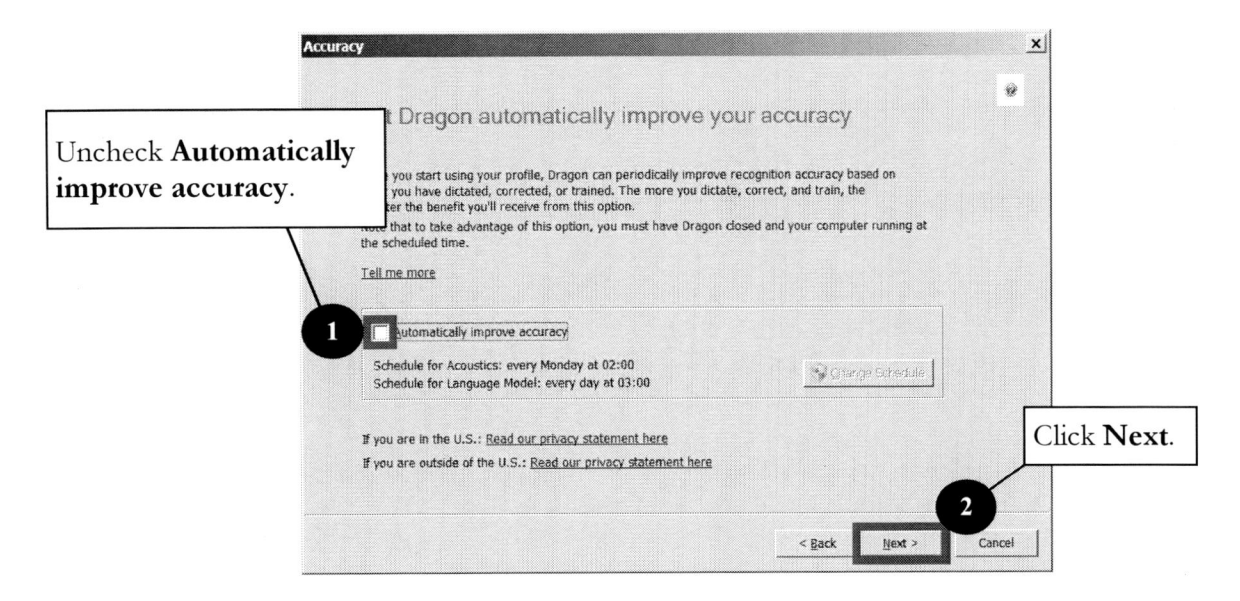

Uncheck **Automatically improve accuracy**.

Click **Next**.

Data Collection

Select **Don't run Data Collection** for this user.

Click **Next**.

105

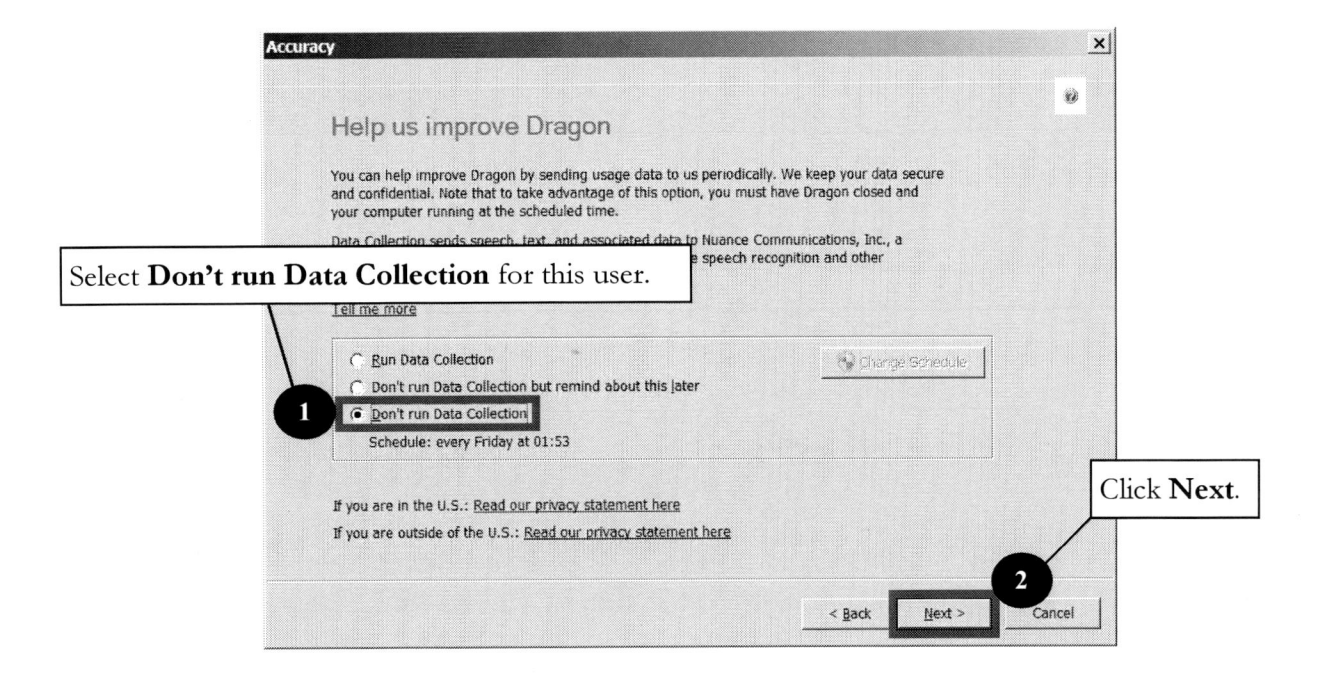

Finish Setup in New User Wizard

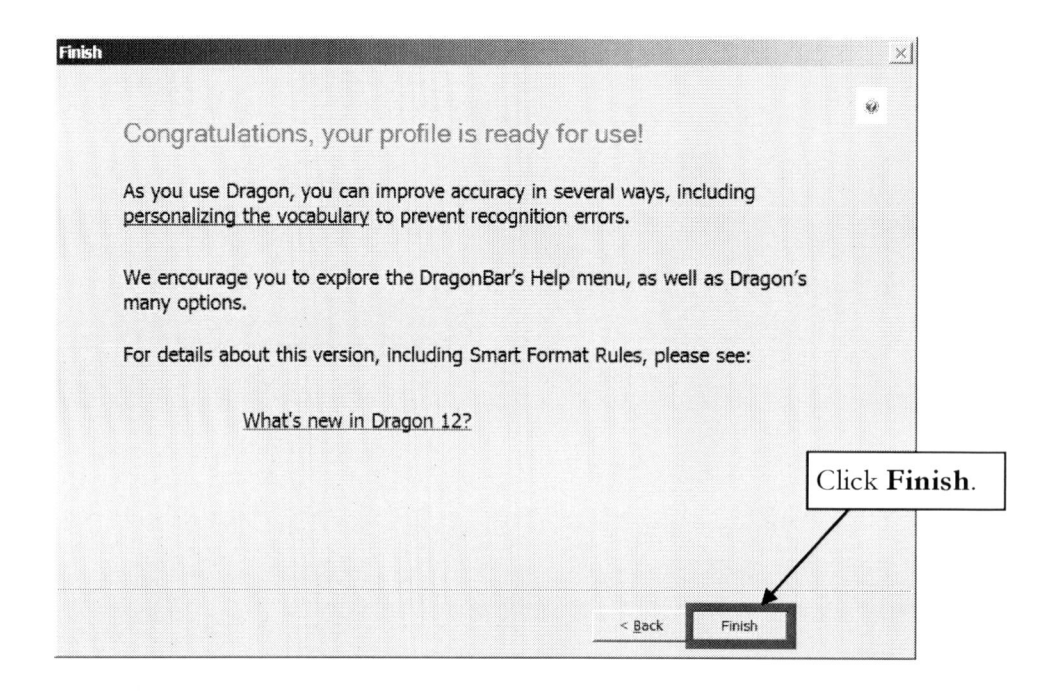

*Click **Finish**.*

Dragon Tutorial

Once your new user is created, Dragon gives you the opportunity to take its tutorial, but it is okay to skip it since this book provides more specific ideas and instructions specific to voice writing careers.

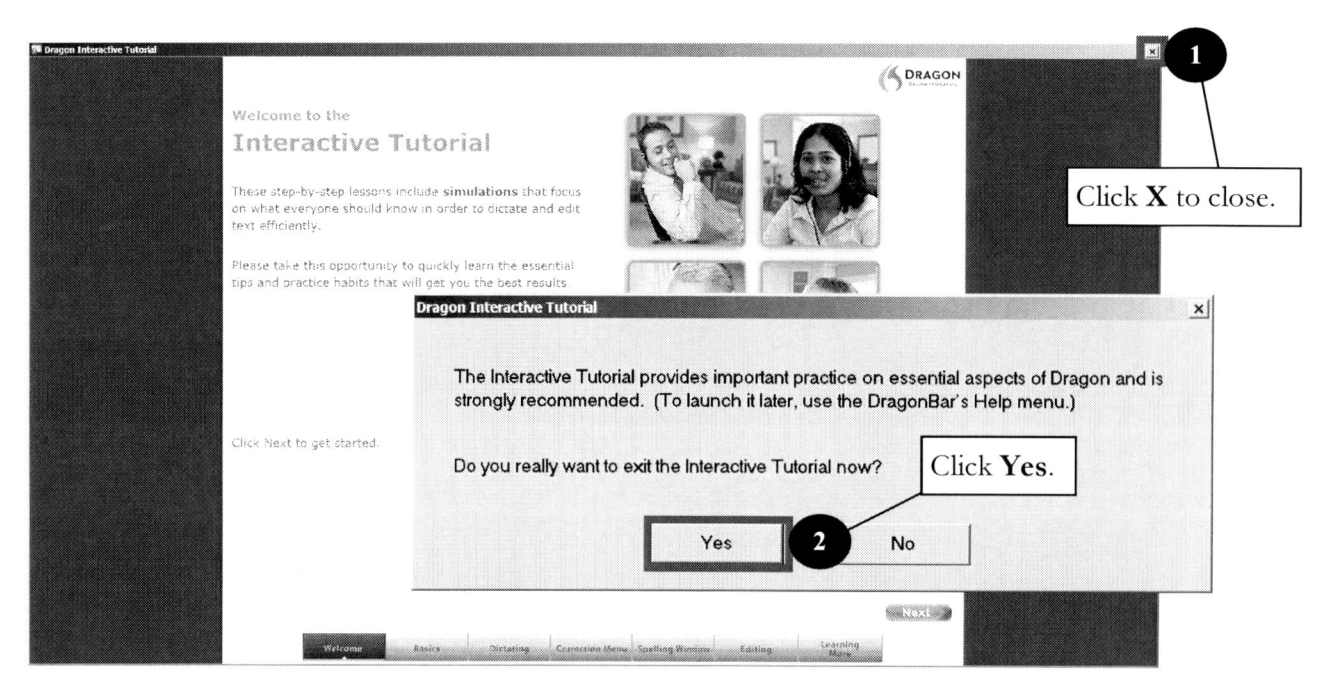

1. Click **X** to close.
2. Click **Yes**.

Chapter 4: Getting Started

Shutting Off Automatic Tips and Sidebar Screens

Here, Dragon presents you with help items intended to make beginning Dragon learners more productive. While this information is beneficial to explore initially, you do not want these screens automatically popping up in the future. Instructing Dragon not to continuously present these screens eliminates the inconvenience of having to manually close them every time you open your Dragon user. This way, you will go straight to performing your dictation work with fewer clicks. You can always access Dragon's Help menu any time you need by pressing F1.

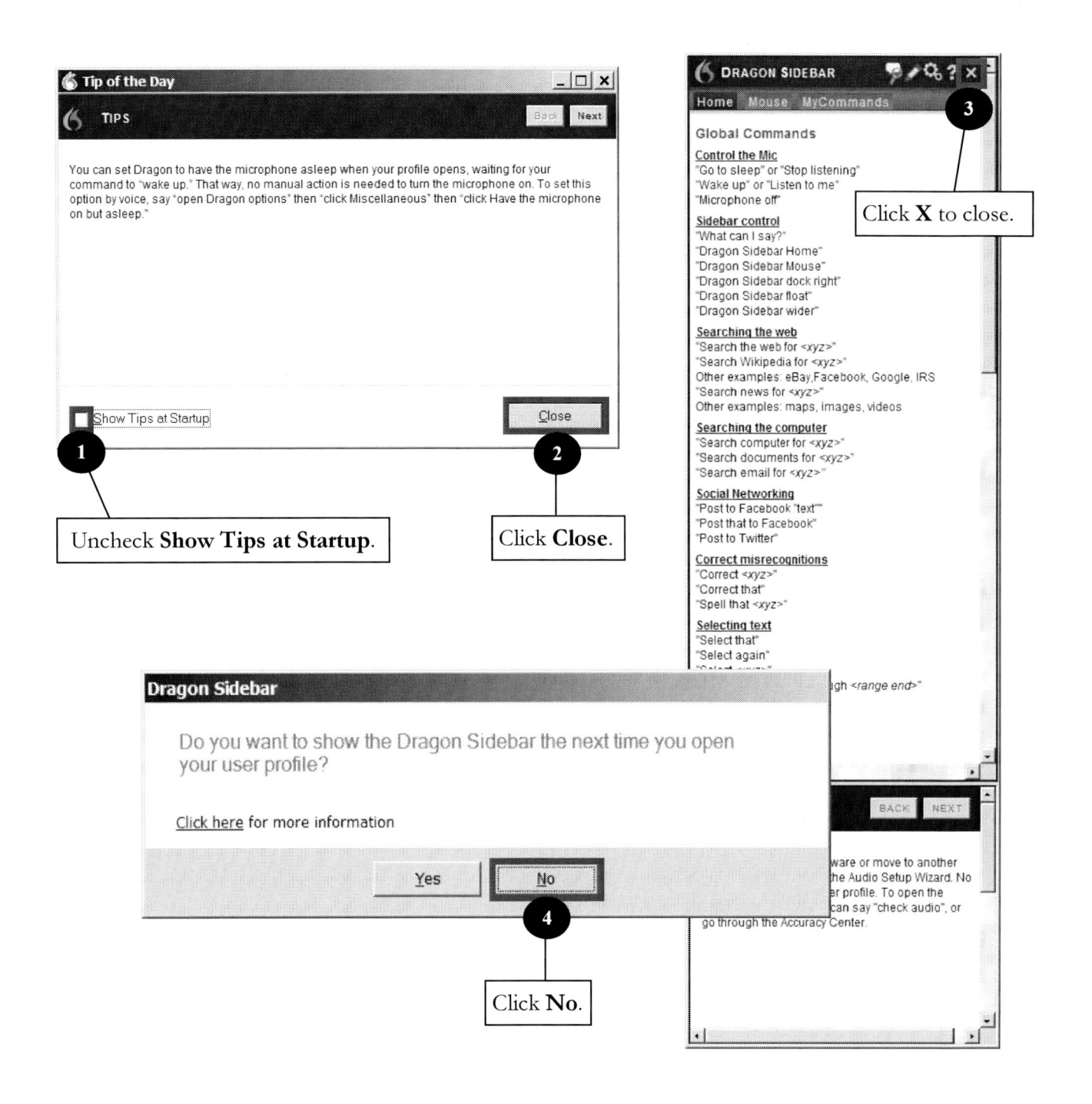

1 Uncheck **Show Tips at Startup**.

2 Click **Close**.

3 Click **X** to close.

4 Click **No**.

107

Performing Additional Voice Training

To continue with the remainder of your general training stories, go to the **DragonBar** > **Audio** menu > **Read text to improve accuracy**. Remember that you only need to read about two more stories to complete your training.

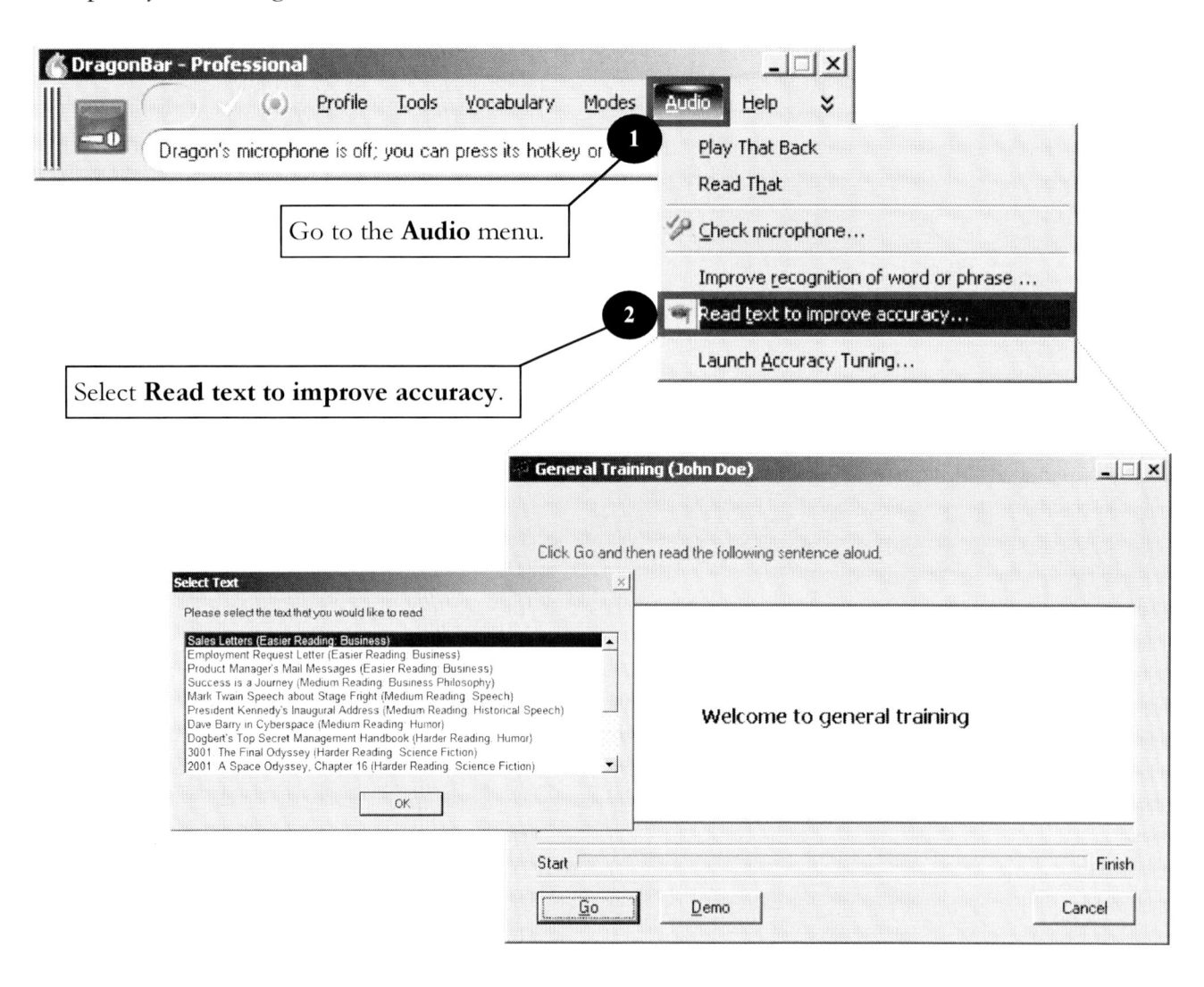

Go to the **Audio** menu.

Select **Read text to improve accuracy**.

After all training is complete, follow the instructions in the next section, *Setting User Options*, then back up and export your user files pursuant to the steps in the *Backing Up/Restoring User Files* and *Exporting/Importing User Files* subsections under the *User File Management* section of Chapter 8.

Chapter 4: Getting Started

Setting User Options

This section will explain how to set your Dragon user options by choosing the most common realtime voice writing preferences for proper formatting of numbers and words and for the use of hot key functions. Since this guide is focused on quickly getting you up and running with realtime, few explanations regarding these settings are given. For complete details about Dragon user options, refer to Dragon's Help menu (F1). If you have CAT software, consult your CAT software manual in case the recommended settings are different.

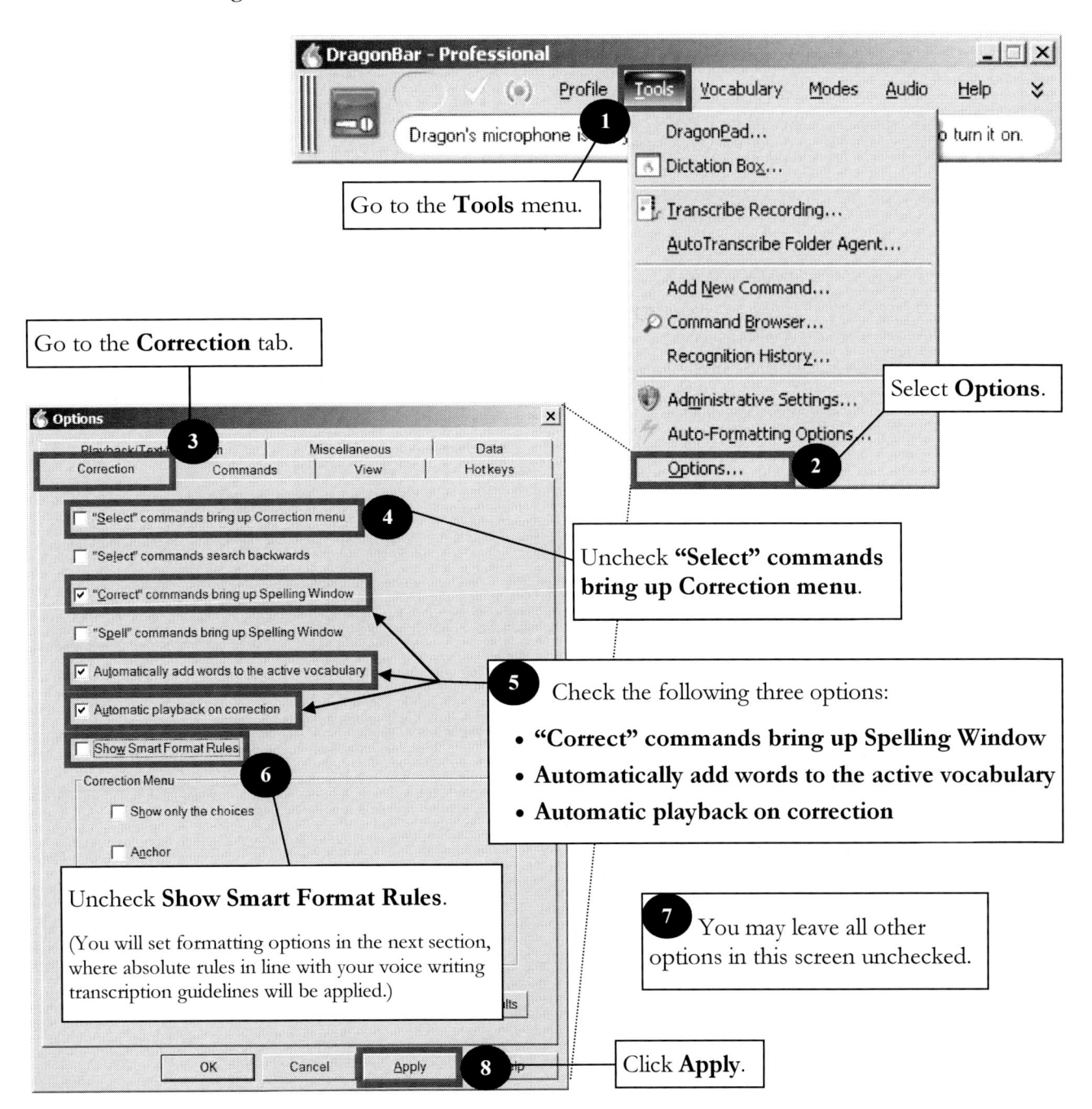

Go to the **Tools** menu.

Select **Options**.

Go to the **Correction** tab.

Uncheck **"Select" commands bring up Correction menu**.

Check the following three options:

- **"Correct" commands bring up Spelling Window**
- **Automatically add words to the active vocabulary**
- **Automatic playback on correction**

Uncheck **Show Smart Format Rules**.

(You will set formatting options in the next section, where absolute rules in line with your voice writing transcription guidelines will be applied.)

You may leave all other options in this screen unchecked.

Click **Apply**.

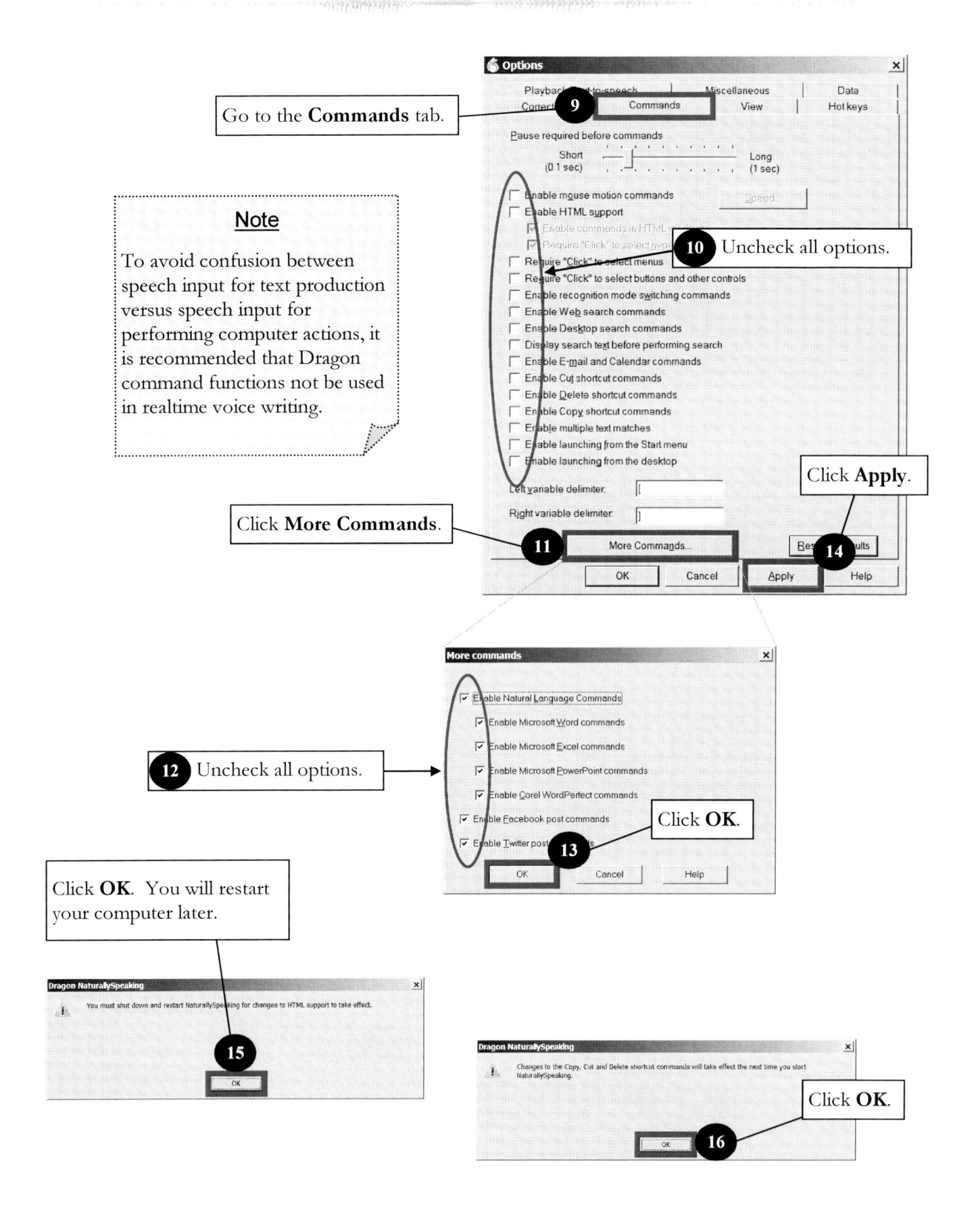

Voice Writing Method • FIFTH EDITION
Dragon NaturallySpeaking® 12

9 Go to the **Commands** tab.

Note

To avoid confusion between speech input for text production versus speech input for performing computer actions, it is recommended that Dragon command functions not be used in realtime voice writing.

10 Uncheck all options.

11 Click **More Commands**.

14 Click **Apply**.

12 Uncheck all options.

13 Click **OK**.

15 Click **OK**. You will restart your computer later.

16 Click **OK**.

110

Chapter 4: Getting Started

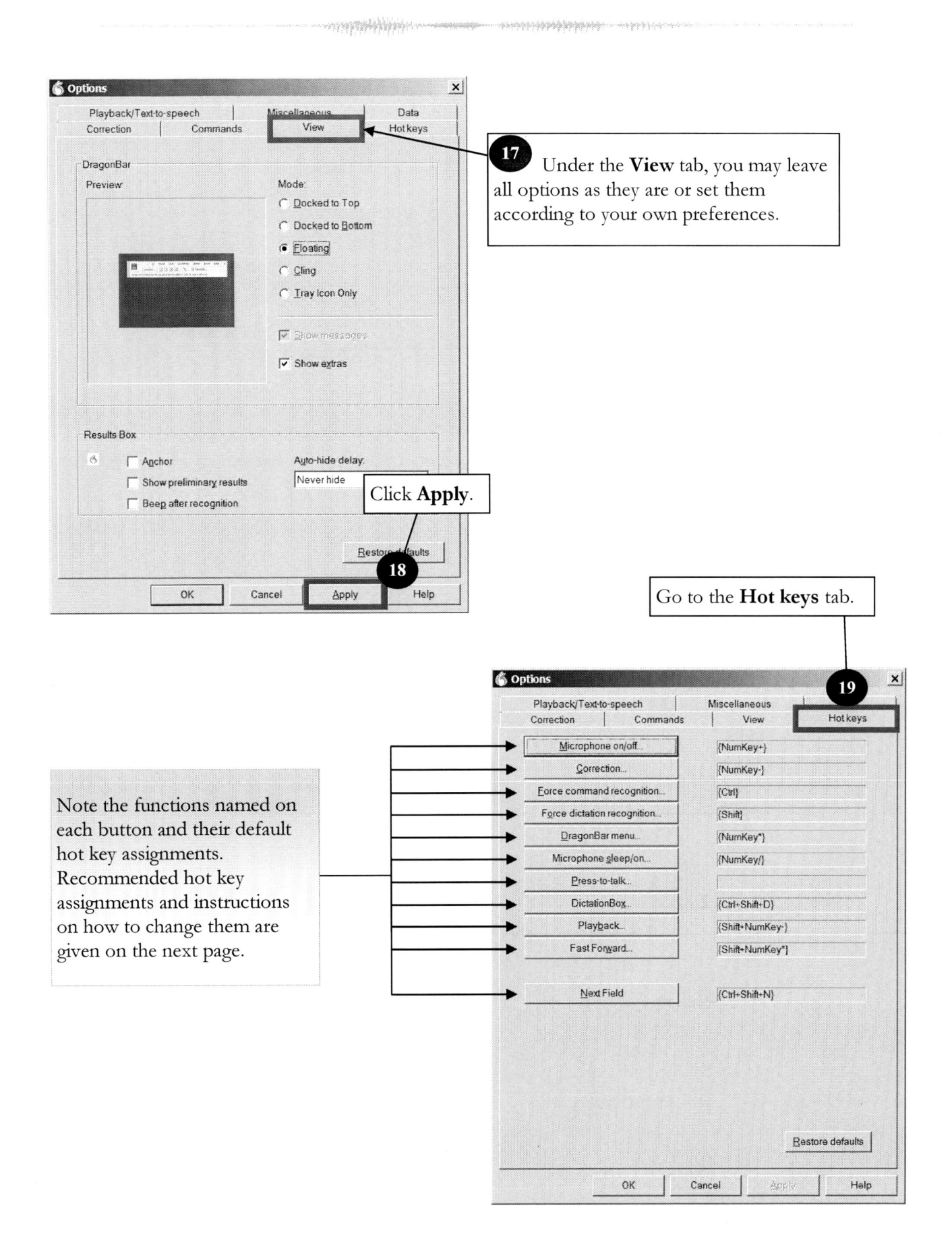

17 Under the **View** tab, you may leave all options as they are or set them according to your own preferences.

Click **Apply**.

18

Go to the **Hot keys** tab.

19

Note the functions named on each button and their default hot key assignments. Recommended hot key assignments and instructions on how to change them are given on the next page.

111

The only hot keys that will be helpful to you for realtime voice writing are for turning the microphone on and off (*Microphone on/off* button), using the correction tool to correct SR errors (*Correction* button), and playing back your dictation (*Playback* button). If the default keys shown do not exist on your keyboard, you will need to change hot key assignments for those functions. Different hot keys may need to be used for performing these functions with CAT software, so refer to your CAT program's manual for that information.

You will not be able to use certain default hot keys that require the use of a number keypad if there isn't one on your keyboard, but the keys recommended below do exist on most keyboards.

To change a hot-key assignment, click the button for the selection you would like to change and press the new key or keystroke combination you prefer to use. Just be sure the keys you're choosing are not commonly used for performing functions in other programs you may work in while operating Dragon.

To remove a hot-key assignment altogether, simply press the *Delete* key after clicking one of these buttons.

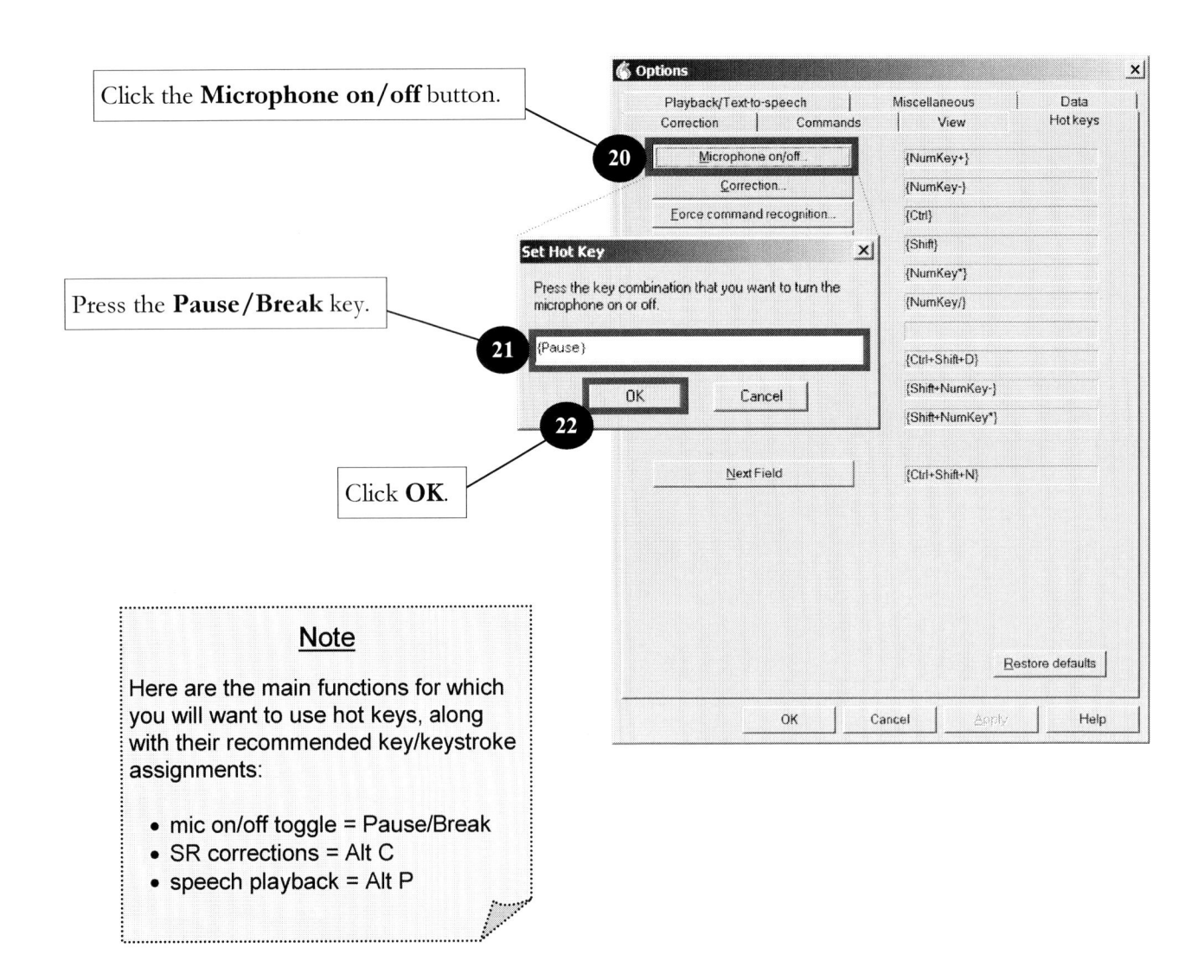

20 Click the **Microphone on/off** button.

21 Press the **Pause/Break** key.

22 Click **OK**.

Note

Here are the main functions for which you will want to use hot keys, along with their recommended key/keystroke assignments:

- mic on/off toggle = Pause/Break
- SR corrections = Alt C
- speech playback = Alt P

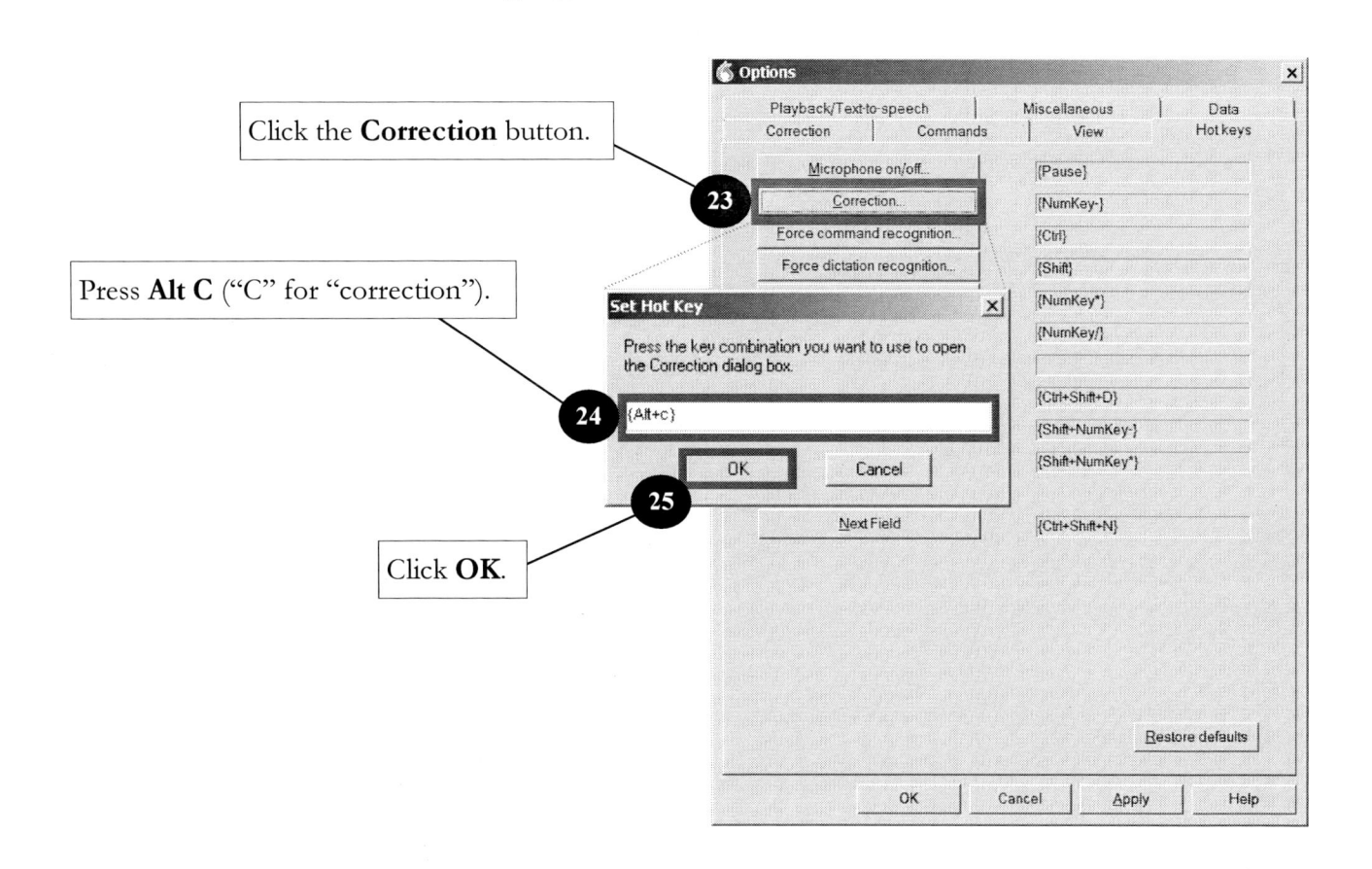

Chapter 4: Getting Started

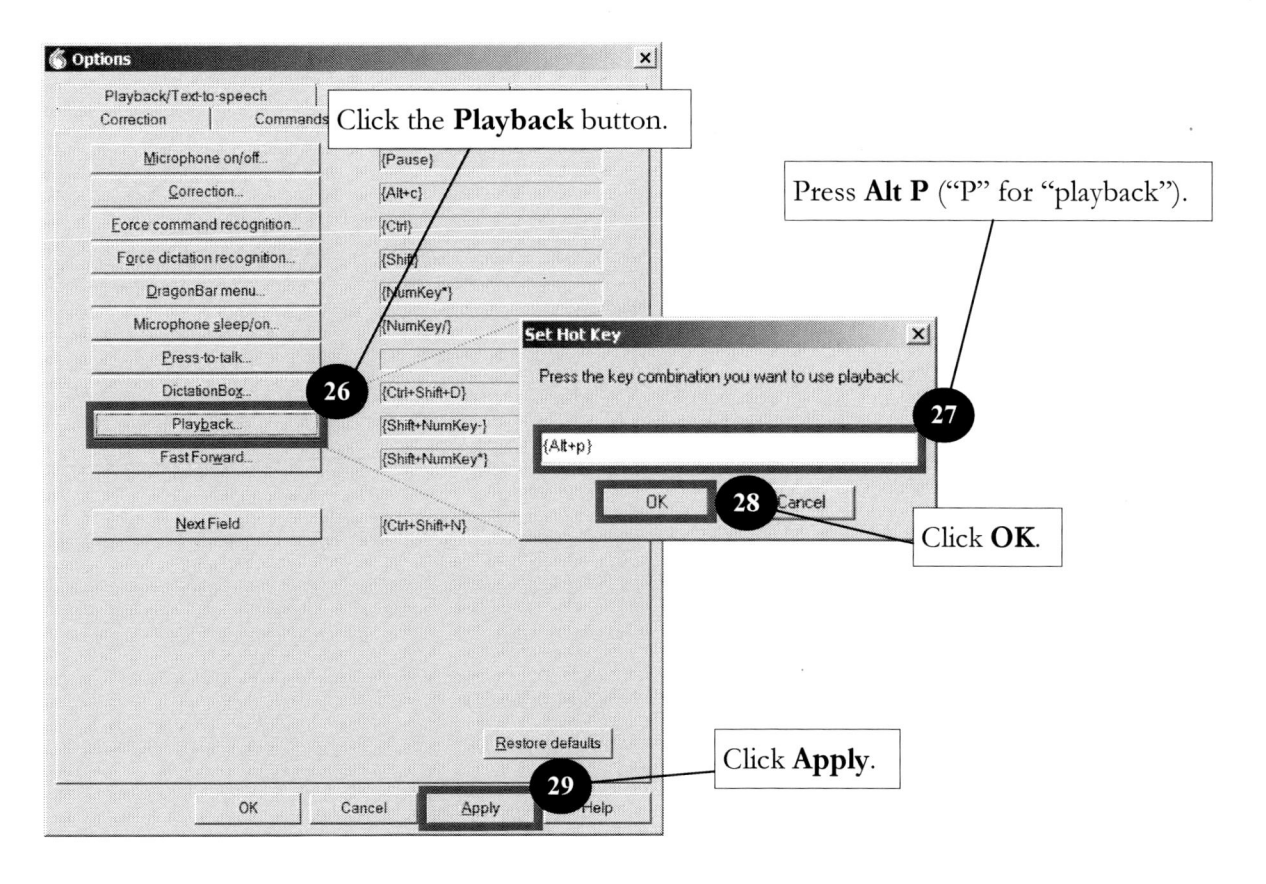

23 Click the **Correction** button.

24 Press **Alt C** ("C" for "correction").

25 Click **OK**.

26 Click the **Playback** button.

27 Press **Alt P** ("P" for "playback").

28 Click **OK**.

29 Click **Apply**.

113

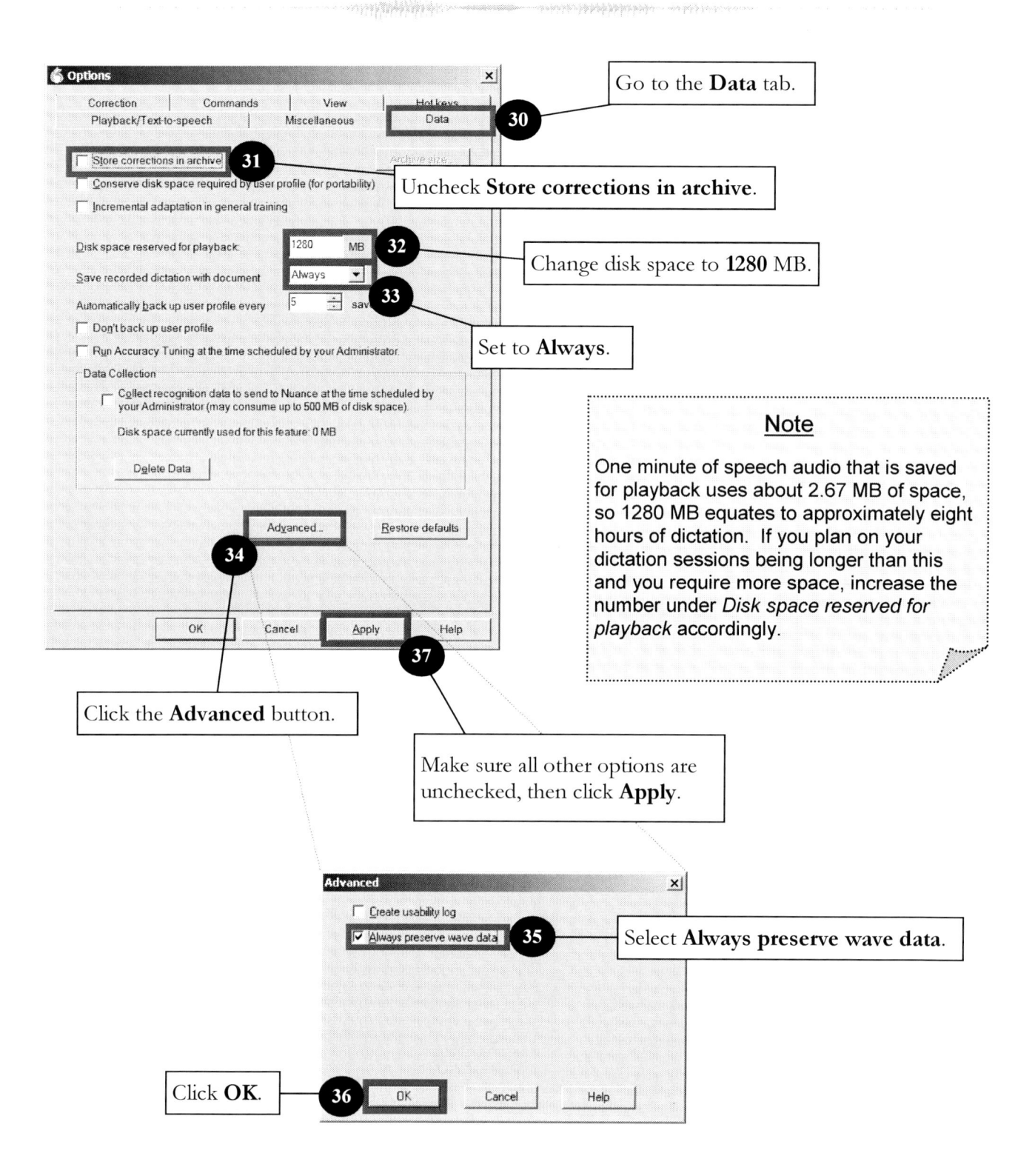

Voice Writing Method • FIFTH EDITION
Dragon NaturallySpeaking® 12

30 Go to the **Data** tab.

31 Uncheck **Store corrections in archive**.

32 Change disk space to **1280** MB.

33 Set to **Always**.

34 Click the **Advanced** button.

Note

One minute of speech audio that is saved for playback uses about 2.67 MB of space, so 1280 MB equates to approximately eight hours of dictation. If you plan on your dictation sessions being longer than this and you require more space, increase the number under *Disk space reserved for playback* accordingly.

37 Make sure all other options are unchecked, then click **Apply**.

35 Select **Always preserve wave data**.

36 Click **OK**.

Chapter 4: Getting Started

Go to the Miscellaneous tab. — 38

Check this option only if you will be using Dragon alone and not dictating into other applications. If checked, Dragon's word-processing screen (DragonPad) automatically opens when you launch Dragon. — 39

Change this setting to 120. This is the maximum number of minutes allowable during silence periods before Dragon will turn off your microphone. — 40

41 Uncheck all other options in this screen.

Click Apply. — 42

Click OK to exit the *Options* screen. — 43

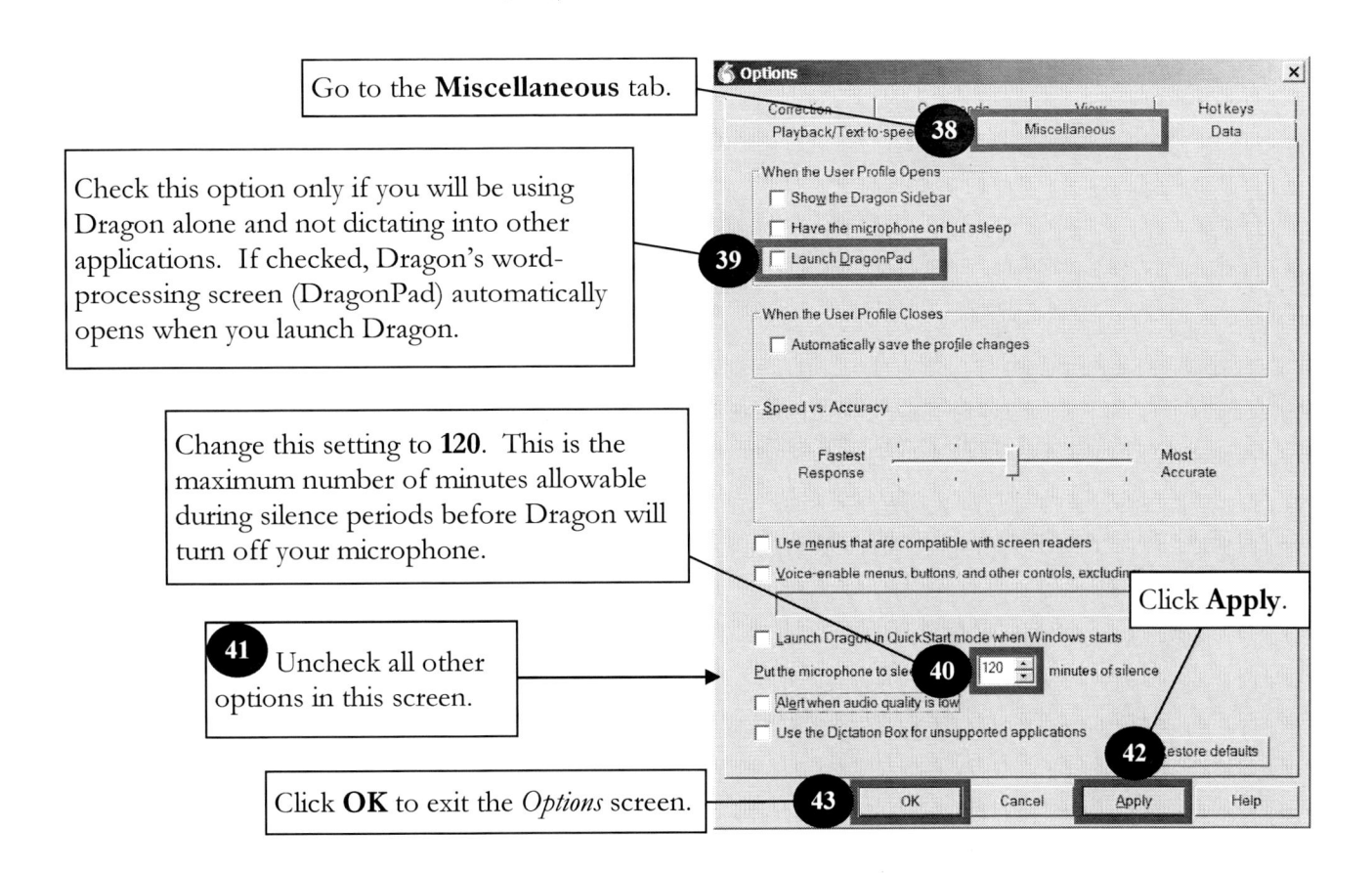

Ideally, you want the fastest text-production speed possible when providing realtime services. Although Dragon continuously lags behind your dictation and often results in delayed text production that is too slow for reporting professions requiring immediate realtime delivery, you should generally leave the *Speed vs. Accuracy* setting at default to allow Dragon to maintain a suitable balance between speed and accuracy. If you are using Dragon without CAT software, you probably won't need to worry about getting the fastest text drop-down anyway.

The only time you really need to be concerned about getting the fastest text speed performance is when sending a realtime feed, in which case you would also need to have the proper type of CAT program that resolves Dragon's text-delay problem. This type of CAT program consistently outputs batches of text onto your computer screen within under a couple of seconds, so there is no need to adjust the response time setting in Dragon. (To view a video demonstration of this type of CAT program so you can see how it works with Dragon to immediately deliver text, go to the www.eclipsevox.com website and follow the EclipseVox™ link.)

If you are using a CAT program that does not overcome Dragon's text-lagging limitations and this is inconvenient to persons relying on your realtime services, be sure to set the slider bar to *Fastest Response* before sending a realtime feed. If your speech files have been adequately trained to produce accuracy results at averages above 96% at 200 words per minute, making this adjustment will not make a noticeable difference in your accuracy results.

Voice Writing Method • FIFTH EDITION
Dragon NaturallySpeaking® 12

Setting Formatting Preferences

The steps below show you how to access Dragon's formatting options and change them, if necessary.

If you have CAT software, obey the instructions provided in your CAT software manual when determining which options to set, since your CAT program may have special options and formatting setup requirements for proper Dragon+CAT compatibility and performance.

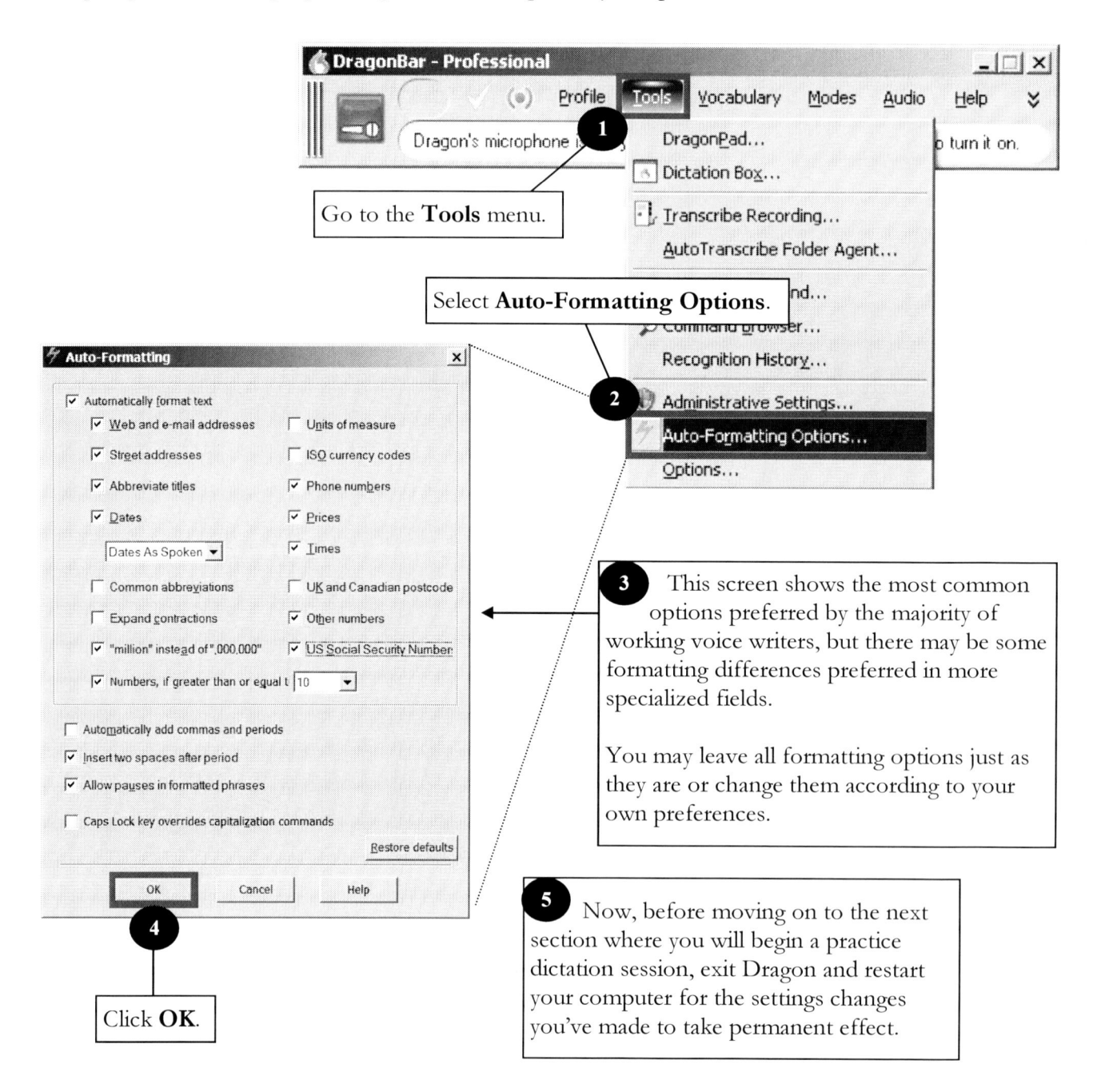

1 Go to the **Tools** menu.

2 Select **Auto-Formatting Options**.

3 This screen shows the most common options preferred by the majority of working voice writers, but there may be some formatting differences preferred in more specialized fields.

You may leave all formatting options just as they are or change them according to your own preferences.

4 Click **OK**.

5 Now, before moving on to the next section where you will begin a practice dictation session, exit Dragon and restart your computer for the settings changes you've made to take permanent effect.

Chapter 4: Getting Started

Performing a Dictation Session

Now you're ready to do some dictation. Make sure your equipment is properly connected and open Dragon before following the steps below.

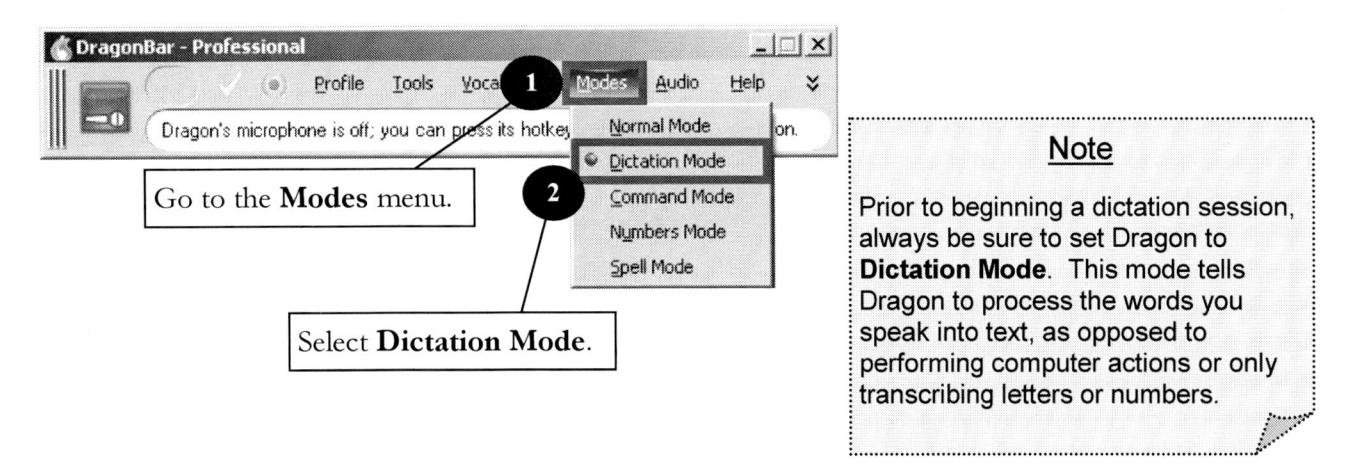

Go to the **Modes** menu.

Select **Dictation Mode**.

Note

Prior to beginning a dictation session, always be sure to set Dragon to **Dictation Mode**. This mode tells Dragon to process the words you speak into text, as opposed to performing computer actions or only transcribing letters or numbers.

The instructions below show you how to use Dragon's document screen, but you can dictate into any application of your choice. If you wanted to use a program such as Microsoft Word, you would simply open Microsoft Word (skipping steps 3 and 4), then turn on Dragon's dictation microphone to begin dictating directly into that application.

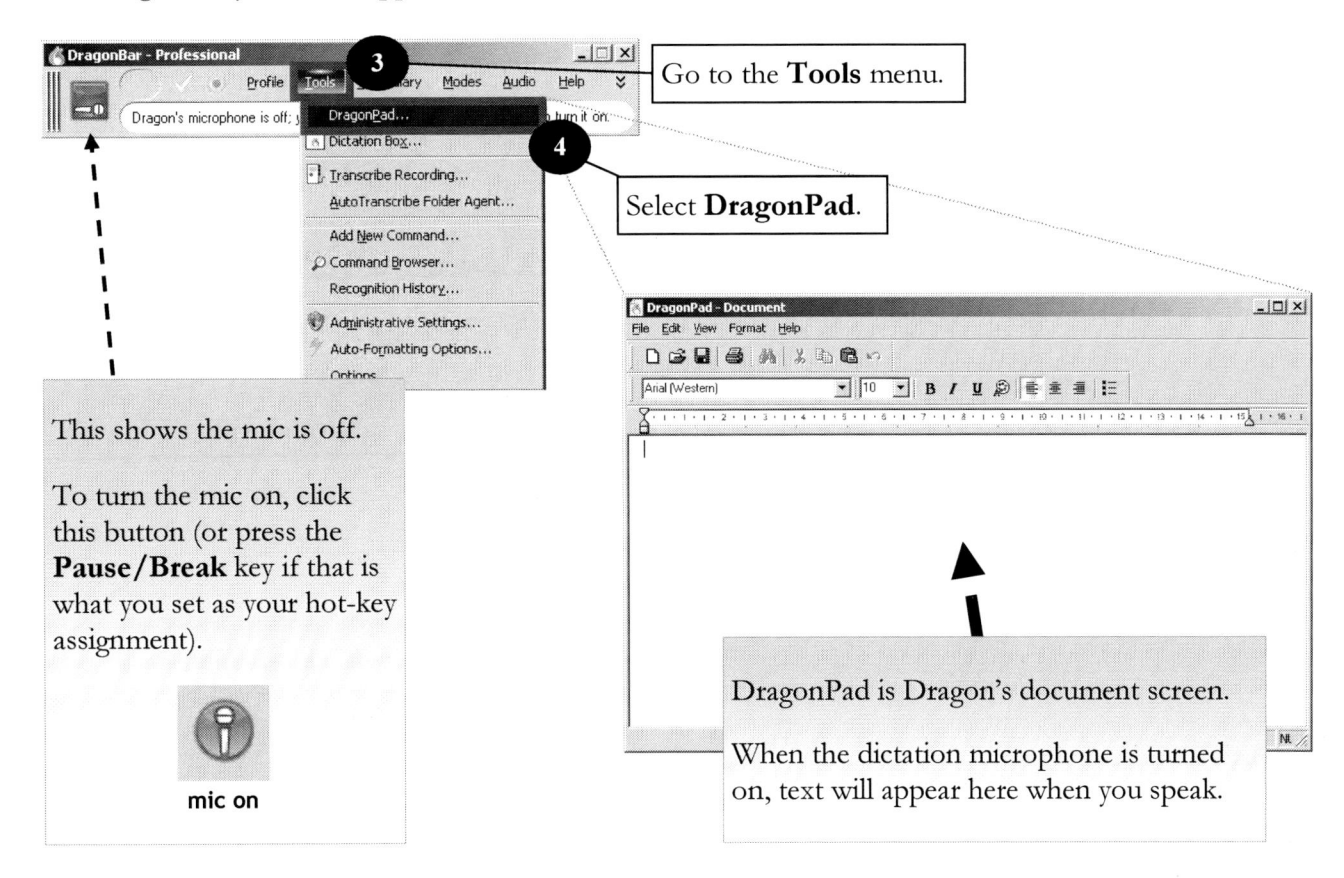

Go to the **Tools** menu.

Select **DragonPad**.

This shows the mic is off.

To turn the mic on, click this button (or press the **Pause/Break** key if that is what you set as your hot-key assignment).

mic on

DragonPad is Dragon's document screen.

When the dictation microphone is turned on, text will appear here when you speak.

117

Voice Writing Method • FIFTH EDITION
Dragon NaturallySpeaking® 12

5 Turn on the dictation microphone.

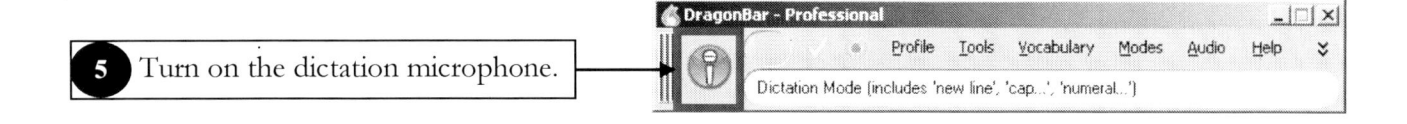

6 Dictate this sample text. Do not dictate any voice codes at this point. Say all words and punctuation marks as you would normally say them, and be sure to speak clearly and evenly.

Story of Sam Walton

(adapted version of Sam Walton's biography taken from www.about.com.)

Sam Walton, the founder of Wal-Mart, was born in Kingfisher, Oklahoma on March 29, 1918. He was raised in Missouri where he worked in his father's store while attending school. This was his first retailing experience and he really enjoyed it. After graduating from the University of Missouri in 1940, he began his own career as a retail merchant when he opened the first of several franchises of the Ben Franklin five-and-dime franchises in Arkansas.

This would lead to bigger and better things and he soon opened his first Wal-Mart store in 1962 in Rogers, Arkansas. Wal-Mart specialized in name brands at low prices and Sam Walton was surprised at the success. Soon a chain of Wal-Mart stores sprang up across rural America.

Walton's management style was popular with employees, and he founded some of the basic concepts of management that are still in use today. After taking the company public in 1970, Walton introduced his profit sharing plan. The profit sharing plan was a plan for Wal-Mart employees to improve their income dependent on the profitability of the store. Sam Walton believed that individuals don't win, teams do. Employees at Wal-Mart stores were offered stock options and store discounts. These benefits are commonplace today, but Walton was among the first to implement them. Walton believed that a happy employee meant happy customers and more sales. Walton believed that by giving employees a part of the company and making their success dependent on the company's success, they would care about the company.

By the 1980s, Wal-Mart had sales of over one billion dollars and over three hundred stores across North America. Wal-Mart's unique decentralized-distribution system, also Walton's idea, created the edge needed to further spur growth in the 1980s amidst growing complaints that the superstore was squelching smaller, traditional mom-and-pop stores. By 1991, Wal-Mart was the largest U.S. retailer with 1,700 stores. Walton remained active in managing the company, as president and CEO until 1988 and chairman until his death. He was awarded the Medal of Freedom shortly before his death.

Walton died in 1992, being the world's second richest man behind Bill Gates. He passed his company down to his three sons, daughter and wife. Wal-Mart Stores Incorporated is also in charge of Sam's Club. Wal-Mart stores now operate in Mexico, Canada, Argentina, Brazil, South Korea, China, and Puerto Rico. Sam Walton's visions were indeed successful.

Chapter 4: Getting Started

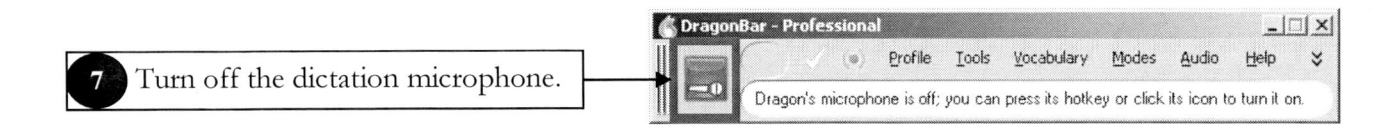

7 Turn off the dictation microphone.

If you encountered any misrecognitions that you would now like to correct, begin by placing your cursor within the text of the DragonPad screen and listening to the playback of your speech using *Alt P* ("P" for playback). As soon as you come across a text error for a word you clearly spoke, press *Alt C* ("C" for correction) to open the correction tool and type in the correct text. Make sure all the text within the correction window matches all audio you hear, then click *OK* (or press *Enter*) to finalize the correction.

If *Alt P* and *Alt C* do not perform the functions mentioned, it is because those keystrokes are not assigned within your Dragon user options. To revisit the instructions for setting those keystroke assignments, go to the *Setting User Options* section earlier in this chapter.

For more details about correcting speech recognition errors, see the *Correcting Misrecognitions* section of Chapter 6.

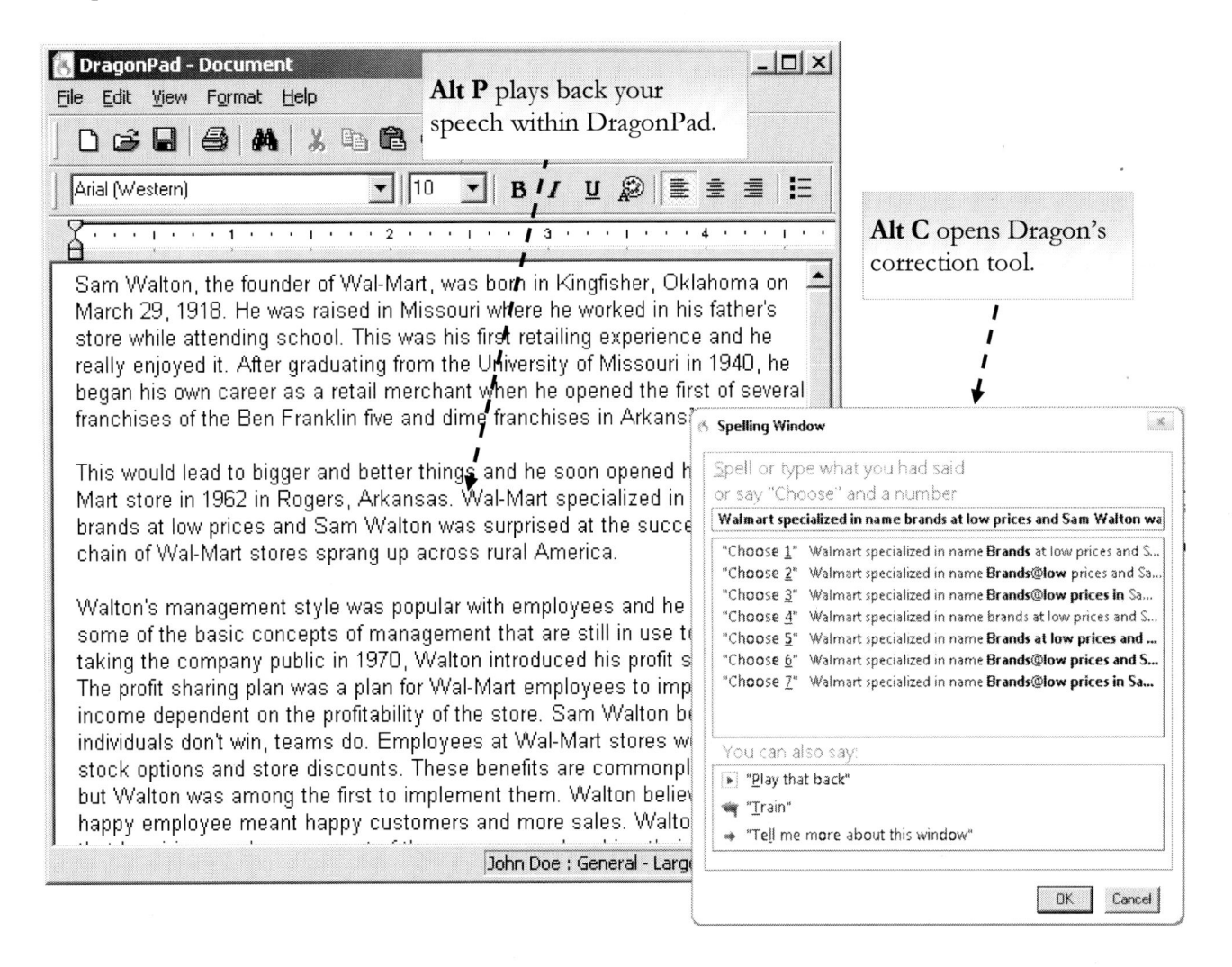

Alt P plays back your speech within DragonPad.

Alt C opens Dragon's correction tool.

119

Saving File of Dictation Session

If you are using the Professional edition of Dragon NaturallySpeaking®, you have the ability to save all the text and speech audio from your dictation session. If you are using the Premium edition, you will only be able to save the text from your file; you will no longer have access to the audio of your dictation once the file has been closed. Since Professional users can always have access to both text and speech audio from a dictation session—before and after closing the file—corrections to misrecognitions may be made at any time. For Premium users, this means corrections to misrecognitions must be made prior to closing the file since speech audio will not be available afterwards.

Follow the steps below to save the file of your dictation session from DragonPad.

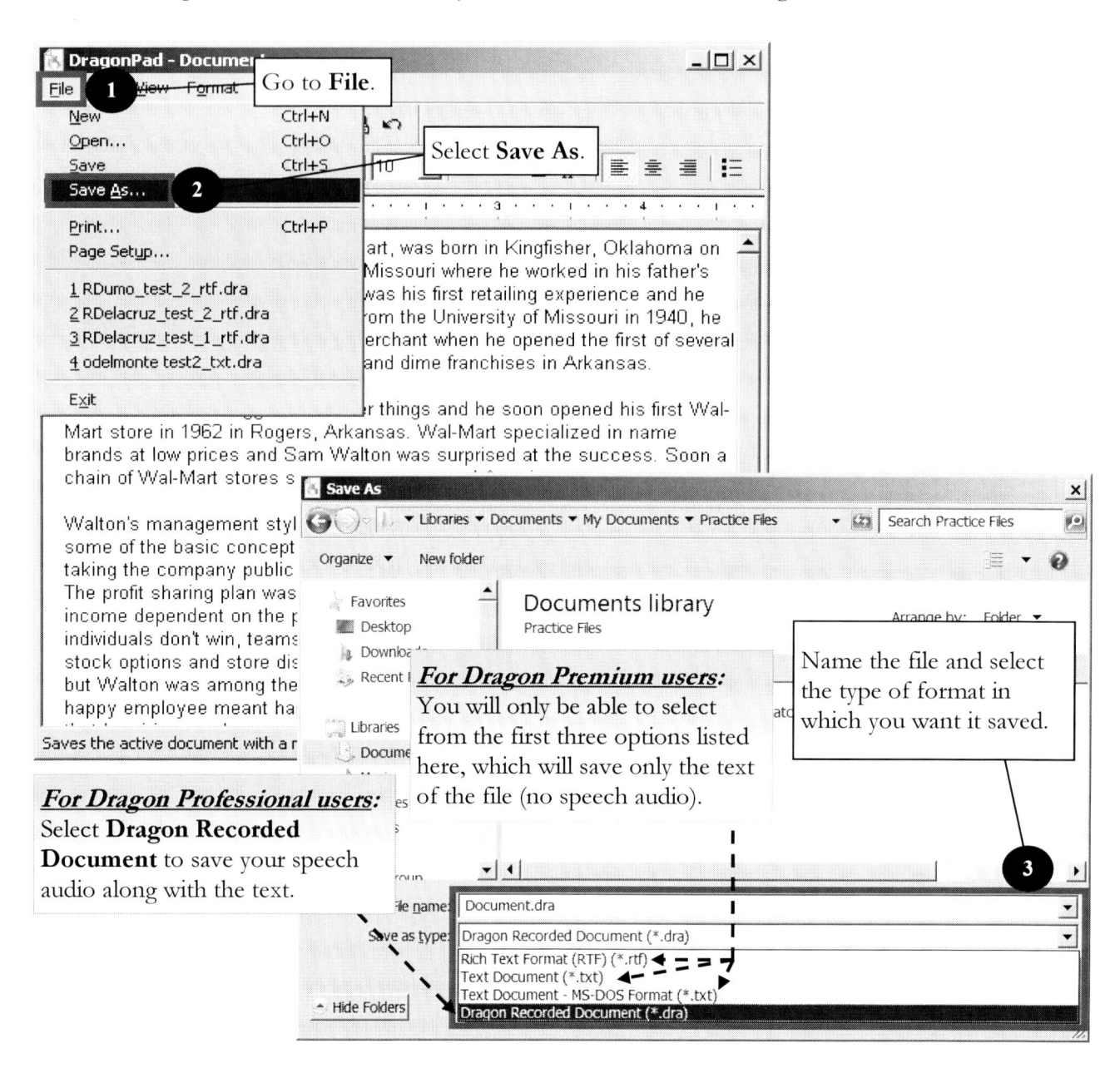

1. Go to **File**.

2. Select **Save As**.

3. Name the file and select the type of format in which you want it saved.

For Dragon Professional users:
Select **Dragon Recorded Document** to save your speech audio along with the text.

For Dragon Premium users:
You will only be able to select from the first three options listed here, which will save only the text of the file (no speech audio).

Chapter 5: Vocabulary Setup and Formatting

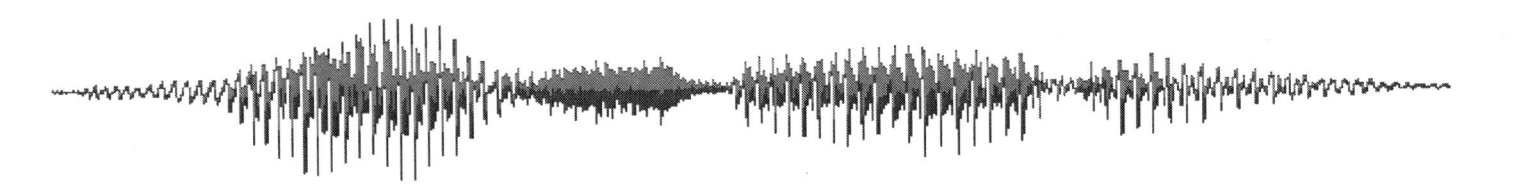

This chapter includes basic must-know information about working with your Dragon vocabulary. You will learn how to make voice writing entries for the voice codes you will be dictating as part of your voice writing theory and how to accomplish all your basic formatting needs using either Dragon and/or Microsoft Word.

If you use CAT software, the mechanisms for manipulating vocabulary input/output are different than when working in Dragon alone or in conjunction with Microsoft Word. Reference diagrams are included near the end of this chapter to give you a general idea of how your vocabulary entries may need to be set up to properly function within the parameters of different types of CAT software. Be aware, however, that the vocabulary and formatting setup procedures vary among CAT programs, so when using these diagrams, refer to your CAT program's user's manual or Help menu for supplemental instructions.

Working with the Vocabulary

Remember that Dragon's base vocabulary already contains most of the words you will encounter in your realtime work. As a result, you do not have to add any common English words. The only types of words you will need to add are unique or specialty words such as for uncommon names, technical terms, and acronyms. The code words that you need to dictate as part of your realtime voice writing language—including punctuation, speaker identifiers, et cetera—fall into the "unique or specialty words" category. Those are the entries you will be setting up throughout this chapter.

The next four sections give you an overview of the way vocabulary items must be spelled and how to add, delete, and modify their formatting properties.

Written- and Spoken-Form Spelling Rules

When making an entry into Dragon's vocabulary, two considerations are made regarding spelling: the written form and the spoken form. The terms "written form" and "spoken form" describe the different categories of spellings that Dragon will need in order to properly recognize a character, word, or phrase when the pronunciation you would expect to give it based upon the way it is spelled does not match the way it is actually pronounced. Thus the **written form** is how a vocabulary item is actually spelled, and the **spoken form** is the spelling for how it is pronounced.

If a pronunciation of a written-form spelling can be made based on rules of English phonetics, it is not necessary to give a spoken-form spelling. A spoken-form spelling only needs to be given if the written form does not match the way it is actually pronounced.

The principles of written- and spoken-form spellings appear in the table below.

Table 40

Spelling Form Principles for Vocabulary Entries

Written Forms	Spoken Forms [1]
1. A written-form spelling must always be given.	1. A spoken-form spelling is only required if a pronunciation cannot be made based on a written-form spelling, and it should be spelled according to rules of English phonetics.
2. A written-form spelling does not have to be spelled according to rules of English phonetics, and any combination of characters may be used (including letters, digits, punctuation marks and other symbols).	2. Except for a hyphen (-) or a single quotation mark ('), a spoken-form spelling should never include symbols, punctuation marks, or any other special characters. Digits are accepted.
3. A written-form spelling is the actual text Dragon will output onto your screen. (If you are using CAT software, this written output may be read by a CAT program to, in turn, output a different final realtime result.)	3. A spoken-form spelling should not include either a stand-alone capital letter or consecutive capital letters unless saying the names of those letters, as you would when reciting letters of the alphabet, is to be made a part of the pronunciation.

1) Whenever intending to use voice codes as spoken-form spellings, be aware that Dragon may not accept them all as proper spoken-form spellings if they conflict with pre-existing vocabulary words (e.g., "ITC," "ITR," and "ITT" already exist in Dragon's vocabulary). In those instances, simply locate and delete the base vocabulary words that pose the conflict (e.g., "ITC," "ITR," and "ITT"); you can then add your vocabulary entries as you originally intended and use those voice codes as spoken-form spellings.

Chapter 5: Vocabulary Setup and Formatting

This table shows some examples of how written forms and spoken forms are used.

Table 41

Examples of Written- and Spoken-Form Spellings

Written-Form Spellings That Match Pronunciations (Requiring Only Written-Form Spellings)		Written-Form Spellings That Do Not Match Pronunciations (Requiring Both Written- and Spoken-Form Spellings)	
Written Form	**Spoken Form**	**Written Form**	**Spoken Form**
MCSE — acronym for "Microsoft Certified Systems Engineer"	(none)	**COTR** — acronym for "contracting officer's technical representative"	cotar
Epzicom — prescription drug	(none)	**Wahde** — proper last name	wah day

To understand why some of the above words require only a written-form spelling while others should include both written and spoken forms, let's address the reasons for all examples given:

- "MCSE" does not require a spoken-form spelling because it is pronounced by saying the letters of the alphabet (*M, C, S, E*).

- "Epzicom" does not require a spoken-form spelling because you can make an accurate vocal pronunciation of it based upon the way it is spelled in its written form.

- It is preferable for "COTR" to include a spoken-form spelling since it is usually pronounced *cotar* and not by reciting the letters of the alphabet (*C, O, T, R*). Alternatively, you could spell the spoken form as "kotar" or any other way that phonetically matches how it is pronounced. If this acronym is sometimes pronounced by reciting letters of the alphabet, then you should make an additional entry in Dragon's vocabulary to account for that other pronunciation also (e.g., written-form spelling of "COTR" with no requirement of a spoken-form spelling).

- "Wahde," a proper last name of Liberian origin, requires a spoken-form spelling because its written form cannot be pronounced according to rules of English phonetics. English speakers simply reading this name on paper would probably pronounce it as either *wod* or *wade*, when it is actually pronounced *wah day*. Even though Dragon would accept your vocal recording of its pronunciation with only a written form entered, you couldn't expect Dragon to adequately recognize this word every time you say it without including a proper spoken-form spelling in this vocabulary entry.

Be aware that if you do not apply the written- and spoken-form spelling rules recommended in this book when making your vocabulary entries, Dragon will not be dependable in recognizing them in your dictation.

Voice Writing Method • FIFTH EDITION
Dragon NaturallySpeaking® 12

Accessing the Vocabulary

Instructions on how to open Dragon's vocabulary and find and select vocabulary items appear below.

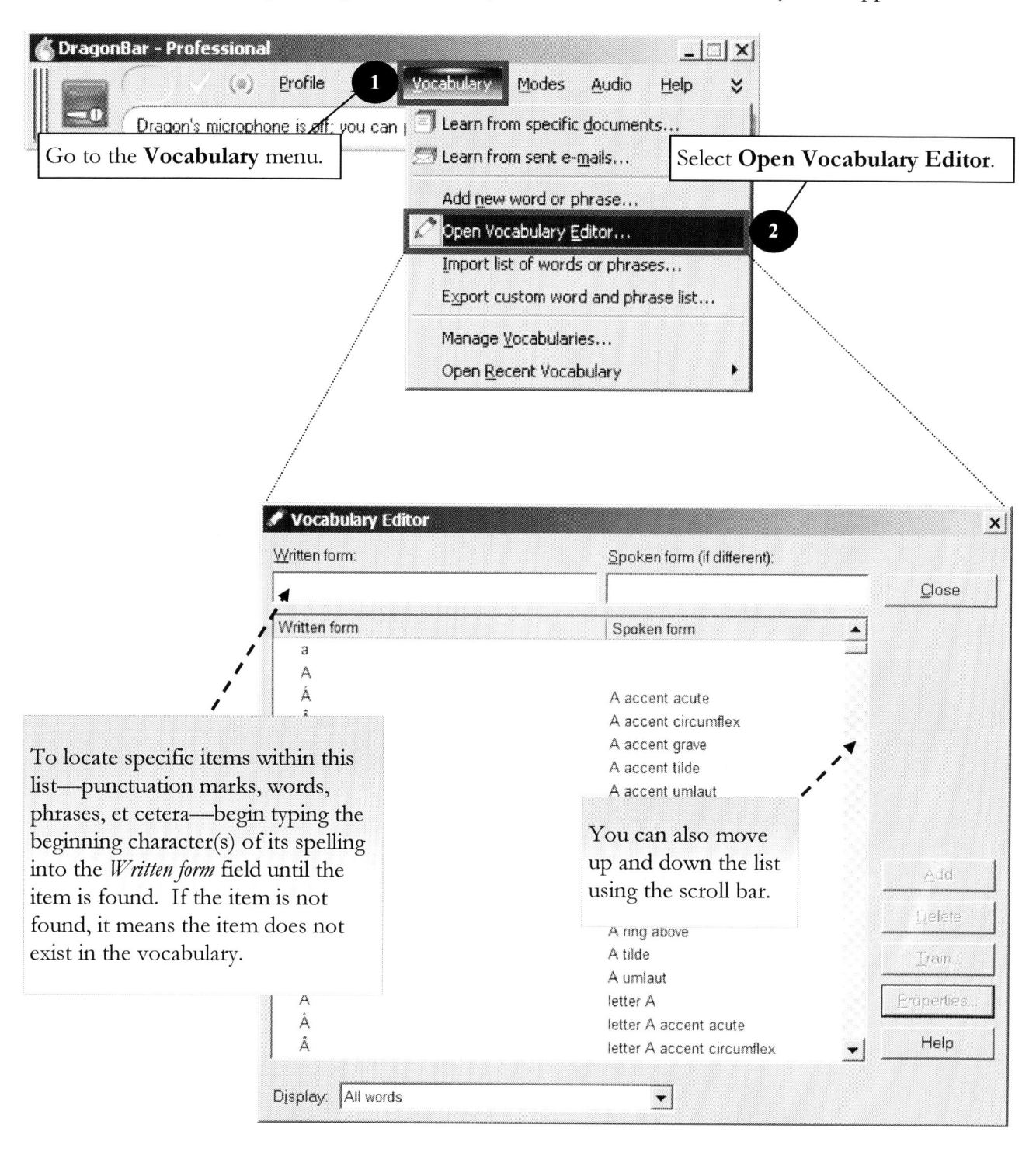

Go to the **Vocabulary** menu.

Select **Open Vocabulary Editor**.

To locate specific items within this list—punctuation marks, words, phrases, et cetera—begin typing the beginning character(s) of its spelling into the *Written form* field until the item is found. If the item is not found, it means the item does not exist in the vocabulary.

You can also move up and down the list using the scroll bar.

Chapter 5: Vocabulary Setup and Formatting

Adding an Entry

To learn how to add items to the vocabulary, follow the steps below for practice.

In the following example, "EBITDAX," an acronym used in financial call reporting, is pronounced *ebidax*. Since the written form of this entry must be spelled with consecutive capital letters, Dragon will assume you will be reciting the letters of the alphabet as part of its pronunciation (*E, B, I, T, D, A, X*) if you do not also give it a spoken-form spelling. Dragon will actually accept a recording for your speech using its regular pronunciation (i.e., *ebidax*) given only its written-form spelling, but Dragon will not recognize it as dependably as you'd like absent a spoken-form spelling matching it's pronunciation.

When making vocabulary entries, remember to always apply the spelling form principles.

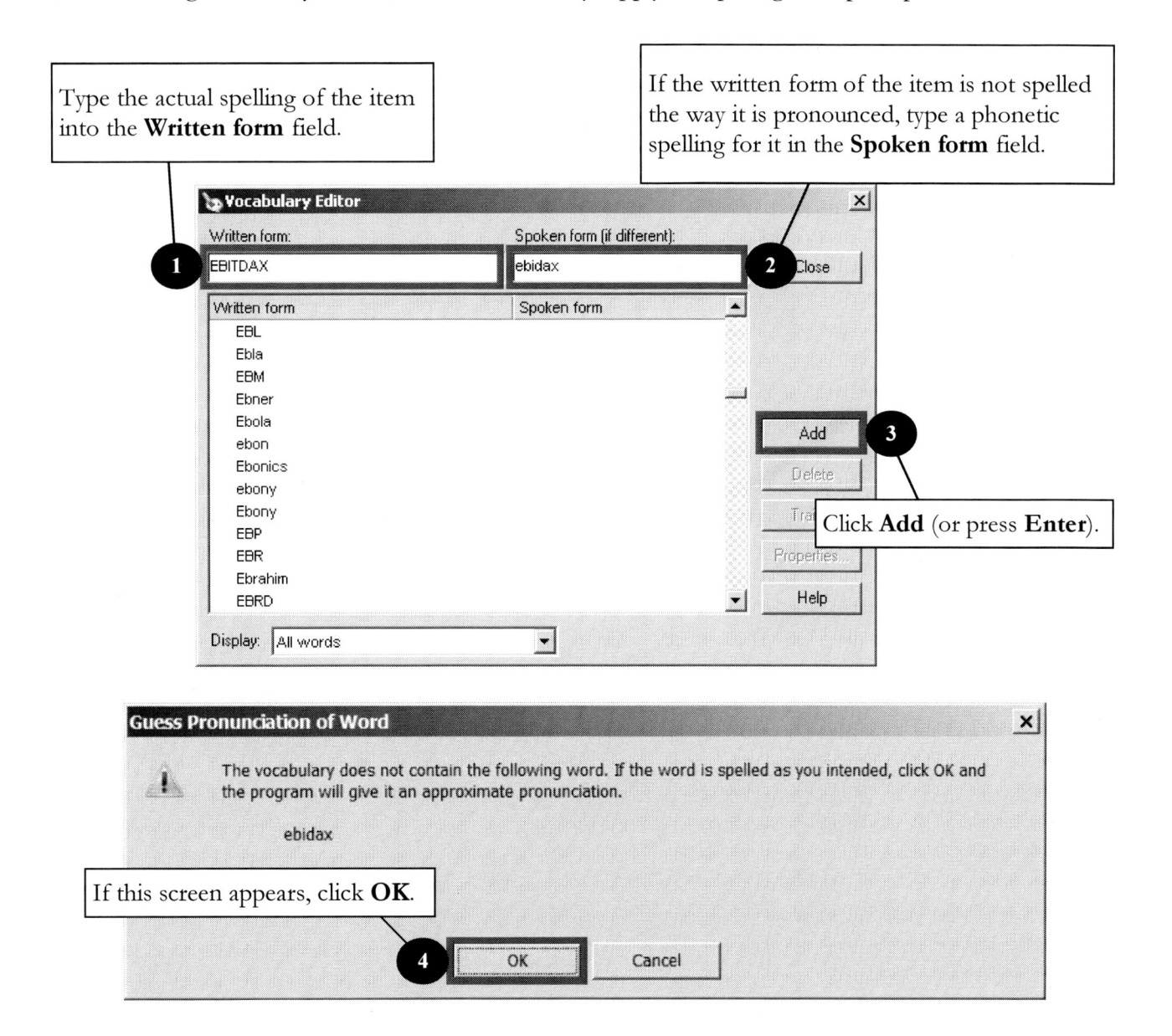

Type the actual spelling of the item into the **Written form** field.

If the written form of the item is not spelled the way it is pronounced, type a phonetic spelling for it in the **Spoken form** field.

Click **Add** (or press **Enter**).

If this screen appears, click **OK**.

125

Voice Writing Method • FIFTH EDITION
Dragon NaturallySpeaking® 12

Your entry will appear in the list. Dragon will have automatically assigned an approximate pronunciation for the entry once it's added; however, it is a good idea to record a vocal pronunciation for it anyway.

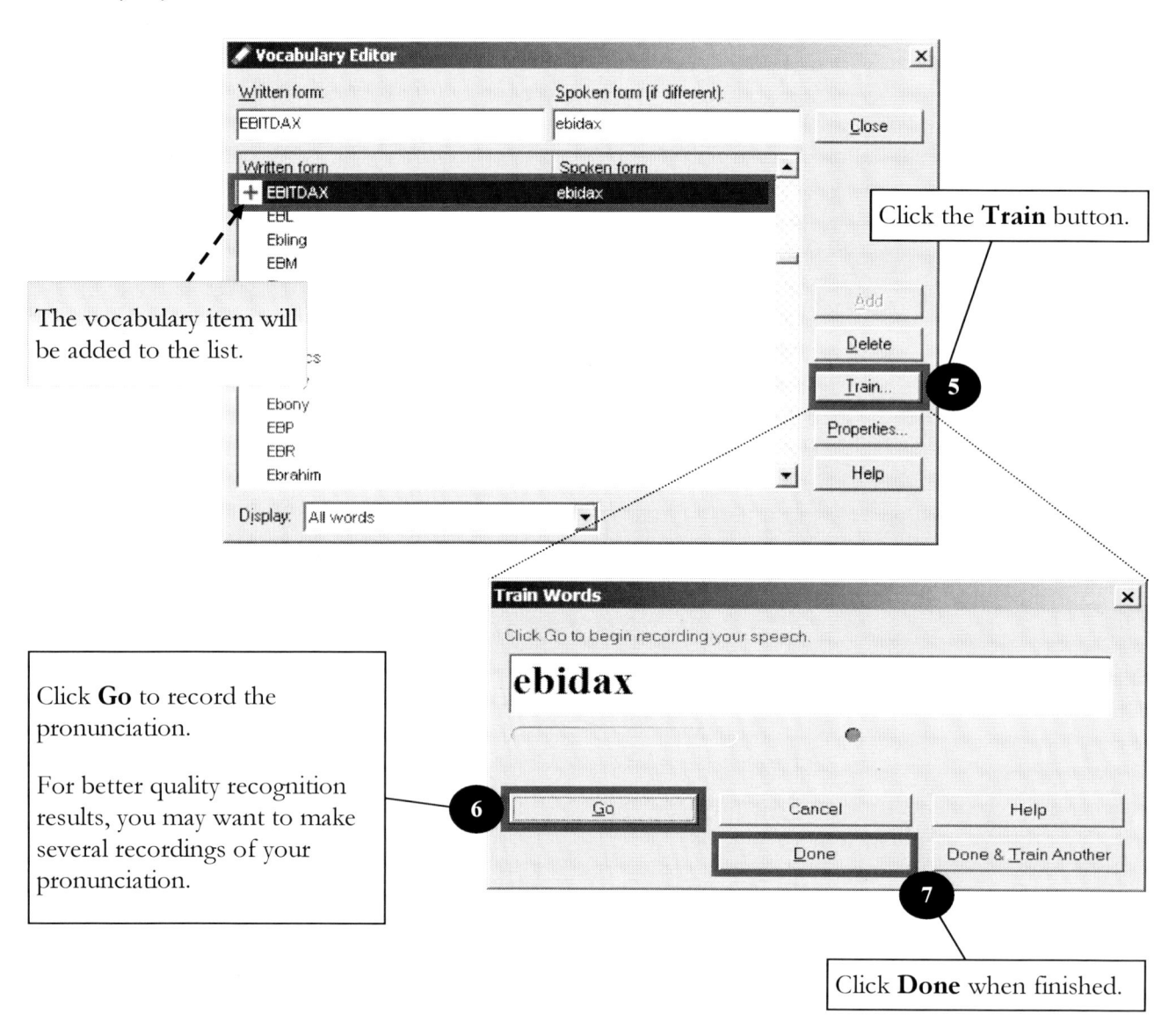

The vocabulary item will be added to the list.

Click the **Train** button.

Click **Go** to record the pronunciation.

For better quality recognition results, you may want to make several recordings of your pronunciation.

Click **Done** when finished.

> **Note**
>
> When adding vocabulary entries that you expect to regularly dictate as part of your voice writing theory—for example, voice codes for punctuation, speaker IDs, et cetera—record several pronunciations for each, saying them at different paces, to give Dragon additional samples of how you will dictate them at varying speeds.

Chapter 5: Vocabulary Setup and Formatting

Deleting an Entry

If you find that Dragon produces words you don't want to come up in your dictation, simply delete them from the vocabulary. As an example, you will regularly dictate the word part "spee" when identifying speakers (e.g., *spee-1*, *spee-2*, *spee-3*, etc.), and Dragon happens to contain the word "Spee" in its vocabulary. If Dragon occasionally recognizes the the word "Spee" instead of these voice codes, you would want to delete "Spee" from the vocabulary. Removing this word will eliminate further conflicts.

Here's how to delete an item from Dragon's vocabulary.

1. Locate the entry.
2. Click the **Delete** button.
3. If this screen appears, click **Yes**.

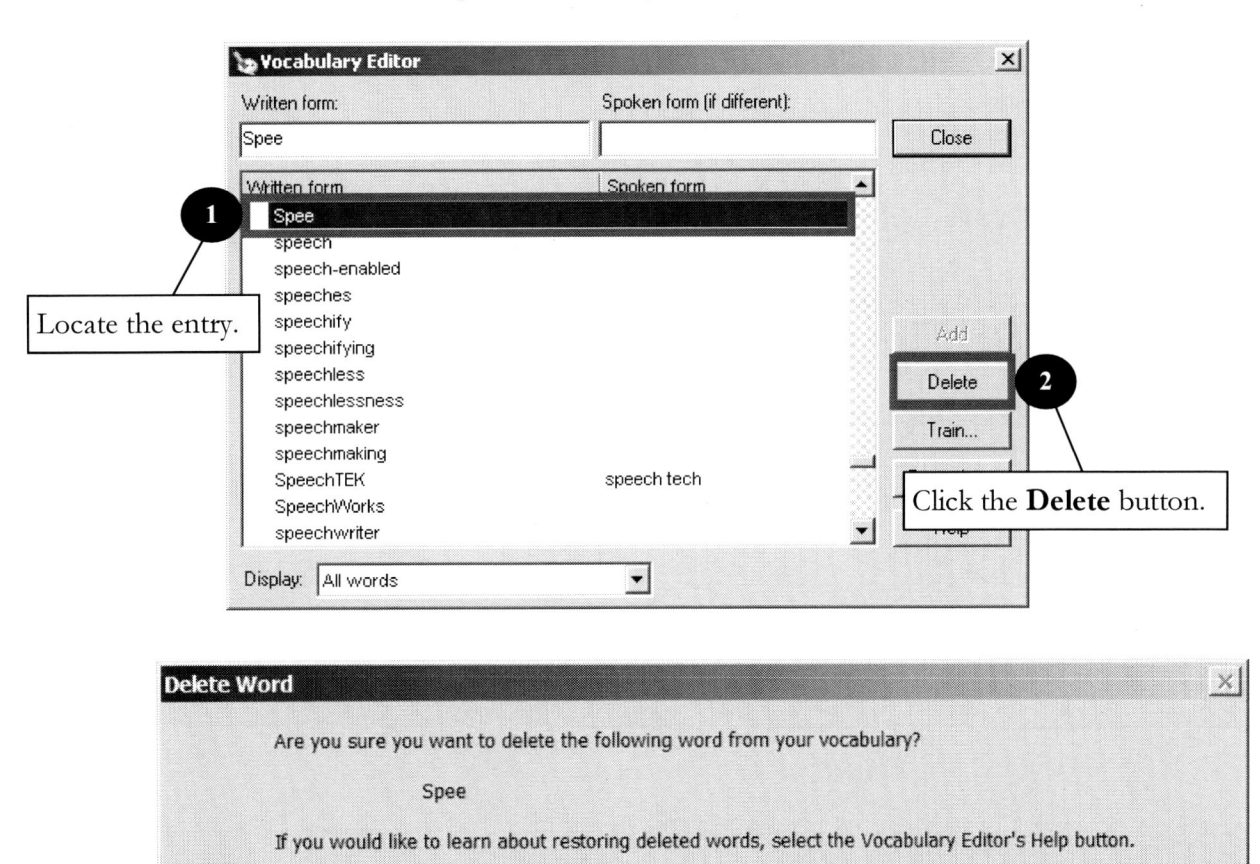

127

Voice Writing Method • FIFTH EDITION
Dragon NaturallySpeaking® 12

Modifying Formatting Properties of an Entry

You may want to modify formatting properties for certain vocabulary entries that are part of your voice writing theory. Suppose, for example, you want to use the voice code "peerk" to produce a period (.). Assuming you've already located that item in the vocabulary, the instructions on how to set the proper spacing and capitalization options are given below.

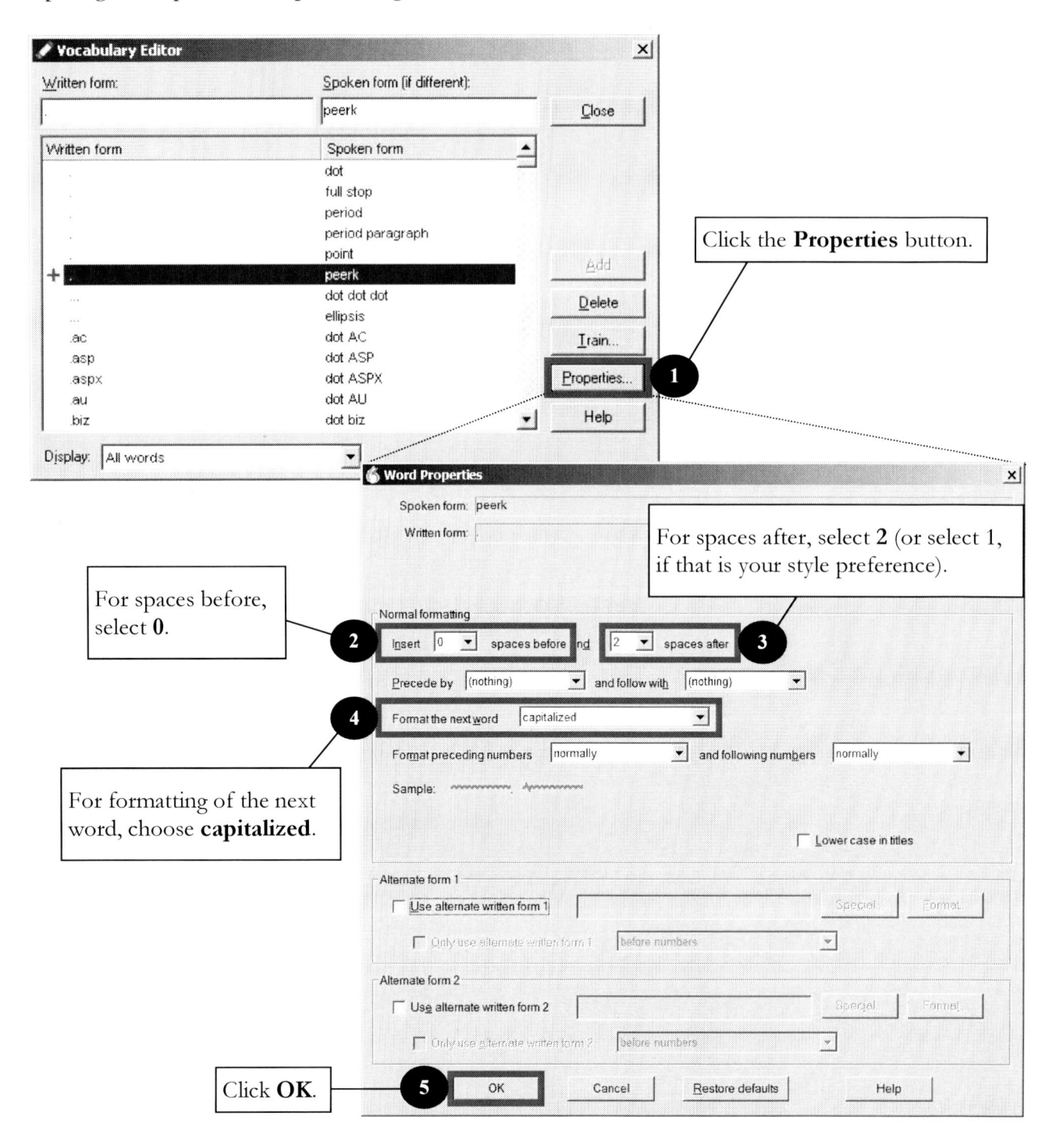

Click the **Properties** button. **1**

For spaces before, select **0**. **2**

For spaces after, select **2** (or select **1**, if that is your style preference). **3**

For formatting of the next word, choose **capitalized**. **4**

Click **OK**. **5**

128

Vocabulary Setup and Formatting Using Dragon Alone

There are two ways you can set up your voice writing entries in Dragon. One is to program them as vocabulary items, and the other is to program them as commands. When setting them up as vocabulary entries, formatting capabilities are limited and you may need to use other software to handle your more complex formatting needs. When setting them up as commands, virtually every type of formatting function can be produced. However, you would need to pause significantly before and after saying a command if you want Dragon to recognize it as an actual command as opposed to vocabulary words. And although your user options may be adjusted to the minimum pause setting required before saying a command, even slight pauses can be too difficult to accomplish during continuous, high-speed dictation. This makes command usage not viable for realtime voice writing. Since pauses are not as necessary for recognition of vocabulary words, it is best to set up your voice writing entries as vocabulary items.

If you have CAT software, you do not need to worry about Dragon's formatting limitations, because all the formatting you will require may be accomplished through your CAT program. Many formatting functions within Dragon are actually ineffective with certain CAT programs anyway. Follow the instructions in your CAT software manual in lieu of the instructions provided herein when setting up your voice writing entries.

If you are using Dragon by itself, the following sections will guide you through the process of setting up your voice writing entries and show you how to achieve proper formatted results, where possible, using only Dragon utilities. To find a way to benefit from more complex formatting functions without CAT software, you can use the automatic text-expansion features found within popular word-processing programs—such as Microsoft Word (*Office* button > *Word Options* button > *Proofing* menu > *AutoCorrect Options* button) or Corel WordPerfect (*Tools* menu > *QuickCorrect* > *QuickWords* tab)—to do the formatting for you. Instead of dictating into Dragon's document screen, DragonPad, you would dictate into your word-processing program's document screen and use its text-expansion utility as a means to convert the regular text output your word-processing software receives from Dragon into formatted text results on your document screen. Instructions on how to handle formatting with Microsoft Word are given near the end of this chapter. To learn how to use such features within a word-processing program other than Microsoft Word, refer to that program's Help menu.

As you set up your voice writing entries in Dragon, realize that you are not limited as to the number of vocabulary items you can add, delete, and modify. However, it is important to know that Dragon does have to maintain a balance of a certain number of entries within its vocabulary. This means that when you add a word, Dragon must take another word away. In total, Dragon contains around 300,000 words. If you created a user pursuant to the steps in Chapter 4 of this book, the base vocabulary that was automatically created for you has approximately 160,000 active words. The rest are stored in Dragon's backup vocabulary. Say you add 1,000 words to your vocabulary. The number of words in your active vocabulary list will still be 160,000; Dragon will have moved 1,000 of its base vocabulary words from the active list over to its backup vocabulary. Dragon does this to make room for the new words you've added. If you do not want Dragon to arbitrarily pick which words to move into its backup list, then you must select and delete words of your own choosing as you add new entries. For this reason, it is recommended that you add and delete words in smaller groups so you can maintain better control over the words moving in and out of the active list.

Punctuation Marks

For any punctuation for which you will be dictating voice codes, it is best to delete any pre-existing punctuation marks from Dragon's base vocabulary to avoid encountering certain conflicts. For example, when you say *time period* and you want your recognition result to be the word "period," Dragon may translate it as "time." (with the punctuation mark). Recognition ambiguities between words and punctuation marks may not seem like much of a concern in the beginning while you are just learning how to do realtime voice writing, but such conflicts will become annoying once you get into professional practice. Besides helping you to avoid recognition conflicts like these, using voice codes for the most commonly used punctuation marks will help you to produce text faster because you will be dictating fewer syllables. But if you prefer, you can just leave the regular base punctuation items as they are and dictate their regular pronunciations. If you want to use Dragon's base punctuation, skip this section.

The example provided here shows you how to modify your vocabulary when setting up a question mark (?) so that it responds to the code-word pronunciation of *kwee*. Refer to the tables in the *Punctuation* section of Chapter 3 to determine which punctuation marks you want to set up in like manner.

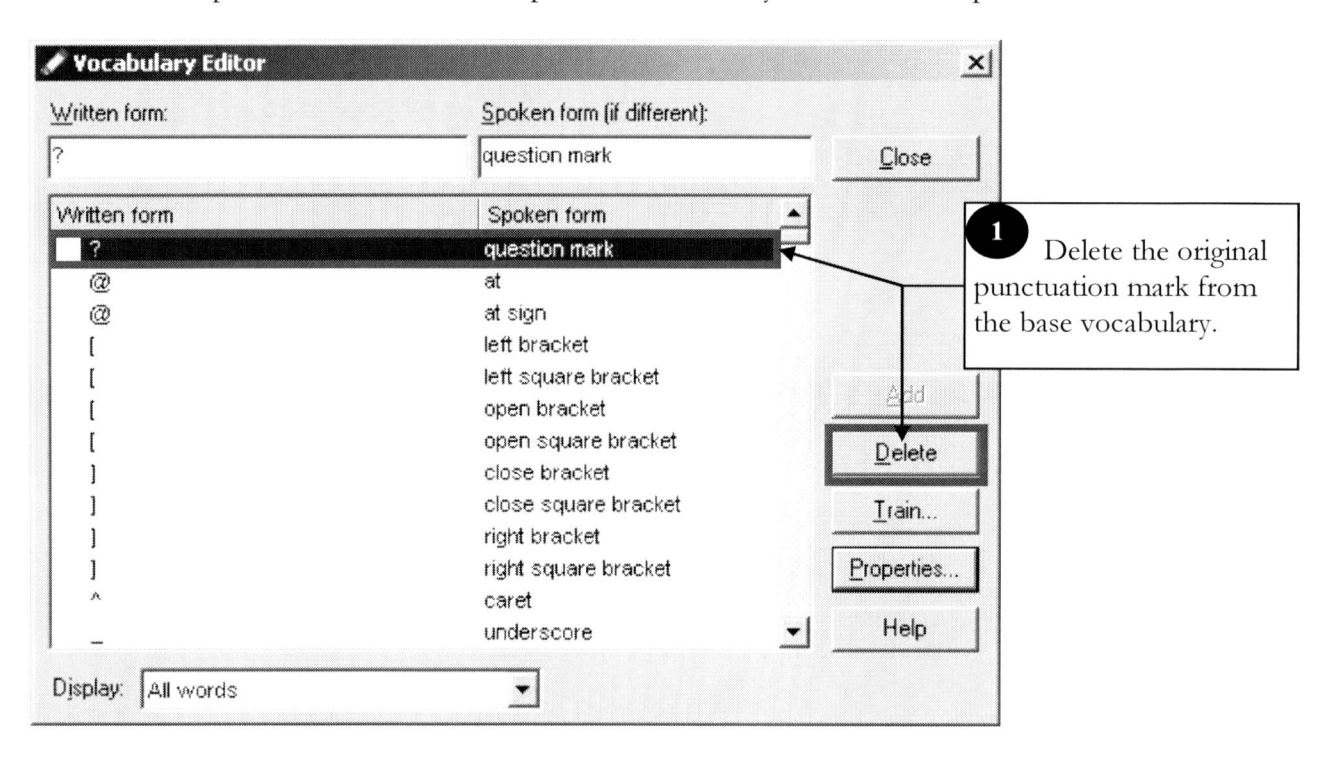

1 Delete the original punctuation mark from the base vocabulary.

You may want to also delete the following base vocabulary punctuation marks: "period" and "full stop" (.); "exclamation mark" and "exclamation point" (!); "colon" and "colon mark" (:) and the base vocabulary word "Colo"; "semicolon" (;); "dash," "double dash," and "N dash" (--); "forward slash" and "slash" (/); "begin double quote," "begin double quotes," "begin quote," "begin quotes," "open double quote," "open double quotes," "open quote," and "open quotes" ("); "close double quote," "close double quotes," "close quote," "close quotes," "end double quote," "end double quotes," "end quote," and "end quotes" ("); "left paren," "left parenthesis," "open paren," and "open parenthesis" ((); and "close paren," "close parenthesis," "right paren," and "right parenthesis" ()).

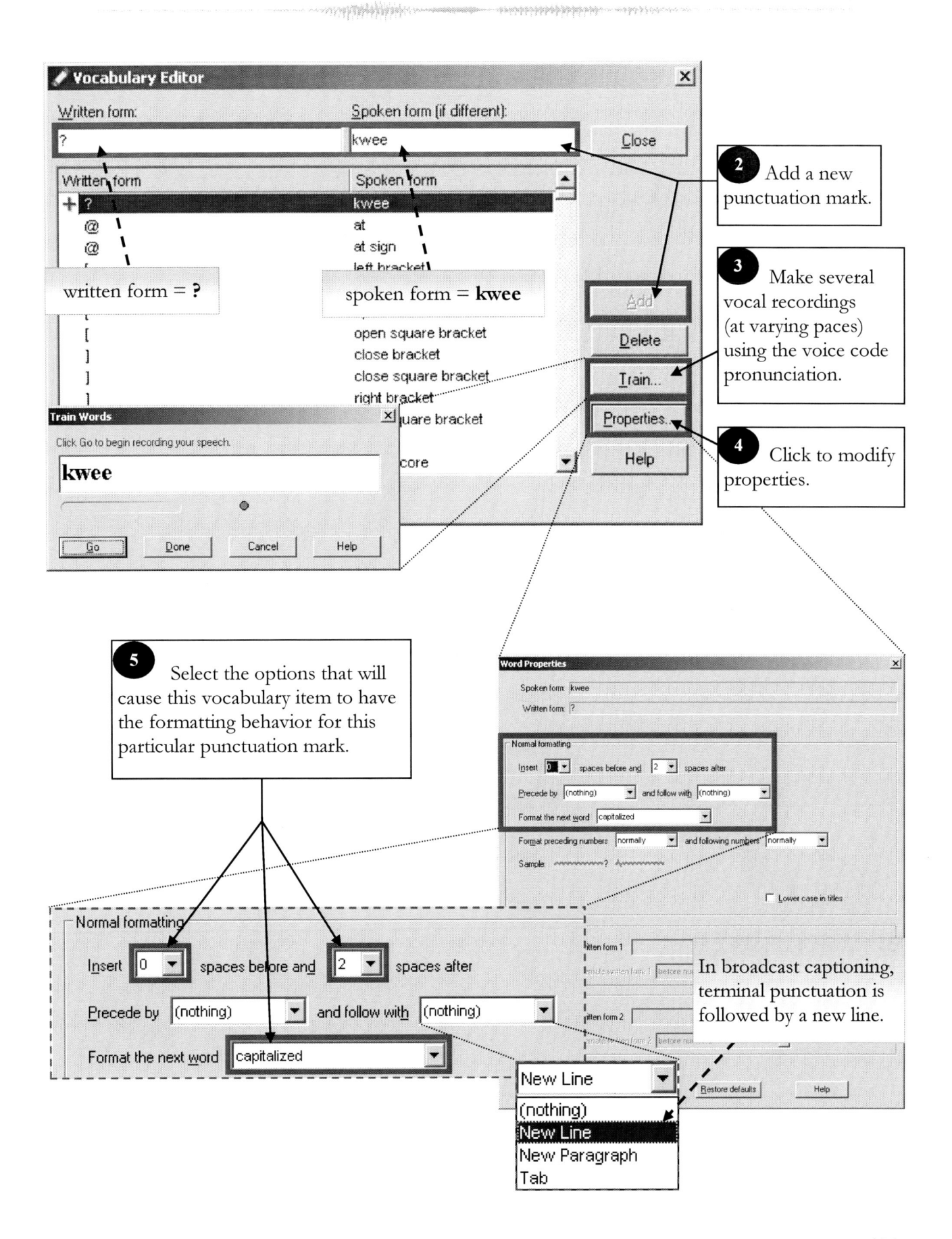

Chapter 5: Vocabulary Setup and Formatting

2 Add a new punctuation mark.

written form = **?**

spoken form = **kwee**

3 Make several vocal recordings (at varying paces) using the voice code pronunciation.

4 Click to modify properties.

5 Select the options that will cause this vocabulary item to have the formatting behavior for this particular punctuation mark.

In broadcast captioning, terminal punctuation is followed by a new line.

131

The table below lists vocabulary setup information for additional punctuation entries you may want to create.

Table 42

Punctuation Entries

New Punctuation Entry		Formatting Properties				
Written Form	Spoken Form	Spaces Before	Spaces After	Format the Next Word	Format Preceding Numbers	Format Following Numbers
.[1]	peerk	0	2 (or 1)	capitalized	normally	normally
![1]	excla	0	2 (or 1)	capitalized	normally	normally
:	colo	0	2 (or 1)	with normal capitalization	normally	normally
;	semco	0	1	with normal capitalization	normally	normally
--	doesh	1	1	with normal capitalization	normally	normally
-	hyka	0	0	with normal capitalization	as numerals	as numerals
/	slok	0	0	with normal capitalization	as numerals	as numerals
"	oco	1	0	with normal capitalization	normally	normally
"	ocap	1	0	capitalized	normally	normally
"	cloco	0	1	with normal capitalization	normally	normally
(	opar	1	0	with normal capitalization	normally	normally
)	clopar	0	1	with normal capitalization	normally	normally
,"	kahcap	0	0	capitalized	normally	normally
,"	kahco	0	1	with normal capitalization	normally	normally
."[1]	peerco	0	2 (or 1)	capitalized	normally	normally
?"[1]	kweeco	0	2 (or 1)	capitalized	normally	normally

1) For broadcast captioning, these terminal punctuation marks should be followed by a new line. In the *Precede by __ and follow with __* section of the *Word Properties* screen, leave *Precede by* set to **(nothing)** and set *and follow with __* to **New Line**.

To ensure that all your formatting has been set up properly, you will want to test your results by dictating these punctuation marks in DragonPad.

Chapter 5: Vocabulary Setup and Formatting

Speaker IDs

Choose the proper speaker IDs from one of the tables in the *Speaker Identification* section of Chapter 3 to determine which ones you need for your particular realtime career specialty.

For speaker identifiers that will always produce the same recurring name or title (e.g., *struc-mac* for "INSTRUCTOR:," *nar-mac* for ">>NARRATOR:" *op-mac* for "Operator," etc.), set them up similar to the example below.

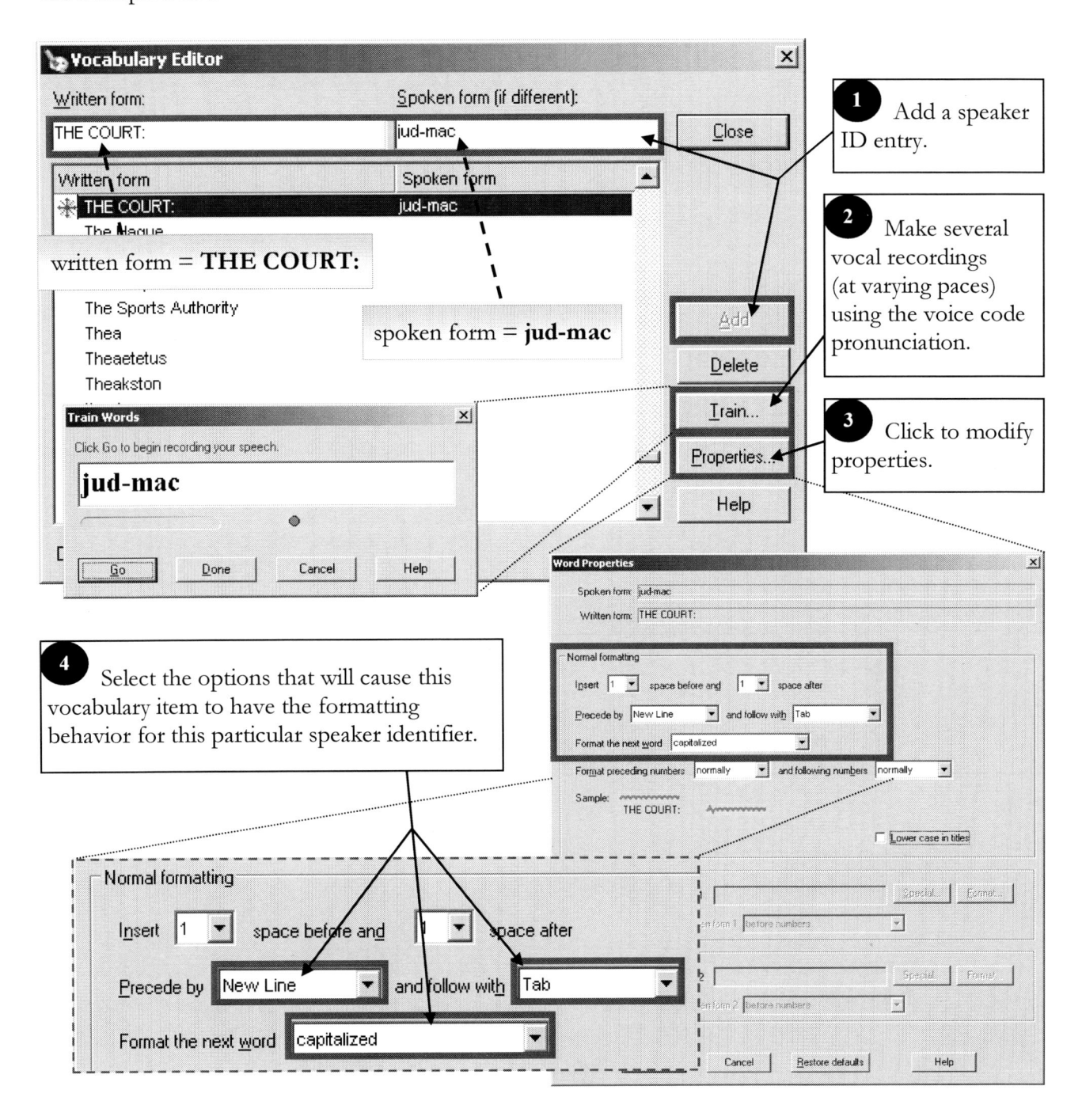

1 Add a speaker ID entry.

written form = **THE COURT:**

spoken form = **jud-mac**

2 Make several vocal recordings (at varying paces) using the voice code pronunciation.

3 Click to modify properties.

4 Select the options that will cause this vocabulary item to have the formatting behavior for this particular speaker identifier.

133

Voice Writing Method • FIFTH EDITION
Dragon NaturallySpeaking® 12

When setting up generic speaker identifiers (e.g., spee-1, spee-2, spee-3, etc.) intended for outputting specific speaker names (e.g., Mr. Smith, Ms. Jones, Mr. Black, etc.), you will use alternate written forms. Alternate written forms can be modified at any time, making it convenient for you to frequently change specific name assignments as you encounter different speakers from one job to the next. An explanation of how to set up such a speaker identifier is given below.

written form = **SPEAKER-1**
(original form only)

spoken form = **spee-1**

1 Add a speaker ID entry.

2 Make several vocal recordings (at varying paces) using the voice code pronunciation.

3 Click to modify properties.

4 Check the box next to **Use alternate written form 1**.

5 Type in the specific name of the speaker.

6 Click to modify the formatting properties for this alternate written form.

(screen shown on next page)

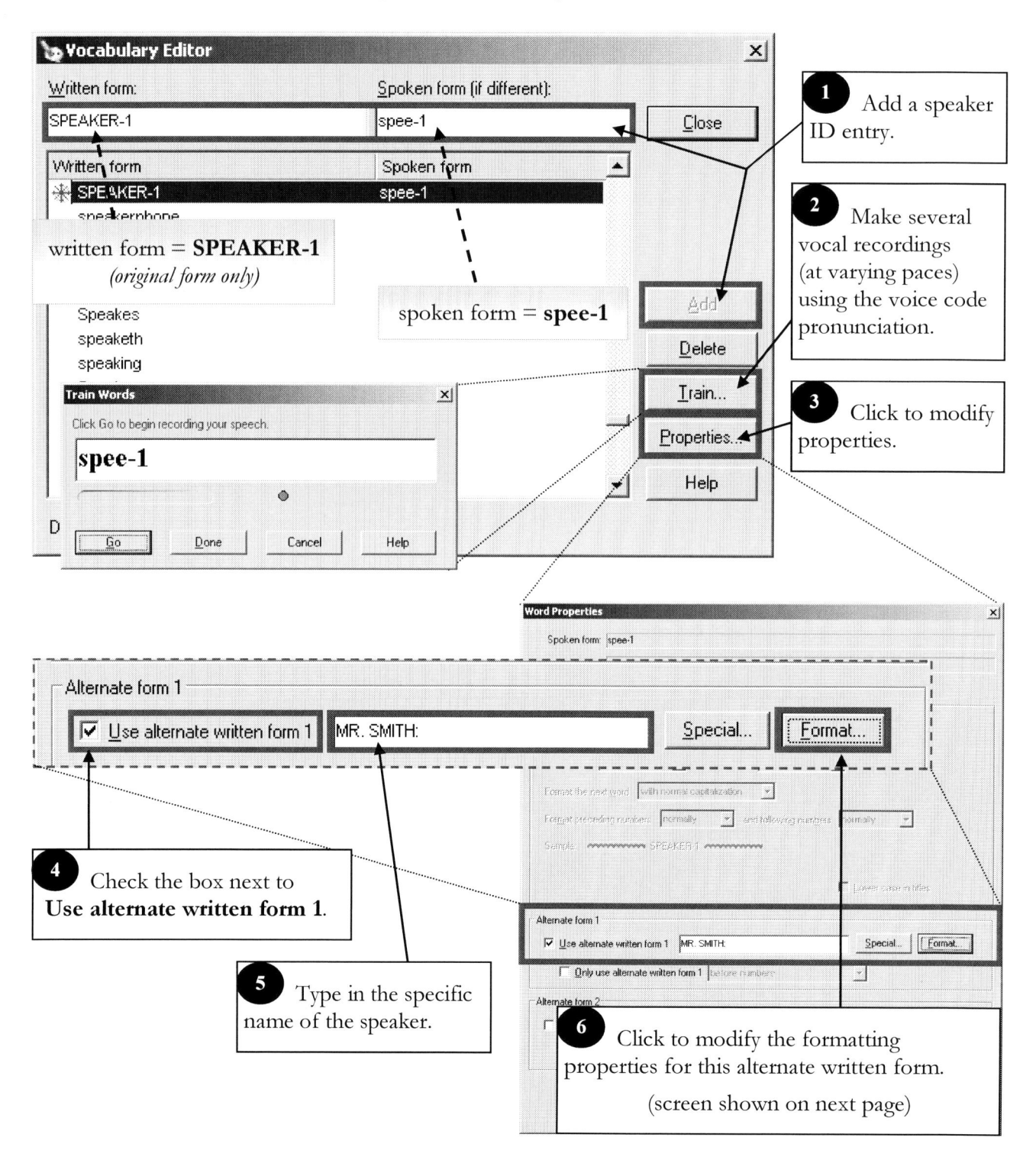

134

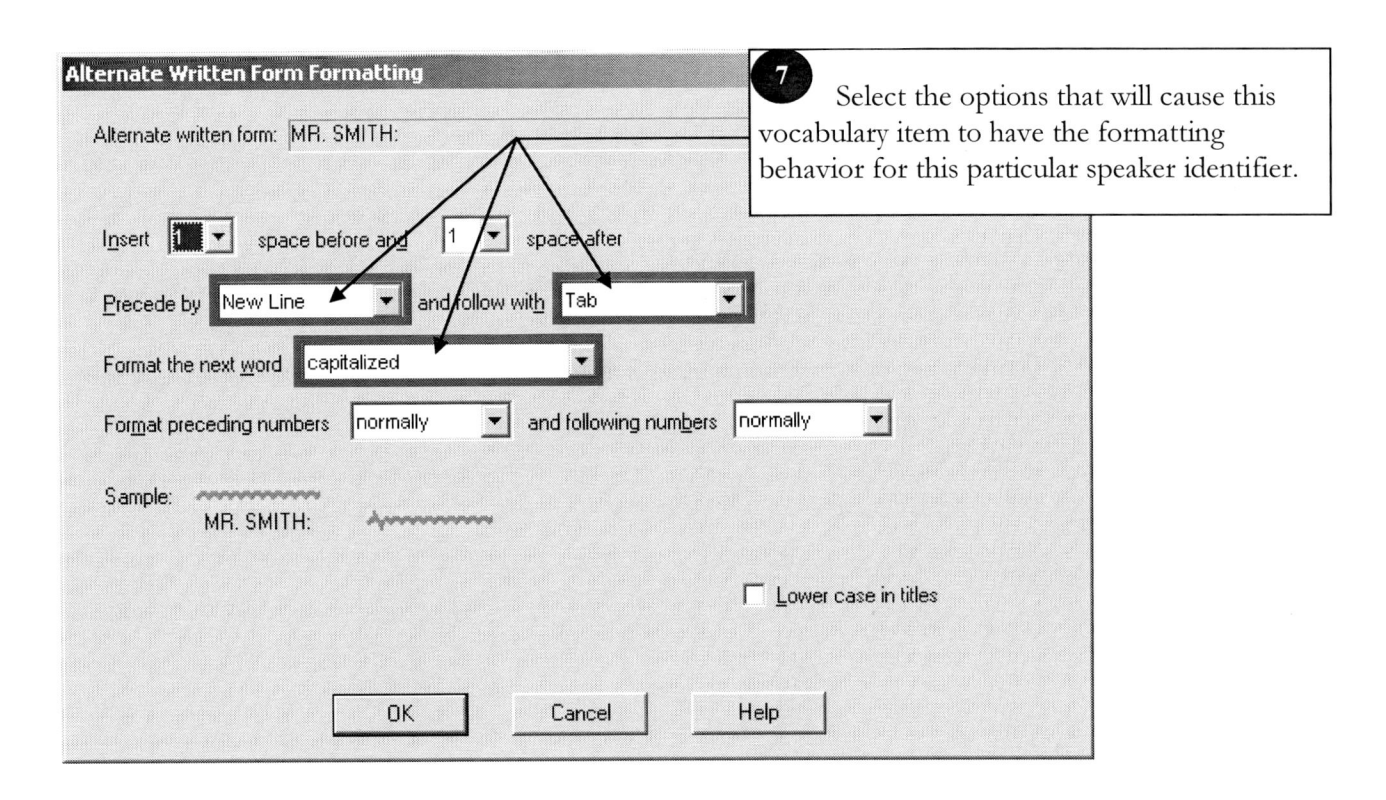

Chapter 5: Vocabulary Setup and Formatting

7 Select the options that will cause this vocabulary item to have the formatting behavior for this particular speaker identifier.

Now when you dictate *spee-1*, the name "MR. SMITH:" will appear on your screen in your chosen format. Set up your remaining name-specific speaker identifiers the same way and test out your results by dictating your speaker voice codes in DragonPad.

When you need to change a speaker name assignment in the future, simply click on the speaker vocabulary item and do the following:

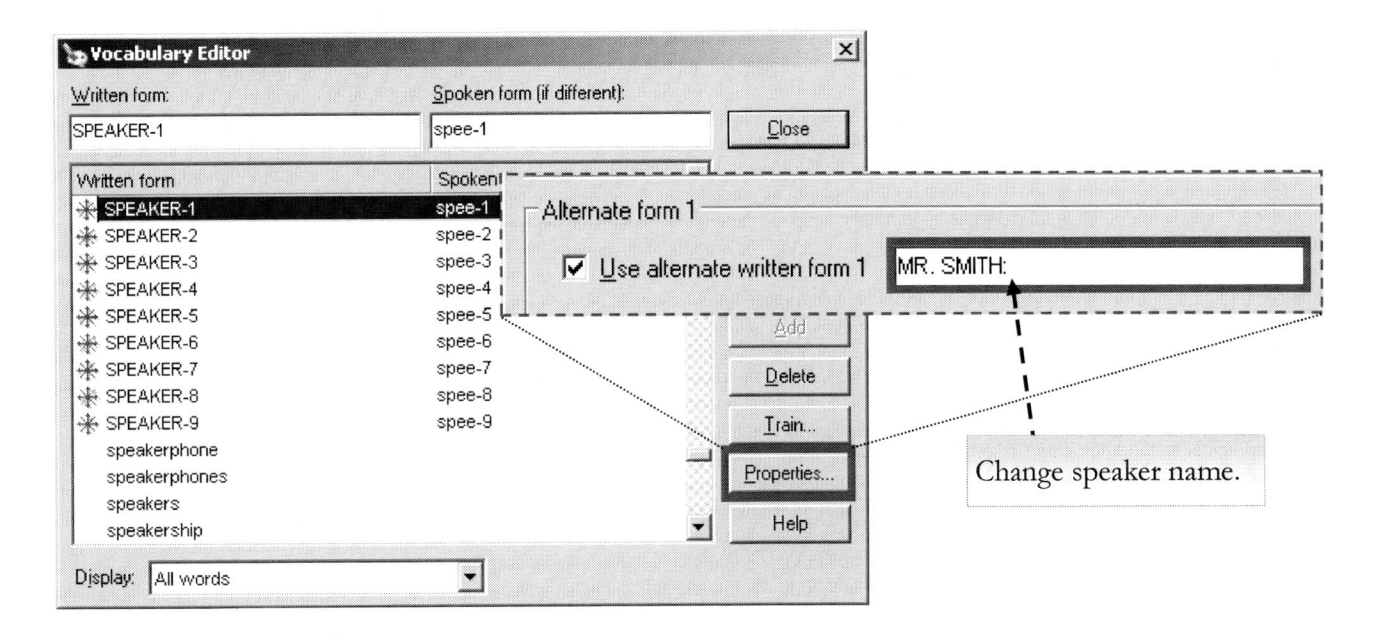

Click the **Properties** button > change the name that appears under **Alternate form 1**

Change speaker name.

135

Voice Writing Method • FIFTH EDITION
Dragon NaturallySpeaking® 12

Q&A Markers

Q&A markers may need to appear in different formats for different realtime career specialties, so follow the guidelines provided to you by your employer or company contractor as you set up your Q's and A's using the examples below.

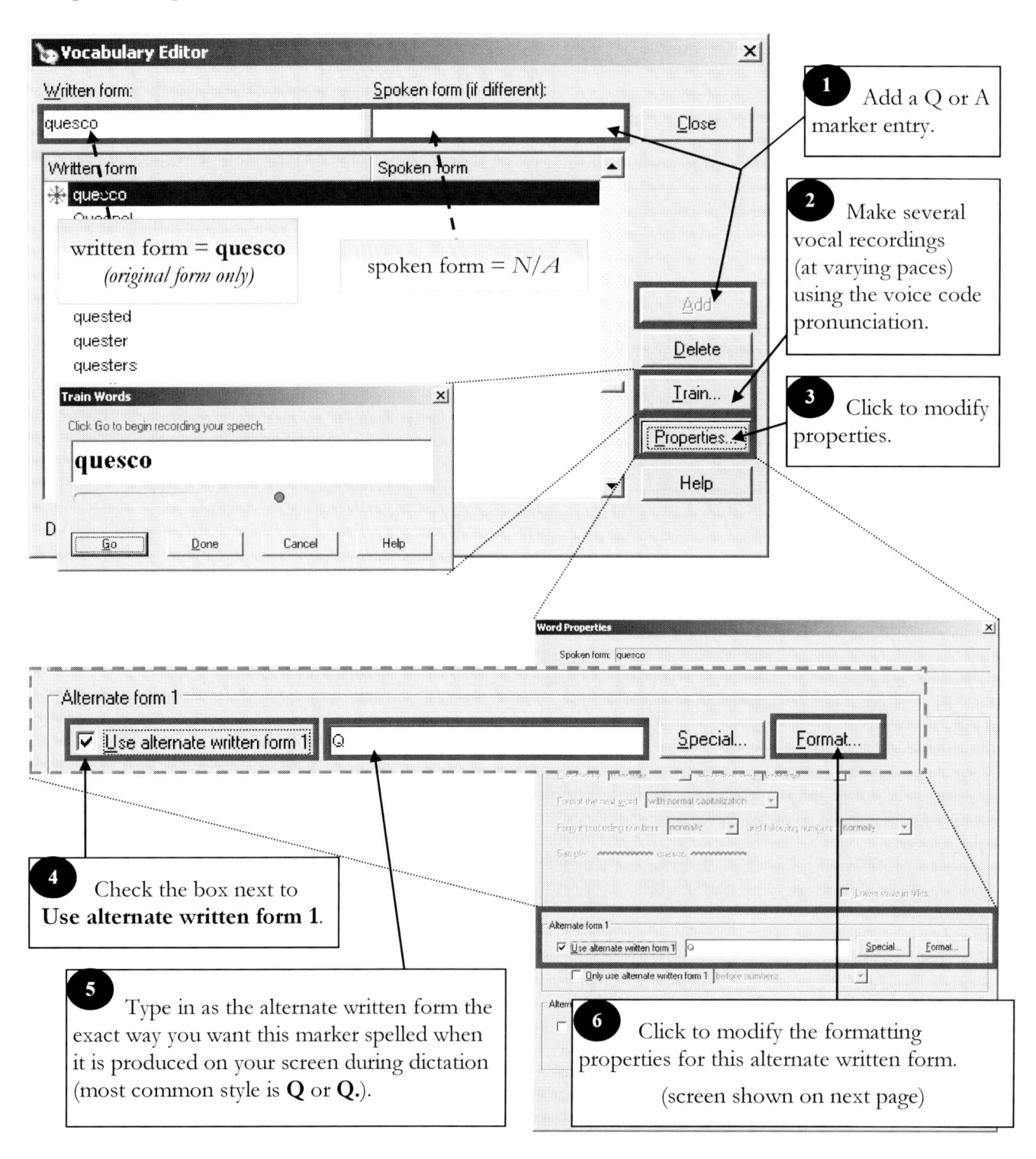

written form = **quesco**
(original form only)

spoken form = N/A

1 Add a Q or A marker entry.

2 Make several vocal recordings (at varying paces) using the voice code pronunciation.

3 Click to modify properties.

4 Check the box next to **Use alternate written form 1**.

5 Type in as the alternate written form the exact way you want this marker spelled when it is produced on your screen during dictation (most common style is **Q** or **Q.**).

6 Click to modify the formatting properties for this alternate written form.

(screen shown on next page)

Chapter 5: Vocabulary Setup and Formatting

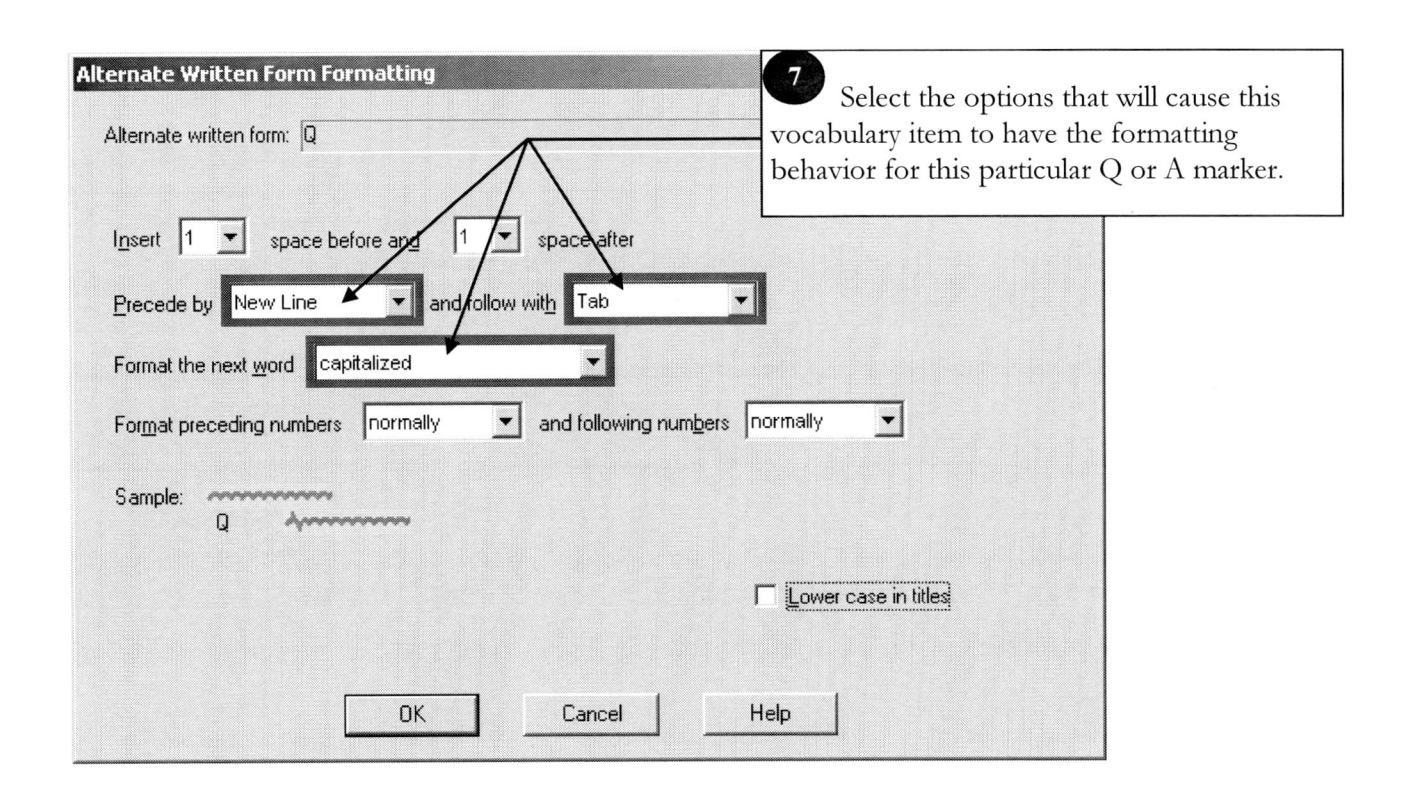

7 Select the options that will cause this vocabulary item to have the formatting behavior for this particular Q or A marker.

Note there is a reason that "Q" was not used as the original written form in this example. Having your question marker entry spelled as "quesco" will rule out misrecognition possibilities between this marker entry and other vocabulary items such as your spelling-technique entry for "Q" (with "Q" being the written form and "Q-mac" being the spoken form) and "Q" which responds to the pronunciations of *letter Q*, *capital Q*, etc. The original written-form spellings for all of your special voice writing entries must be unique and not match the spellings of any other vocabulary items. This will become very important later on when you learn about the topic of vocabulary building involving the customization of Dragon's grammatical model. (See the *Vocabulary Building* section in Chapter 6.) Because Dragon will be analyzing contexts of words as they are written only, you will need to custom-prepare documents to reflect the actual written-form identities of all the words you speak (their original written forms) so that Dragon can discern the difference amongst them as it learns about their contexts.

Now, set up your answer marker in like manner, using "respo" as the original written-form spelling and "A" as the alternate written-form spelling.

When finished, test your formatted results by dictating your Q&A markers into DragonPad.

All Other Voice Writing Entries

Refer to all the tables in Chapter 3 to determine which other voice writing entries need to be set up. Remember when setting up the rest of your entries that you should use voice codes as the spoken-form spellings. But if you have CAT software, instead comply with your user's manual instructions as opposed to the ones provided in this book, because those steps may be different.

Since the majority of your other voice writing entries do not require complex formatting, Dragon's vocabulary formatting options should work well in most cases. If not, see the next section which explains how to use Microsoft Word to accomplish more defined formatted results.

Vocabulary Setup and Formatting Using Dragon with Other Applications

The next three sections address how the voice writing entries in your Dragon vocabulary may need to be set up to work in conjunction with Microsoft Word and CAT applications to achieve formatted results within those applications.

Microsoft Word

If you find that Dragon's formatting options are too limited to handle your formatting needs for certain voice writing entries and you do not have CAT software, you may obtain more complex formatting results using Microsoft Word's text-expansion feature. To derive the most benefit from this feature, the only formatting option you want to use within Dragon's vocabulary properties is capitalization; Microsoft Word will handle all the remaining formatting functions for you (e.g., tabs, indents, new lines, centered text, underlined text, etc.).

To illustrate how Microsoft Word's word-expansion feature will convert the words it receives from Dragon into formatted text during your dictation, let's begin with a simple example: an examination line for a court reporting transcript. Depending on the transcript formatting guidelines used by particular court systems or freelance reporting firms, formats of examination lines in court reporting transcripts vary. An examination line can either be left-justified or centered and may either appear on a line by itself or on the same line as a speaker's name. Many court reporters like to underline their examination lines so they stand out better on the page. Dragon does not provide options to center and underline vocabulary items. And because you need to be working in Dragon's *Dictation Mode* to prevent commands from mistakenly being recognized during realtime transcription, neither will you be able to use Dragon's commands to center and underline text. Assuming you want to create an examination line that will be centered and underlined, a Microsoft Word text-replacement entry would need to be set up as a corresponding entry to your Dragon vocabulary entry for that examination line. Setup instructions appear on the next page.

Chapter 5: Vocabulary Setup and Formatting

written form = **EXAMINATION**

spoken form = **exi exi**

1 Add the entry for an examination line.

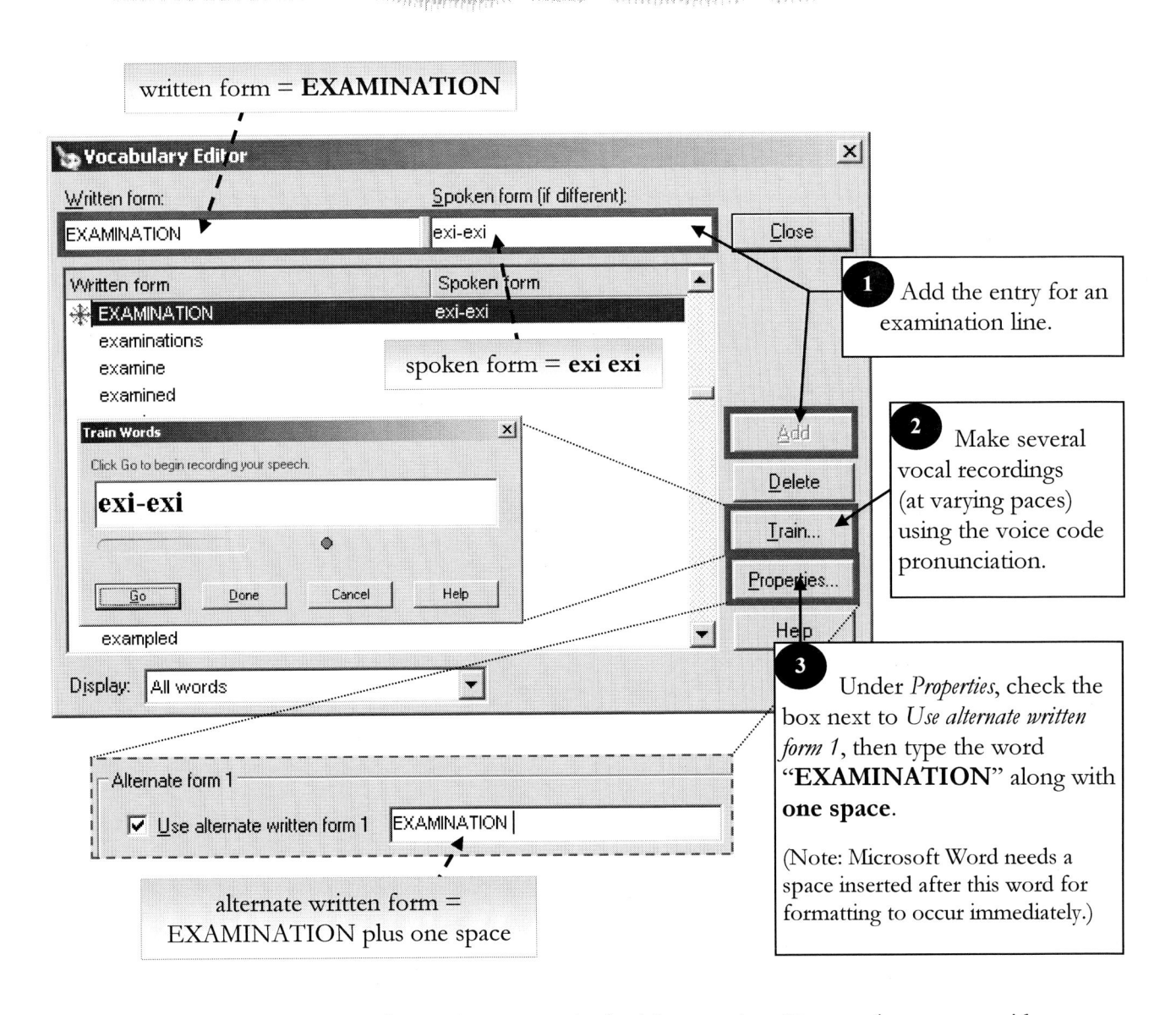

2 Make several vocal recordings (at varying paces) using the voice code pronunciation.

3 Under *Properties*, check the box next to *Use alternate written form 1*, then type the word "**EXAMINATION**" along with **one space**.

(Note: Microsoft Word needs a space inserted after this word for formatting to occur immediately.)

alternate written form = EXAMINATION plus one space

You will not need to select any formatting properties in this case, since Dragon does not provide you with the options of underlining or centering text. At this point when you dictate *exi-exi*, Dragon will simply output the text "EXAMINATION" along with one space.

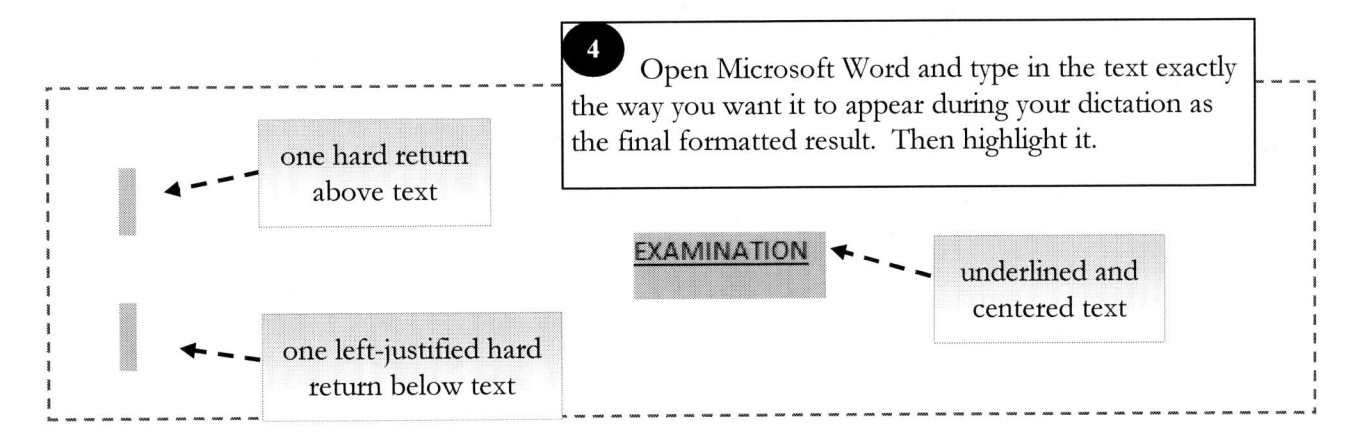

4 Open Microsoft Word and type in the text exactly the way you want it to appear during your dictation as the final formatted result. Then highlight it.

one hard return above text

underlined and centered text

one left-justified hard return below text

139

Voice Writing Method • FIFTH EDITION
Dragon NaturallySpeaking® 12

Next, you would open the *AutoCorrect* table in Microsoft Word. The way to access this utility is different in different versions of Microsoft Word. Generally, you would open *AutoCorrect* using one of the following paths:

Tools menu > *AutoCorrect*

or

Office button > *Word Options* button > *Proofing* > *AutoCorrect Options* button > *AutoCorrect* tab

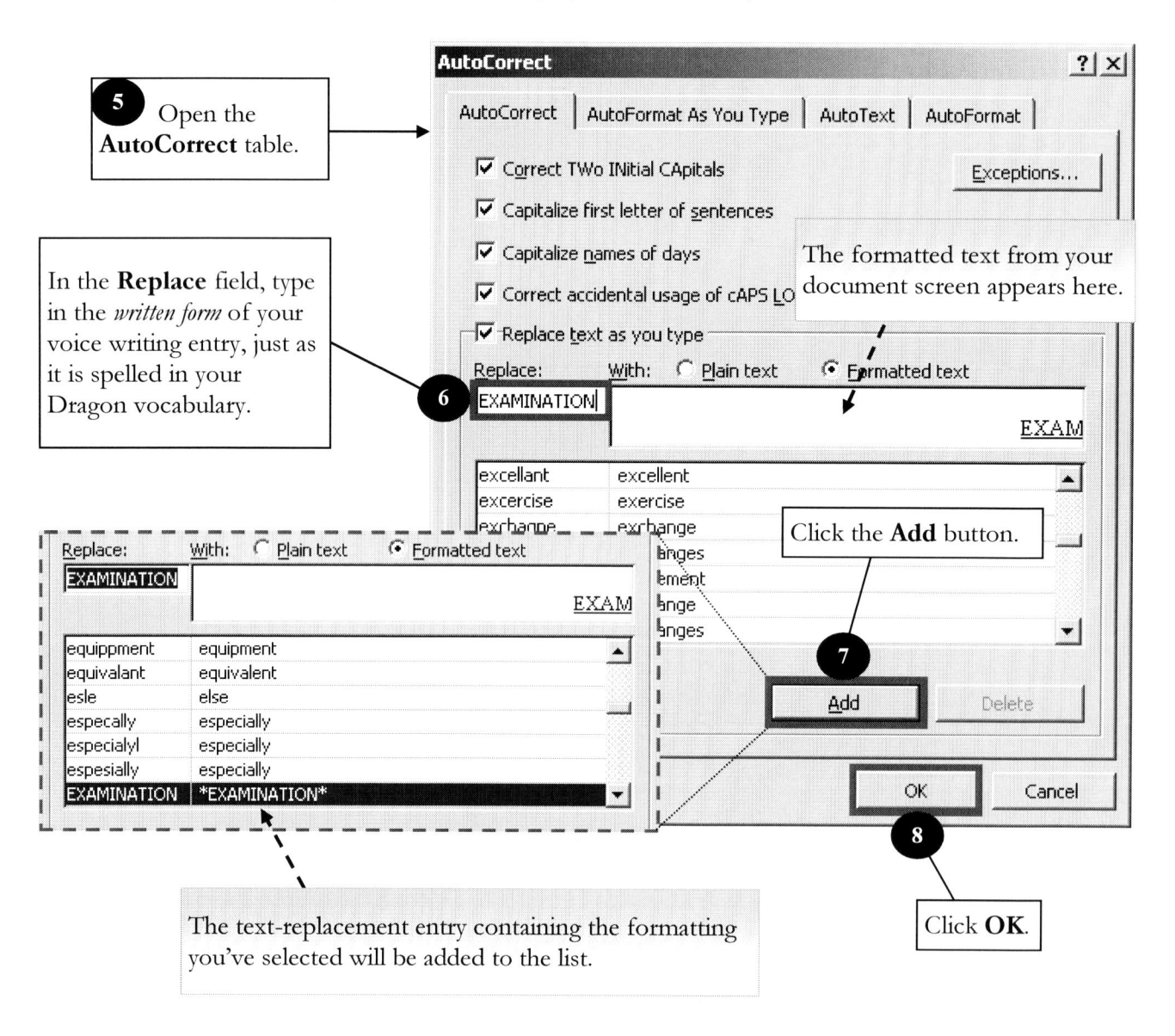

5 Open the **AutoCorrect** table.

In the **Replace** field, type in the *written form* of your voice writing entry, just as it is spelled in your Dragon vocabulary.

6

The formatted text from your document screen appears here.

Click the **Add** button.

7

8 Click **OK**.

The text-replacement entry containing the formatting you've selected will be added to the list.

To test out your formatted results, turn on Dragon's dictation microphone and dictate the voice code into your Microsoft Word document screen.

(Note: If the *AutoCorrect* table is not correctly producing your formatted results, see Microsoft Word's Help menu or user's manual.)

140

Chapter 5: Vocabulary Setup and Formatting

CAT Program with a Dictionary

If you have CAT software with a dictionary database, that is likely the utility you will use to manage your voice writing entries. The purpose for such a database is to store input and output instructions for your CAT program to read so it can properly format the text it receives from Dragon during your dictation. The diagram below illustrates the general concept of how your voice writing entries would be set up in both your Dragon vocabulary and CAT dictionary for effective communication to occur between those two programs. The reference to voice writing entries here means the special words—your voice codes—that you will be dictating as part of your realtime voice writing language, such as punctuation marks, speaker identifiers, Q&A markers, parentheticals and other indications, et cetera. For all other words you will dictate to produce plain English text, it is not necessary to create entries for them in your CAT dictionary unless you want that text to appear in special formats and/or you intend to create conflict-choice options from which to pick in order to train your CAT system to learn the difference between which words to output in different contexts.

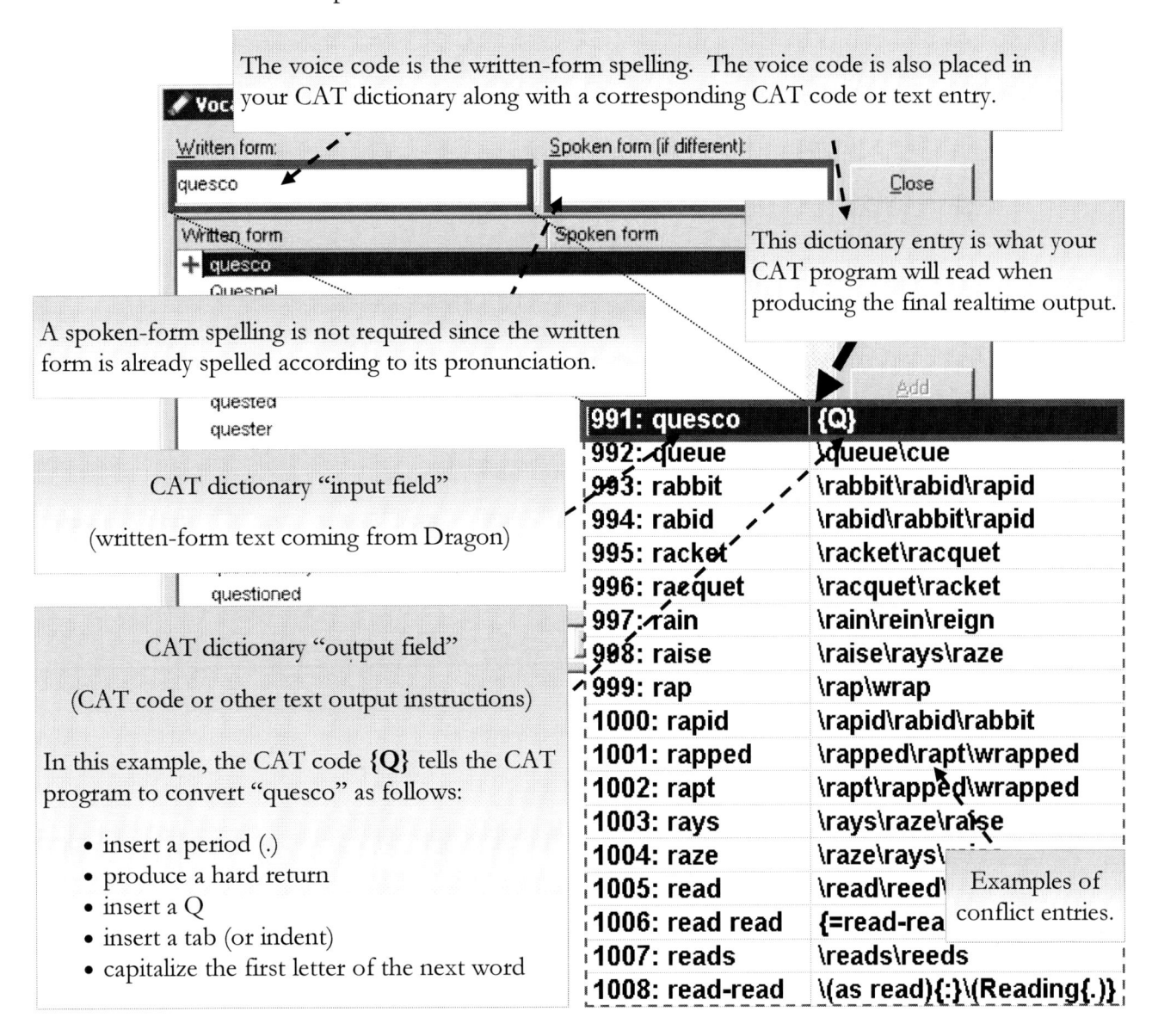

The voice code is the written-form spelling. The voice code is also placed in your CAT dictionary along with a corresponding CAT code or text entry.

This dictionary entry is what your CAT program will read when producing the final realtime output.

A spoken-form spelling is not required since the written form is already spelled according to its pronunciation.

CAT dictionary "input field"

(written-form text coming from Dragon)

CAT dictionary "output field"

(CAT code or other text output instructions)

In this example, the CAT code **{Q}** tells the CAT program to convert "quesco" as follows:

- insert a period (.)
- produce a hard return
- insert a Q
- insert a tab (or indent)
- capitalize the first letter of the next word

991: quesco	{Q}	
992: queue	\queue\cue	
993: rabbit	\rabbit\rabid\rapid	
994: rabid	\rabid\rabbit\rapid	
995: racket	\racket\racquet	
996: racquet	\racquet\racket	
997: rain	\rain\rein\reign	
998: raise	\raise\rays\raze	
999: rap	\rap\wrap	
1000: rapid	\rapid\rabid\rabbit	
1001: rapped	\rapped\rapt\wrapped	
1002: rapt	\rapt\rapped\wrapped	
1003: rays	\rays\raze\raise	
1004: raze	\raze\rays\	
1005: read	\read\reed\	Examples of conflict entries.
1006: read read	{=read-rea	
1007: reads	\reads\reeds	
1008: read-read	\(as read){:}\(Reading{.}}	

141

After setting up your voice writing entries in your Dragon vocabulary and CAT dictionary, begin a realtime translation session in your CAT program and dictate the words you've entered to ensure that your output results appear on your screen in their proper format.

(Note: Be sure to refer to the user's manual of your CAT program for exact Dragon vocabulary and CAT dictionary setup instructions. All CAT programs are designed differently, so these steps will vary.)

CAT Program without a Dictionary

If you have CAT software without a dictionary database, it is likely that you would manage all your voice writing entries only through Dragon's vocabulary. This type of CAT program would still require you to use CAT codes for your CAT program to be able to read your special output instructions in order to format text; however, the CAT code would be entered as the written form of your Dragon vocabulary entry. You would type in your voice code as the spoken-form spelling since, in that case, the written-form spelling would not match the pronunciation. Since this type of CAT program does not provide a dictionary, formatting functions are restricted to the capabilities of your CAT program.

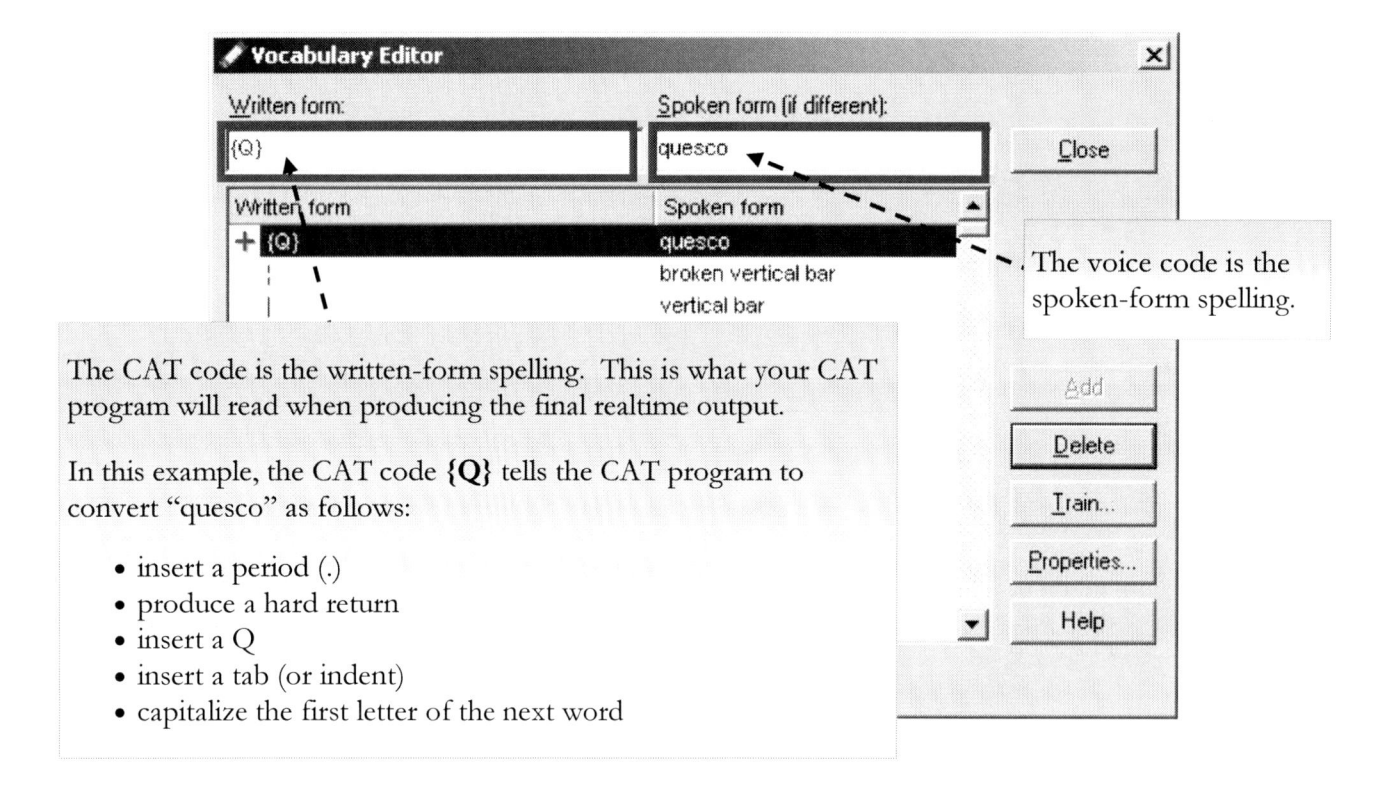

The voice code is the spoken-form spelling.

The CAT code is the written-form spelling. This is what your CAT program will read when producing the final realtime output.

In this example, the CAT code **{Q}** tells the CAT program to convert "quesco" as follows:

- insert a period (.)
- produce a hard return
- insert a Q
- insert a tab (or indent)
- capitalize the first letter of the next word

After setting up your voice writing entries in your Dragon vocabulary, begin a realtime translation session in your CAT program and dictate the words you've entered to ensure that your output results appear on your screen in their proper format.

(Note: Be sure to refer to the user's manual of your CAT program for exact Dragon vocabulary setup instructions. All CAT programs are designed differently, so these steps will vary.)

Chapter 5: Vocabulary Setup and Formatting

Practice Dictation to Test Formatted Results

Before beginning a dictation session, follow all the steps under the *Computer and Equipment Setup* sections of Chapter 4 and remember to set Dragon to *Dictation Mode*.

If you are dictating in Microsoft Word, here are some tips that will help you free up memory for better speech recognition performance:

- Do not run other applications besides Dragon and Microsoft Word.

- Disable any automatic features of Microsoft Word that are not necessary for you to use during dictation, such as spell check, grammar check, et cetera.

If you are dictating in a CAT program, follow the instructions in your user's manual.

Now do a practice dictation session in whatever program you will be using—whether it be Dragon alone or Dragon in conjunction with another application— and read from any material you'd like in order to test out your formatted results.

When you receive speech recognition errors, use the written forms exactly as they appear in your Dragon vocabulary when typing in corrections. If you elected to use alternate written forms, type in the alternate written forms when making corrections. Do not be alarmed if you receive quite a few errors when dictating your voice codes at this point. Continue to make corrections and re-dictate until you obtain correct recognition results.

Understand that making corrections to misrecognitions is just the first step in improving accuracy. This work applies only to the acoustic model and vocabulary, two of the three main design elements of your speech recognition engine. Beyond making corrections to speech recognition errors, you still have more accuracy-improvement work to do. This involves the process of vocabulary building to customize the grammatical model, which is the third and final design element, in order to produce proper recognition based on context. We will take a more comprehensive approach to accuracy improvement that considers all three of these design elements—the acoustic model, the vocabulary, and the grammatical model—in the next chapter.

Chapter 6:
Improving Accuracy

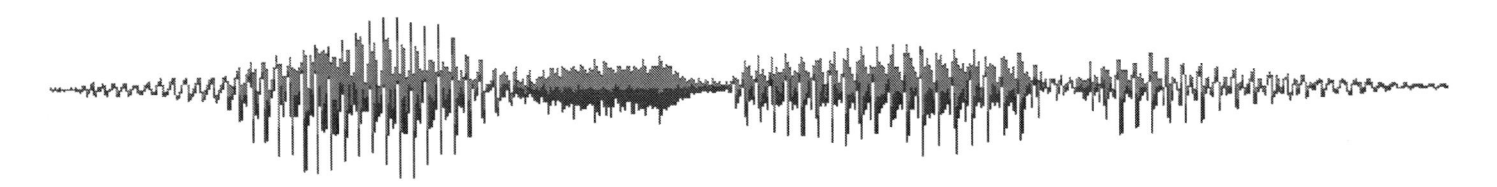

This chapter focuses on increasing your overall SR performance.

First, you will learn how to adjust audio levels for your speech input. This allows Dragon to make accurate acoustic measurements for the volume level and quality of your speech prior to beginning a dictation session. This will also enhance Dragon's filtering capabilities in properly determining the difference between the sounds coming from your speech input versus other sounds it hears.

Next, you will learn why certain speech recognition errors occur, whether or not they warrant correcting, and how to correct them when necessary. Through the correction process, accuracy is improved by adapting your vocabulary and acoustic model. This is how Dragon learns more specific information about your speech, including not only the way you say words and word combinations, but the nuances and pattern inconsistencies associated with the various ways you pronounce them at fluctuating rates. After all, human speech is not perfect, and rarely do we say words in the same way twice.

Other measures taken for accuracy improvement are adding words and phrases. You will learn how to add words individually and automatically from documents. Adding common phrases will help increase your speech recognition accuracy by giving Dragon specific samples of how you will say those exact words, possibly running them together when your enunciation becomes less clear during high-speed dictation.

Another important topic addressed in this chapter is vocabulary building. This is where you will gather documents representative of the type of transcripts or text material you will be producing in your dictation, prepare them in the special format needed for Dragon to understand the syntax of the words contained in them, and run them through Dragon's vocabulary-building utility for contextual analysis. In essence, you will be customizing Dragon's grammatical model so the special words you say as part of your realtime dictation language will be contextually recognizable.

Applying all the concepts presented in this chapter will ensure you have given Dragon the specific information it needs to enforce proper recognition involving each main design element: the acoustic model, the vocabulary, and the grammatical model.

Adjusting Audio Levels

Audio settings play an important role in Dragon's ability to accurately process your speech into text based on acoustics. Your dictation voice may be affected by outside influences as you go from one working environment to the next. If you are dictating in a quiet courtroom or conference room, your voice will be much softer than if you are dictating in a classroom setting that has lots of background noise. In some rooms, you may be able to hear a pin drop; yet in others, you have to strain your ears. You will only dictate as loudly as comfortable for the environment.

If you are using an open-mic headset, the qualities and levels of extraneous sound in the room may change, affecting Dragon's noise-filtering capability. Even when you are using a speech silencer in very noisy environments, though you have formed a good seal onto your face, your silencer's microphone may still pick up extraneous noise outside your silencer. To remedy these concerns, you should adjust the audio levels each time you enter a new working atmosphere. If you do not adjust them, your accuracy will suffer.

This section will show you how to adjust audio levels to account for acoustic changes that take place during the day and when moving to different environments. If you are working in an area where the background sound level remains the same day after day, you may not need to adjust the audio settings as often. You will still want to readjust them occasionally, however, if your voice becomes affected by other factors such as fatigue, stress, or respiratory problems.

Before adjusting audio levels, realize that the acoustic measurements made by Dragon during your audio adjustment will not be accurate if the quality of your speech input changes significantly when you begin dictating. To ensure consistent recognition results, continue to speak in the same volume during your actual dictation that you use during your audio setup.

Remember, if using CAT software, to refer to your user's manual for instruction on how to access Dragon's audio setup tool from within that program since steps will be different.

Now make sure your dictation input devices are properly connected and that the USB sound card is plugged into one dedicated USB port of your computer. Then perform the steps below.

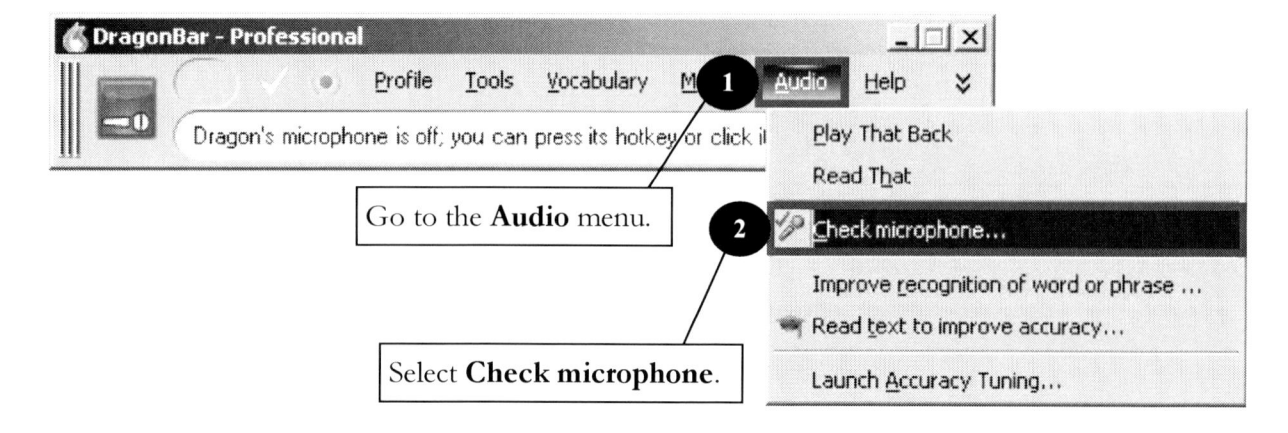

Go to the **Audio** menu.

Select **Check microphone**.

Chapter 6: Improving Accuracy

If you are using an open-mic headset, position it properly, as shown, and remember to keep it in that same position during dictation. If you are using a speech silencer, make sure it forms a secure enough seal around your mouth so as to prevent the internal microphone from detecting extraneous noises coming from the outer environment.

Open-Mic Headset Positioning    *Speech Silencer Positioning*

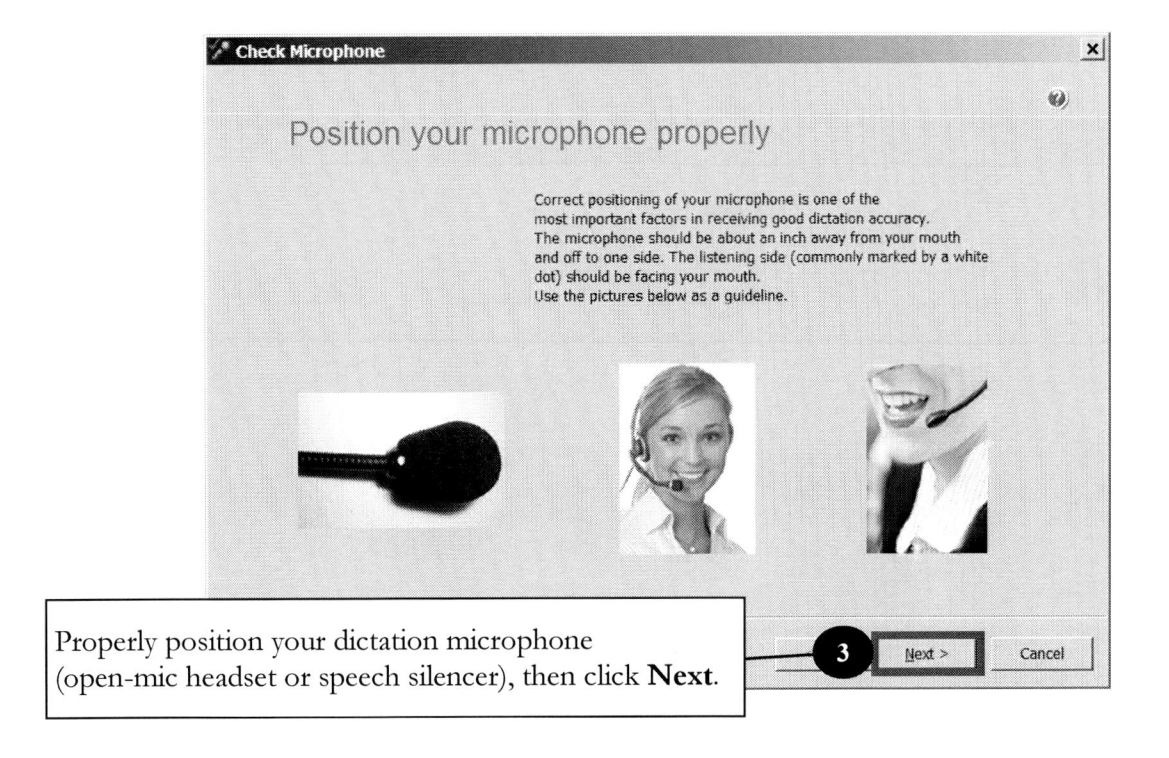

Properly position your dictation microphone (open-mic headset or speech silencer), then click **Next**.

Here, the volume level of your speech input will be measured. Remember to speak in the same volume that you intend to use later while dictating.

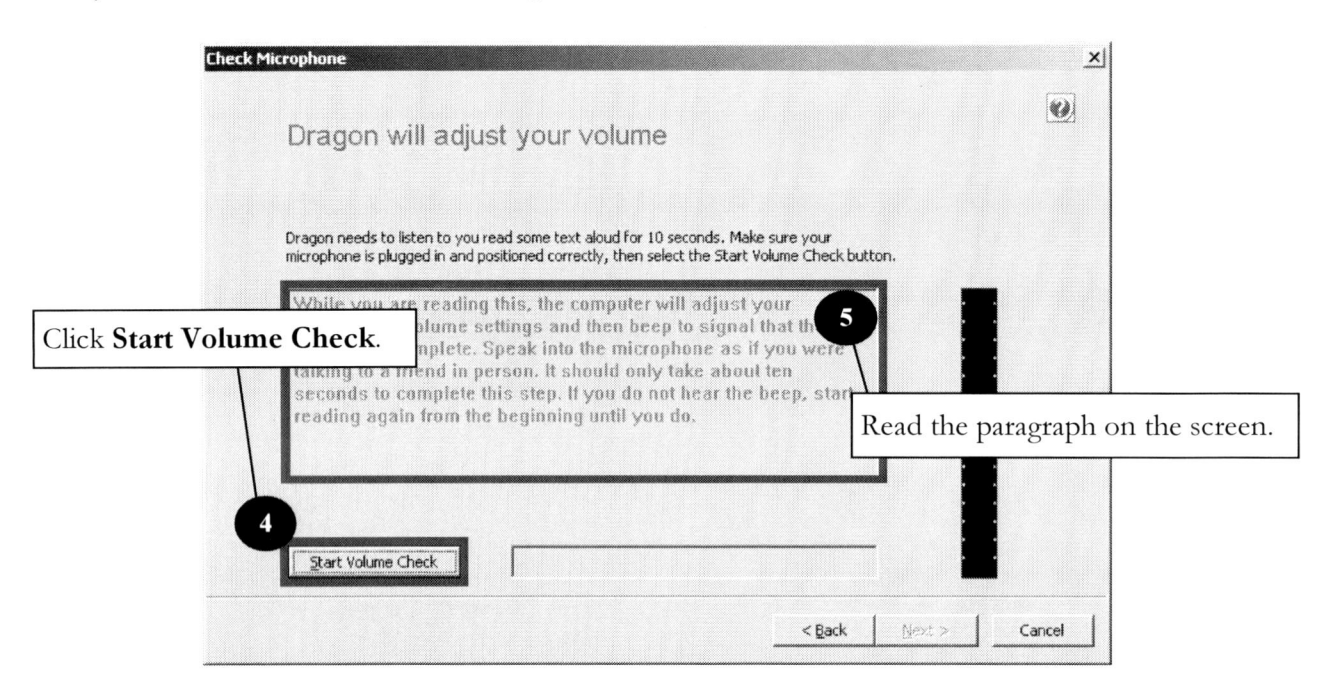

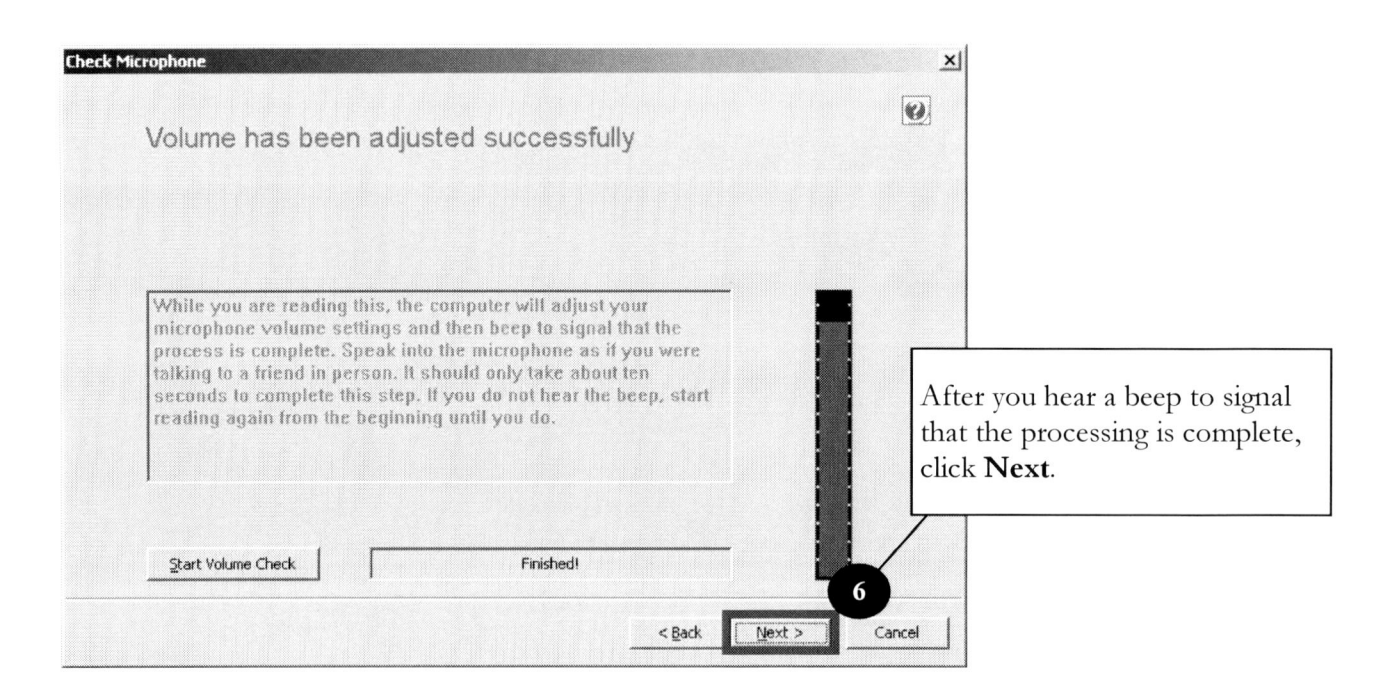

Click **Start Volume Check**.

Read the paragraph on the screen.

After you hear a beep to signal that the processing is complete, click **Next**.

Chapter 6: Improving Accuracy

Dragon will now measure your speech input against other sounds that are extraneous or which come from the background. If you use a speech silencer, it adjusts the audio levels for the acoustic environment within the silencer. More specifically, what is being measured is the difference between the sound of your *speech input* versus *extraneous noise* versus *silence*. Therefore, to assist Dragon in its sound-filtering capabilities so it will better recognize your speech amidst all other sounds it hears, it is important not to speak continuously without pausing. Rather, pause at the end of each sentence so Dragon will hear silence and background noise. To prevent Dragon from translating your breath sounds into words, it also helps to produce a couple of breath sounds during those pauses so Dragon will know to filter out those sounds as well during your dictation.

Click **Start Quality Check**.

Read the paragraph on the screen, using natural pauses.

After you hear the beep to signal that processing is complete, make sure you get a *Passed* result. If not, repeat this step.

Click **Finish**.

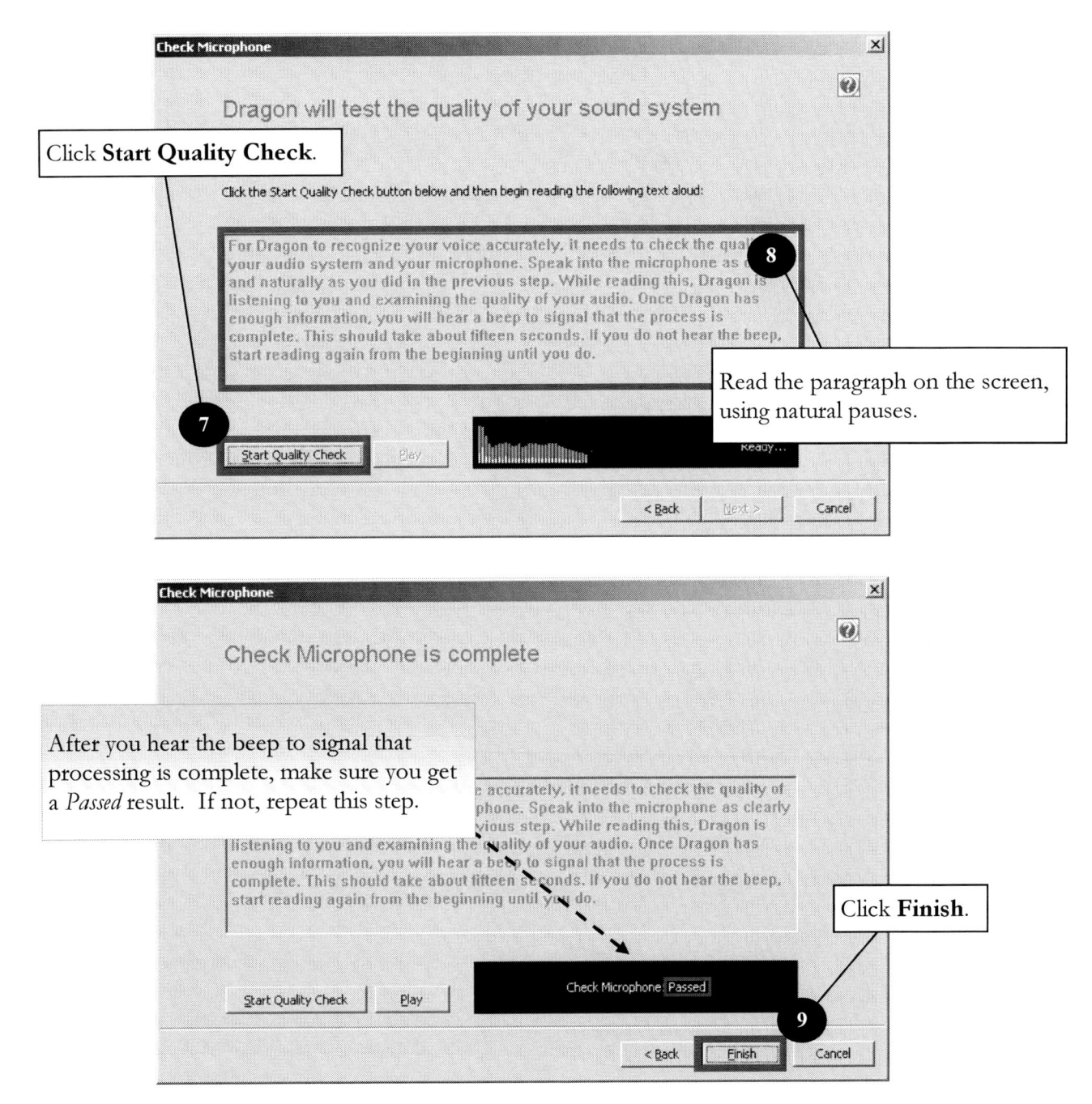

149

Correcting Misrecognitions

This section addresses the technique of improving SR accuracy by making corrections to misrecognitions in Dragon. If you are using Dragon alone and set your Dragon user options according to the steps outlined in Chapter 4 of this book, the keystroke you will use to open the correction tool is *Alt C* ("C" for correction); *Alt P* ("P" for playback) begins your speech playback. If you have CAT software, refer to your software manual for instructions on accessing this tool and performing corrections, since steps will be different from how they are described here.

Before you can increase speech recognition accuracy by making corrections to SR errors through listening-and-repeating practice, you should first correct misrecognitions that occur when you simply "read" text. The idea behind this is to learn how to dictate properly for speech recognition.

Our tendency is to dictate every word we hear, because that is the standard expected within the reporting industry. But Dragon does not respond well when the urgency to keep up with speakers causes us to lose control of our dictation performance, resulting in poor clarity of input. The mental coordination efforts involved in the process of listening and repeating can scatter the focus necessary for personal linguistic programming. If we have poor speech skills at the time we practice listening and repeating, the combined effects of untrained speech with an untrained voice model will compound recognition troubles, making it difficult to discern which misrecognitions are due to human speech errors versus valid computer errors which require correction. This is not an effective way to determine accuracy, because it prevents us from objectively measuring Dragon's true recognition capabilities, unencumbered by human performance flaws. To effectively measure computer performance levels involving human-to-computer interactions, we must remove from the equation as many human performance flaw factors as possible.

You do not want to waste time introducing to your voice model dictation discrepancies (i.e., poor enunciation, lack of context, and/or inconsistent speaking patterns) that will prevent Dragon from making a well-informed best-match guess. You can spend all day correcting misrecognitions of improper speech, but the errors will return unless you change your dictation habits. For example, if, in an effort to keep up with the pace of fast speakers, you slurred the words "in her muscles" and Dragon produced "inner muscles" because that's what it sounded like you said, then correcting the recognition of your slurred speech will defeat the purpose of effectively attuning a voice model; you would actually end up decreasing Dragon's ability to recognize the sound of "inner muscles" according to proper speech, and errors would still result from the slurred pronunciation.

Total dedication should be given first to how we dictate. The best method of working on proper speech development is to focus on dictation (without listening and repeating) by reading text aloud and forcing yourself to apply the dictation techniques described in Chapter 2 until they become second nature. You can do this using the SpeedMaster™ software program that accompanies this book (see Chapter 7 for details). Then when Dragon receives proper input, you will know that any errors produced are attributable to Dragon, enhancing your ability to determine what adjustments to make during the correction process. If you enunciate clearly, speak at a steady pace, and insert the punctuation required for SR contextual judgment, you will know that word misrecognitions are due to a lack of acoustic

Chapter 6: Improving Accuracy

and/or vocabulary information within Dragon when improving your voice model. This is the most efficient way to perform voice model improvement work, as you will reach realtime accuracy faster when the corrections you spend time making are isolated to SR errors.

To prepare for the dictation and correction process, gather texts that are representative of the type of subject matter you will encounter in your realtime work. Dictate into Dragon by reading from that material in two-page increments, pausing to correct errors. Repeat this process until you have reached your minimum goal of 96% accuracy at standard realtime speeds. Improving your voice model by working in smaller segments will save you the time of having to repetitively correct a higher number of those same errors in larger segments. For example, improving your voice model would take you 20 times longer in 40 pages' worth of material if the recognition errors you corrected were redundant occurrences, when you could have fixed those same errors in less time by correcting them in the first two pages.

Once your voice model is replete with adjustments based on proper speech styles and your accuracy has surpassed 96% at a sustained dictation rate of 180 words per minute, you can turn your attention to merging the skill of dictating properly with the skill of listening and repeating. You will then be able to focus on improving Dragon's ability to recognize the way you speak during the listening-and-repeating process, and it will be easier to determine whether further adaptations need to be made to the voice model or to your speech.

To learn how to correct different types of misrecognitions as they are encountered, you must first do a practice dictation session so that you have errors to work with. To assist you with this exercise, provided on the next page is sample text for you to read.

Remember to always set Dragon to *Dictation Mode* prior to dictating so that Dragon will process all the words you speak into text rather than performing computer actions or only transcribing letters or numbers.

Follow the steps below to begin your dictation practice session.

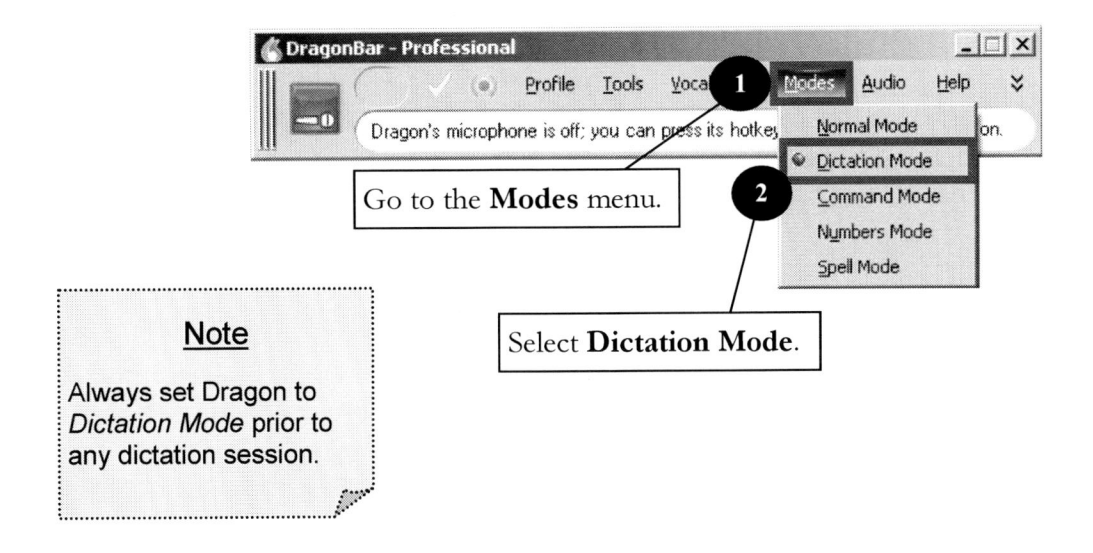

Go to the **Modes** menu.

Select **Dictation Mode**.

Note

Always set Dragon to *Dictation Mode* prior to any dictation session.

Voice Writing Method • FIFTH EDITION
Dragon NaturallySpeaking® 12

3 Go to the **Tools** menu.

4 Select **DragonPad**.

5 Turn on the dictation microphone.

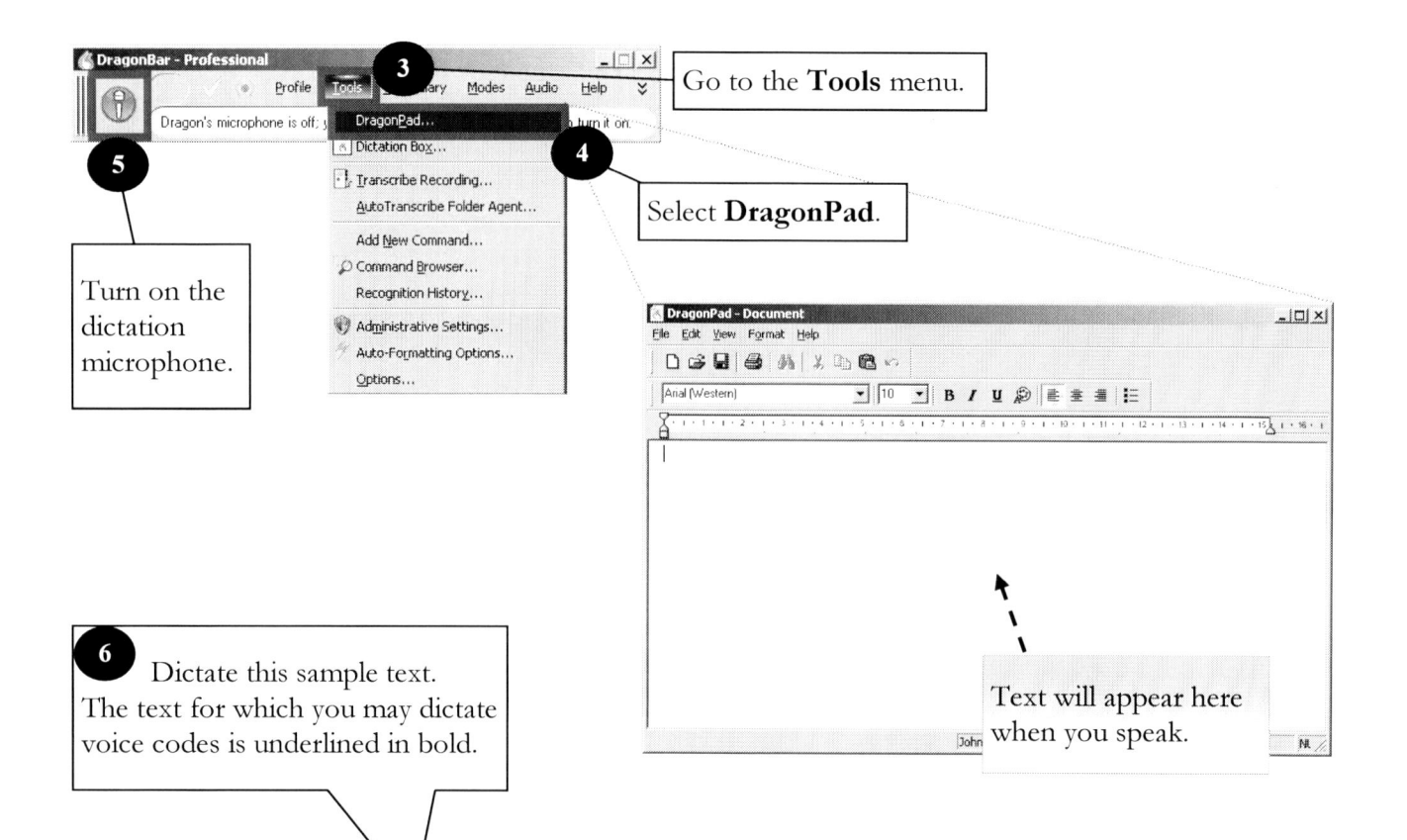

6 Dictate this sample text. The text for which you may dictate voice codes is underlined in bold.

Text will appear here when you speak.

Sample Dictation Text

Using the voice writing techniques outlined in this book, you can control the way you produce numbers and letters, such as **"2"** versus **"two"** and **"B"** versus **"P."** You can also produce parentheticals such as **"(indicating)"** and **"(demonstrating)"** or a verbal indication like **"uh-huh"** using special pronunciations as part of your dictation method**.** Assigning special pronunciations to certain punctuation marks, you can produce a word like **"period"** without running into conflicts with the period symbol**.** Even though you must consistently dictate from 180 to 200 words per minute, in instances where you will say the word **"to"** between numbers, you should pause slightly before and after saying **"to"** or use a different pronunciation for this word **(**since "to" can easily be recognized as **"2"** while dictating a sequence of numbers**).** If you do not pause, you make it an error. Other errors may occur simply because the word you dictated is not in the vocabulary**.** When adding words with uncommon pronunciations **(**for example, the last name **"Kuhlhaenk")**, you should use a phonetic spelling for its spoken form in order to achieve proper and consistent recognition**.**

Chapter 6: Improving Accuracy

This is an example of the *Sample Text* having been dictated in DragonPad. Although your accuracy results will probably be different than what appears here, note the errors indicated and read the subsequent explanations offered. Table 43 on the next page provides information about how and why these errors occurred and the corrective actions that should be taken. This will help you determine how to improve your own accuracy.

Using the voice writing techniques outlined in this book, you can control the way you produce numbers and letters, such as "2" versus "two" and "B" versus "P." You can also produce parentheticals such as "(indicating)" and "(demonstrating)" or a verbal indication like EAA using special pronunciations as part of your dictation method. By assigning special pronunciations to certain punctuation marks, you can produce a word like "period" without running into conflicts with the period symbol. **errors produced** you must consistently dictate from 182 200 words per minute, in which case you will say the word "to" between numbers, you should pause slightly before and after staying "to" or use a different pronunciation for this word (since to can easily be recognized as "2" while dictating a sequence of numbers). If you do not pause, you make it an error. Other errors may occur simply because the word you dictated is not in the vocabulary. When adding words with uncommon pronunciations (for example, the last name chronic), you should use a phonetic spelling for its spoken form in order to achieve proper and consistent recognition peer

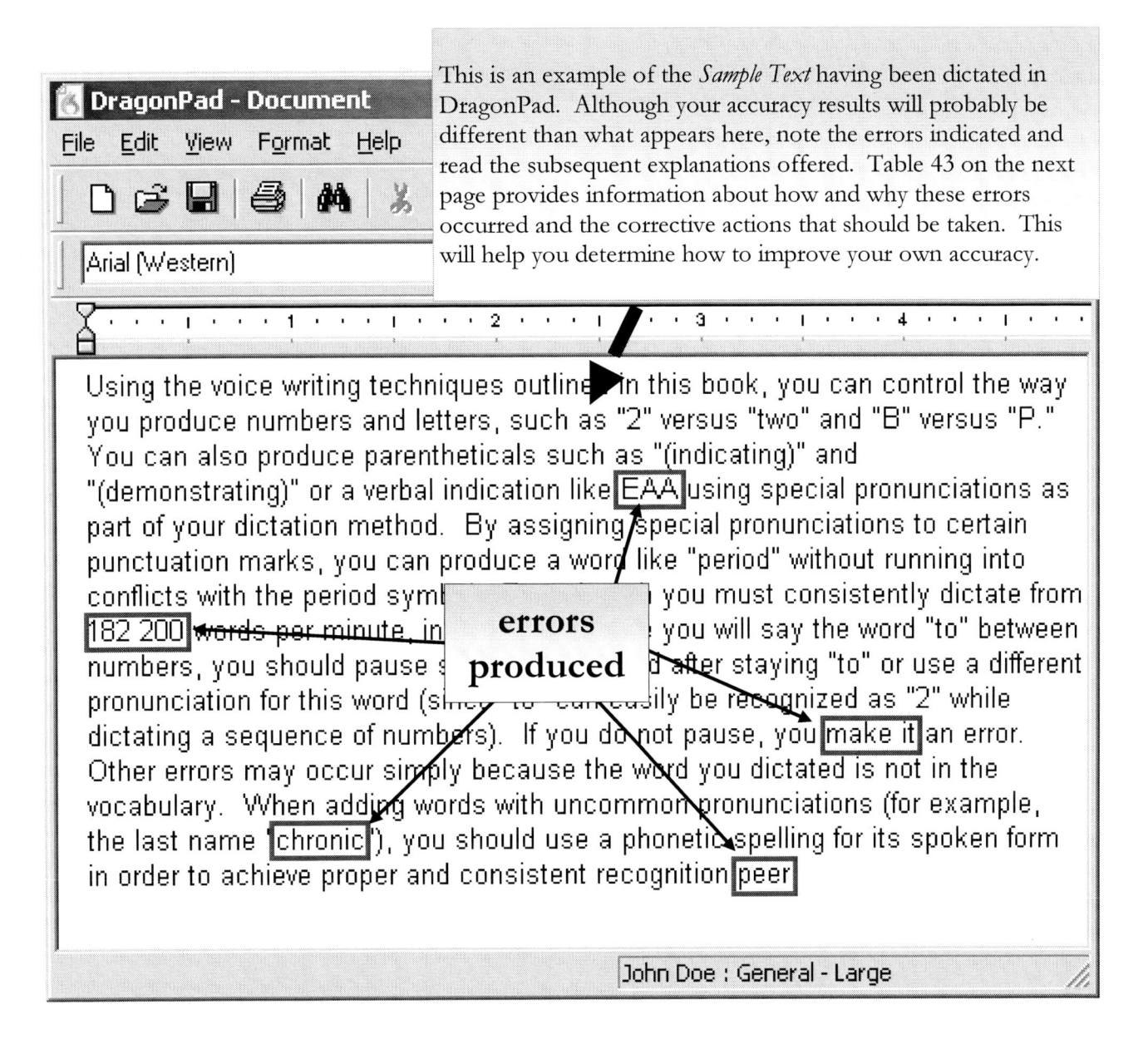

John Doe : General - Large

Table 42

Examples of Errors		
Spoken During Dictation	*Actual SR Result*	*Intended SR Result*
yay-yay	EAA	uh-huh
one hundred eighty to two hundred	182 200	180 to 200
may get	make it	may get
Kuhlhaenk	chronic	Kuhlhaenk
peerk	peer	.

Table 43

How to Treat Errors

Error	Cause of Error and Corrective Action to Take
Actual result... EAA ———— Intended result... uh-huh	This pertains to the voice writing entry that was made in Dragon's vocabulary for "uh-huh" as the written form and "yay-yay" as the spoken form. Even though this entry exists in the vocabulary and a pronunciation for it has already been recorded, you will still experience these types of misrecognitions with a new set of speech files. If your speech was clear, you should correct this error using the correction tool (Alt C). Always use written-form spellings when making corrections. Also, you would want to click the *Train* button within the correction window to record a vocal pronunciation upon making a correction for this, since it is a special item you may end up regularly dictating.
Actual result... 182 200 ———— Intended result... 180 to 200	While dictating numbers, no pause was made before or after saying the word "to," so it was recognized as part of the number. Alternatively, you may use the voice code "tah tah" instead of pausing in order to produce the word "to." For this type of error, do not make corrections in Dragon. Simply correct how/what you speak.
Actual result... make it ———— Intended result... may get	With any new voice model, Dragon will have trouble making distinctions between certain common English words that you haven't yet had the opportunity to correct at least once. When you make corrections, Dragon will learn the different nuances in how you say these words versus others that may sound similar. In these cases, Dragon makes a best-guess match and chooses another word that sounds pretty close to what you said. This type of recognition error will be encountered frequently in the beginning, especially with small words such as "and" versus "in" and phrases such as "make it" versus "may get." If your speech was clear, you should correct this error using the correction tool (Alt C). Always use written-form spellings when making corrections.

Chapter 6: Improving Accuracy

Table 43 (Continued)

How to Treat Errors

Error	Cause of Error and Corrective Action to Take
Actual result. . . **chronic** (pronounced "kahonnick") --- *Intended result. . .* **Kuhlhaenk**	This is an uncommon proper name which does not exist in Dragon's base vocabulary, and it must only be added if you anticipate regularly encountering it in your future realtime work. If adding this word to the vocabulary manually, you should give it two different spellings: 1.) a written-form spelling of "Kuhlhaenk" and 2.) a spoken-form spelling of "kahonnick" or other phonetic spelling to represent its actual pronunciation (*kahonnick*). Note that you do not have to open the vocabulary to add this word; it will be added automatically through the correction process. If you expect to encounter this word again in the future and your speech was clear, you should correct this error using the correction tool (Alt C). Always use written-form spellings when making corrections.
Actual result. . . **peer** --- *Intended result. . .* **.** (punctuation symbol)	Since this is a custom word that was recently added to Dragon's vocabulary and no corrections have yet been made for Dragon to learn the difference between how you say your code words versus ordinary words, Dragon may favor choosing more common English words that sound similar. This type of error may be encountered regularly in the beginning, and accurate recognition will be achieved and maintained once you've made enough corrections for Dragon to decipher the difference between these pronunciations. If your speech was clear, you should correct this error using the correction tool (Alt C). Always use written-form spellings when making corrections.

When you see errors on your screen, it is important not to just assume they are actually speech recognition errors. They may or may not be. It all depends on whether your speech was clear enough to be deciphered by Dragon in the first place. If your speech is unclear, it is not Dragon's fault for mistranslating what you said. Make sure the sound of your speech is clear before you decide to treat errors as misrecognitions.

See the next page to learn how to make corrections.

155

Voice Writing Method • FIFTH EDITION
Dragon NaturallySpeaking® 12

To properly identify all the errors that need to be corrected, it is best to listen to the entire playback of your speech. Begin by placing your cursor within the text of the DragonPad screen and listen to the playback by pressing *Alt P* ("P" for playback). As you come across a text error for a word you clearly spoke, press *Alt C* ("C" for correction) to open the correction tool.

1a Place your cursor at the beginning of the document and press **Alt P** to play back your speech audio.

When you come across an error, press **Alt C** to correct it.

1b If you already know which error you want to correct, simply highlight it and press **Alt C** to isolate that particular portion of text when correcting the error.

2 If the correct option appears in the list, select it. If not, type in the correction manually.

The **Play that back** option replays your speech audio.

The **Train** option lets you record your speech for corrected text.

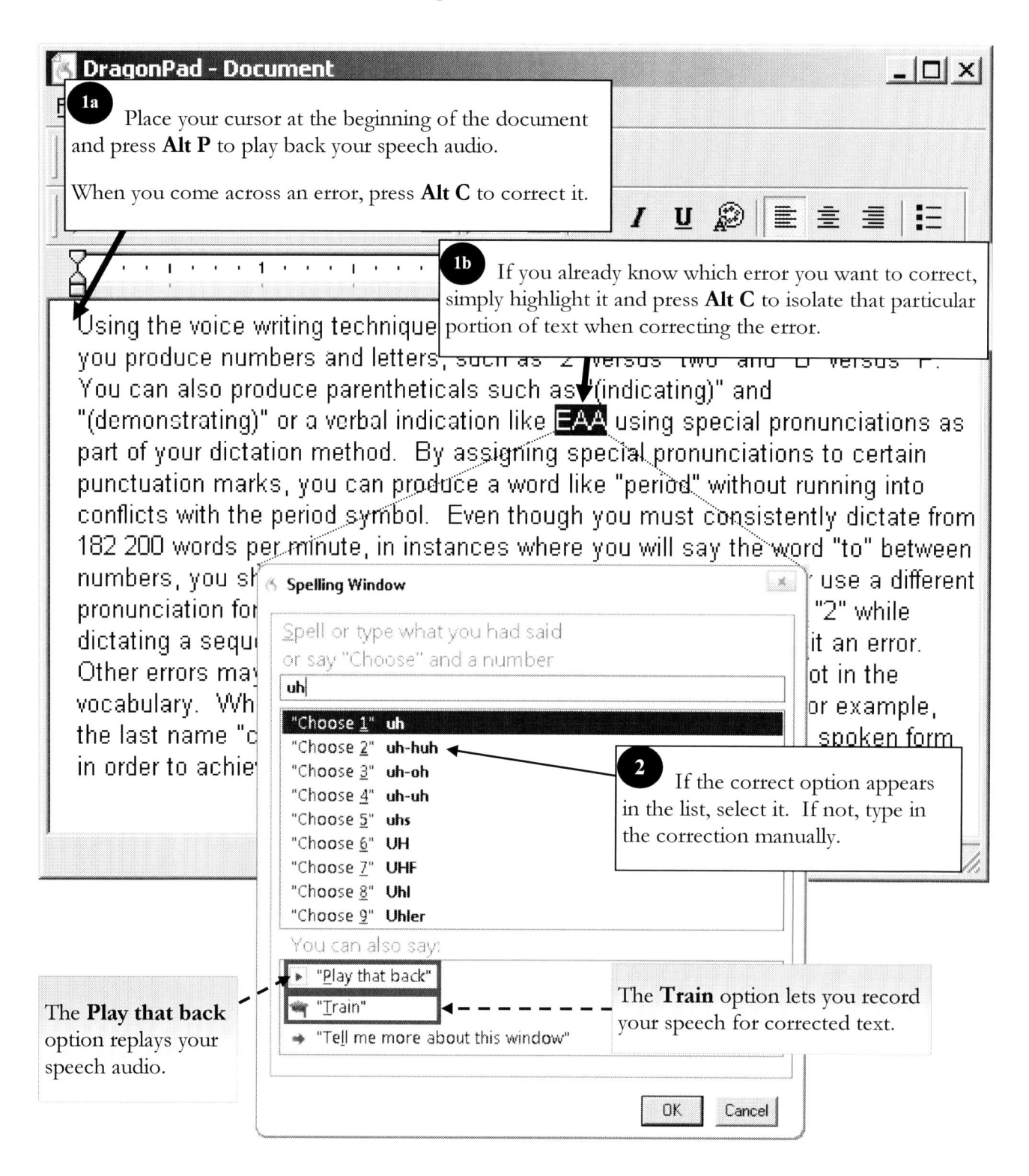

156

Chapter 6: Improving Accuracy

If the correct choice appears as an option within the list, you can select it without having to type it manually. If it does not appear within the list, type it in manually and click *OK* (or press *Enter*).

You will notice as you type the correction that the list of choices changes with each new character or letter you type. If the word exists within the vocabulary but was not one of the original options, eventually it will appear as you continue typing. If the word is not in the vocabulary, the word will be added upon making the correction.

If your enunciation is clear, then make the correction. If you're not sure whether it was clear or not and want to hear the playback again, click the *Play that back* option. If it sounds like you either mispronounced word(s) or mumbled or slurred your speech in any way, then don't correct Dragon; correct yourself in the way you speak by dictating it more clearly next time.

Before finalizing a correction, make sure that, upon your speech playback, all of the text within the correction window matches all audio you hear. Sometimes the beginning and/or ending speech audio for a misrecognized word cannot be heard because some of its sound parts are connected to other words occurring either before or after the word you are attempting to correct. If this is the case, you will have to perform the correction of the misrecognized word in conjunction with other words surrounding it. Close the correction window and begin the correction over again. Include as many words as necessary until all sounds completely match the text you've selected.

When you make corrections, Dragon does not automatically update your speech files. It only temporarily stores your correction work until you elect to permanently save that data. Any time you want to manually save your work during a correction session, go to the *Profile* menu and select *Save User Profile*. But if you forget to save your user files manually, do not worry because Dragon will ask you if you want to save them when you exit your dictation session. It is always a good idea to remember to save your user files manually anyway, just in case your computer becomes unstable and unexpectedly shuts down Dragon while you are in the middle of making corrections. As a general rule, you should periodically save your user files after every 20-or-so corrections.

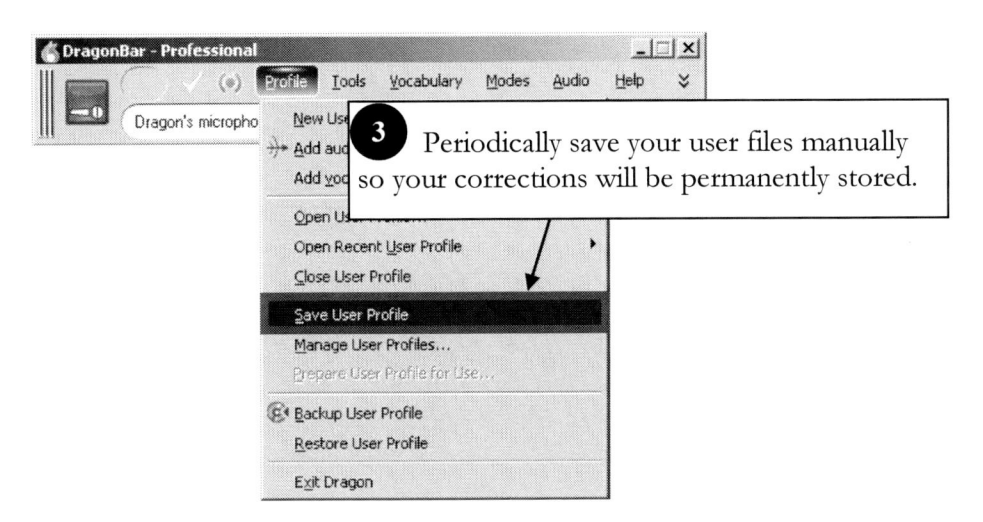

3 Periodically save your user files manually so your corrections will be permanently stored.

Your corrections will have been applied to your voice model and any words that were not already contained in Dragon's vocabulary will have been added.

157

Adding New Words

As part of your preparation for a realtime session, you will want to ensure that the words you will be dictating exist in Dragon's vocabulary. Since Dragon already contains thousands of common English words within its base vocabulary, generally the only words you will have to add are uncommon technical words and terms, common words with uncommon spellings, unique proper names and acronyms, and foreign words. This section focuses on how to add words individually or automatically from a list, but words may also be added to the vocabulary either directly (see *Adding an Entry* under the *Working with the Vocabulary* section of Chapter 5), through the correction process (described in the previous section), or during the vocabulary-building process (see the *Vocabulary Building* section of this chapter).

To quickly add a single word to the vocabulary, follow the steps below.

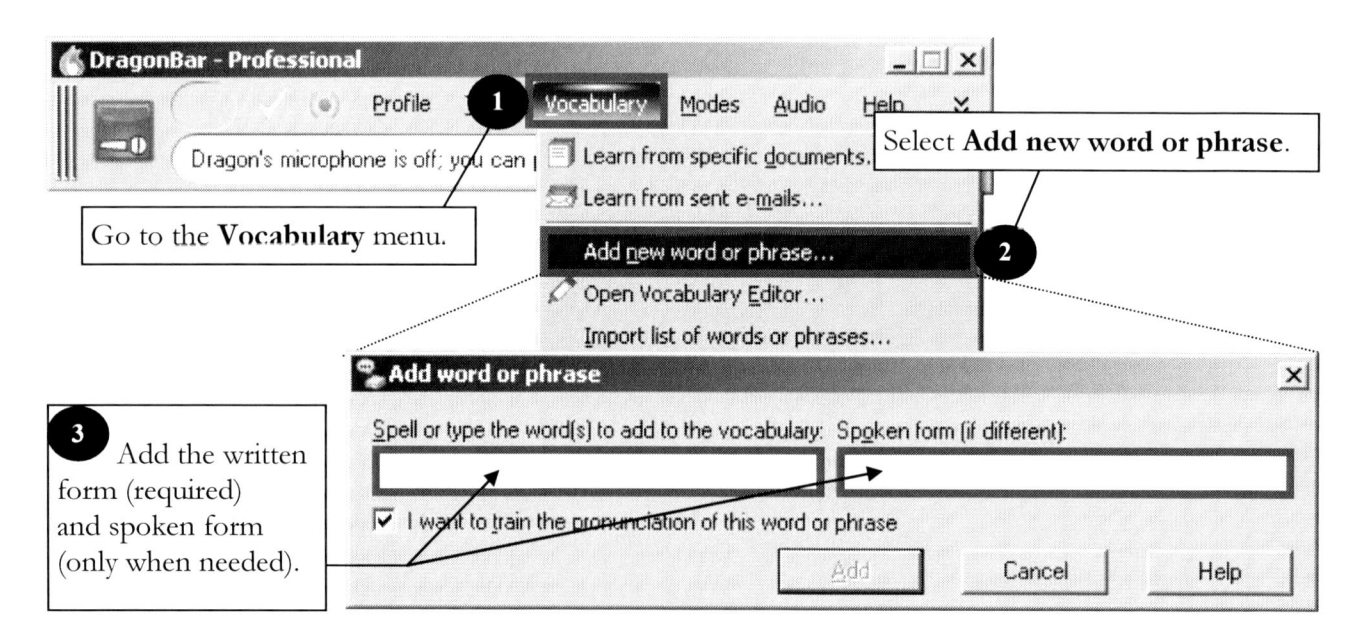

Go to the **Vocabulary** menu.

Select **Add new word or phrase**.

③ Add the written form (required) and spoken form (only when needed).

If you have any lists or text files of documents containing special words that you expect to encounter during your realtime session, you can process those files through Dragon so any new words found will be automatically added to your vocabulary. You should save any documents like these in ASCII format (with a .txt extension) before running them through Dragon to avoid file-format incompatibilities when Dragon reads your document.

If some of the words you need to add require spoken-form spellings in addition to their written forms, you should manually prepare your word list by typing in the written form along with a backslash (\) and then the spoken form. Each new written- and spoken-form combination must appear on a new line in order for Dragon to read them as individual entries that must be added as separate items to the vocabulary. Note that the written form is listed first, followed by a backslash (\), then the spoken form. When you create a document this way, Dragon will read each line that appears as a separate entry, with a backslash (\) to indicate the written-form spelling is on the left and the spoken-form spelling is on the right. If a spoken-form spelling is not required, simply type in the written form by itself without a backslash (\).

Chapter 6: Improving Accuracy

For example, your word list document would appear as follows:

EBITDAX\ebidax
non-GAAP\non-gap
CapEx

When you have finished creating your document, remember to save it in an ASCII text (.txt) file format.

To have Dragon automatically find and add new words from documents, whether they be word lists or entire transcripts, follow the steps below:

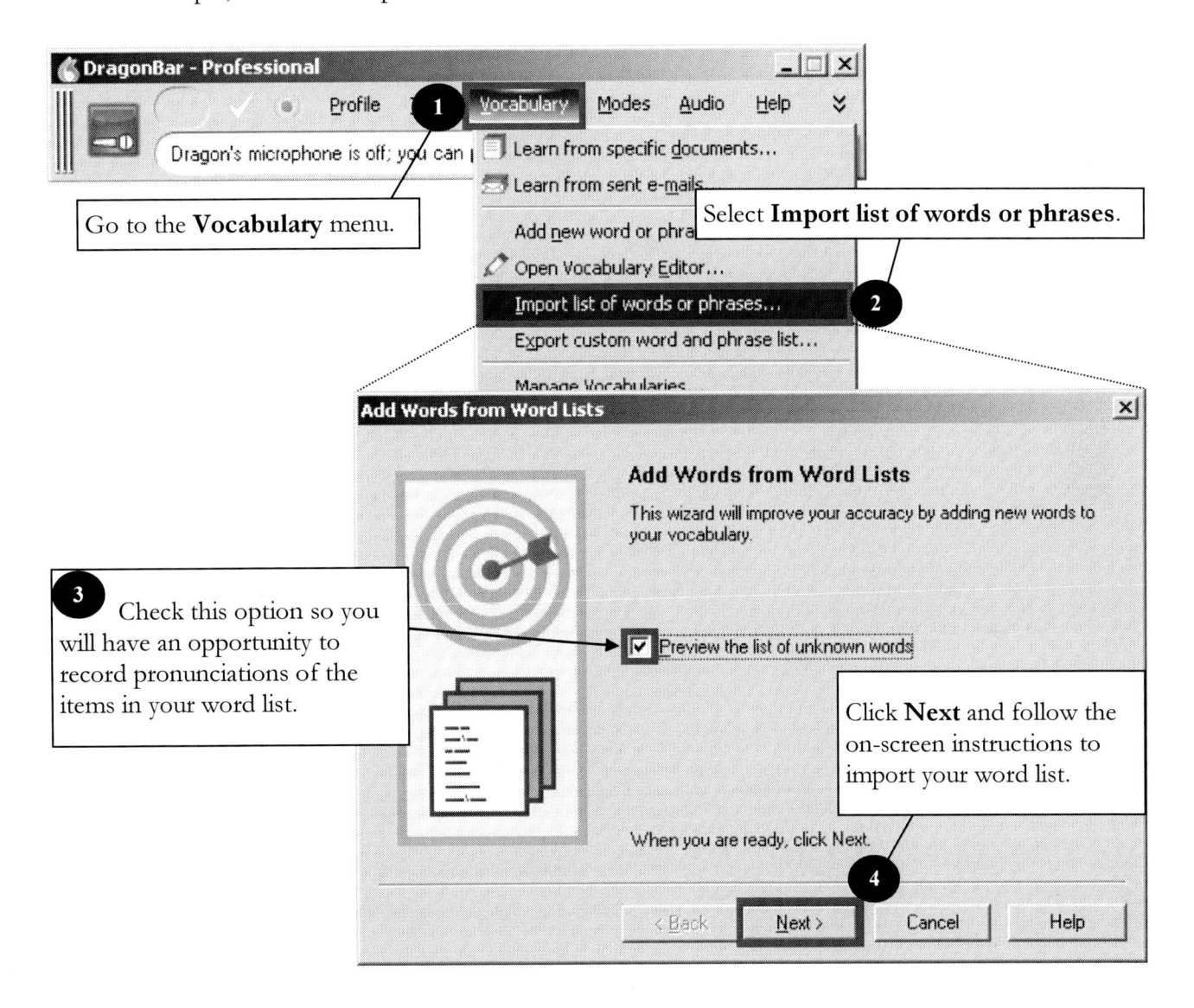

1. Go to the **Vocabulary** menu.

2. Select **Import list of words or phrases**.

3. Check this option so you will have an opportunity to record pronunciations of the items in your word list.

4. Click **Next** and follow the on-screen instructions to import your word list.

After your words have been added, remember to save your user files by going to the *Profile* menu and select *Save User Profile*.

159

Adding Phrases

For quickly determining which phrases to add to your vocabulary, you can use the N-Gram Phrase Extractor, a state-of-the-art data-mining tool that will automate the process of finding commonly-occurring word groups contained within documents. This program can be accessed on the Internet by going to the following Web address, then clicking the *Links* button and the *N-Gram Phrase Extractor* hyperlink:

http://www.voicewritingmethod.com

To utilize this tool, you must have documents for it to analyze. The documents you obtain should be representative of the type of material you will be dictating in your realtime work and they should be saved in an ASCII file format (any plain text format including a .txt extension). Depending on the options you select, the N-Gram Phrase Extractor will identify the most-common phrases in two- to five-word combinations, sorted according to their highest occurrences.

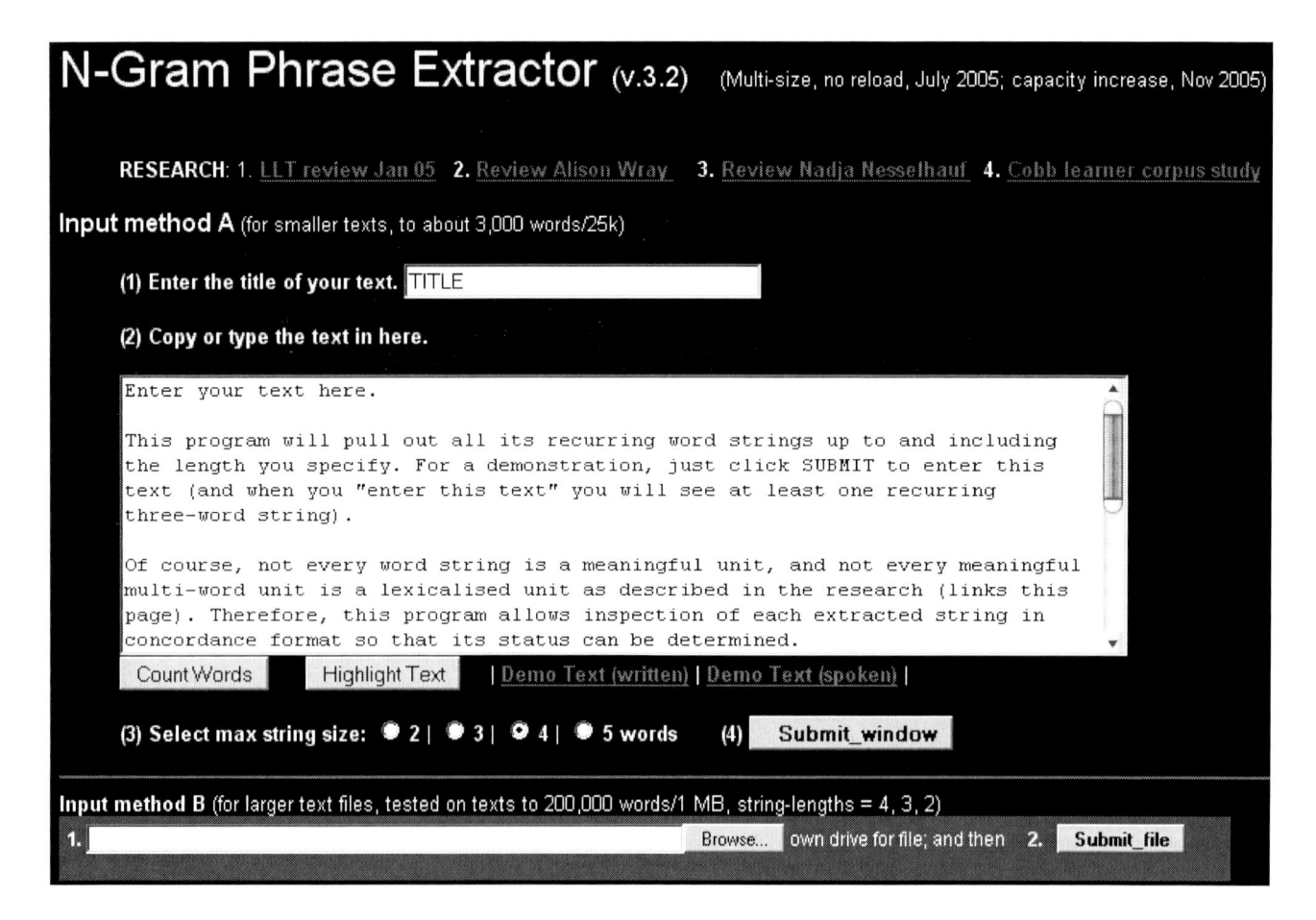

Chapter 6: Improving Accuracy

If transcript files are not available, then you will have to identify phrases individually as you encounter them through your dictation work.

The phrases you should focus on adding first are common, small-word combinations. The primary purpose in identifying phrases that include small words is to improve Dragon's ability to recognize them in a multisyllabic context, because Dragon has trouble deciphering sounds of smaller words individually and apart from other words as human speech becomes less perfect during high-speed dictation.

You may want to identify other common phrases so you will be able to create brief forms for them. Dictating brief forms will help speed up your dictation pace by reducing the need to always dictate punctuation and/or for minimizing the number of syllables that would ordinarily be required to produce common multiple-word groups that are high in syllabic density.

Whether you want to enter an individual phrase or a list of multiple phrases, you can add them to the vocabulary by the same methods used to add words. (See the previous section of this chapter on *Adding New Words* for more details.)

When adding a list of phrases requiring written- and spoken-form spellings, remember to type them in their proper format (written-form spelling\spoken-form spelling) and separate each vocabulary item by a new line. If a spoken-form spelling is not required, simply type in the written form all by itself. For example, your document would appear as follows:

>   if, however,\if however
>   do you know
>   ; isn't that correct\INTC
>   do you actually recall\doyactal

If you create a phrase list, remember to save it in an ASCII text (.txt) file format. Then to process the file in Dragon, go to the *Vocabulary* menu and select *Import list of words or phrases*. Follow the onscreen instructions to add your list of phrases.

After your phrases have been added, remember to save your user files by going to the *Profile* menu and select *Save User Profile*.

Vocabulary Building

When you performed your practice dictation sessions prior to reaching this point, you may have noticed that you had to speak more slowly than you would like in order to produce accurate results, especially when dictating your voice codes. You may even have had to use significant pauses, which is not an acceptable level of performance when you're expected to provide realtime services at high speeds. This is because Dragon is expecting you to use words in contexts that are based on proper English grammar styles and not based on any custom language styles for the way we speak using our realtime dictation theory.

To increase Dragon's ability to recognize you during high-speed dictation, you must customize Dragon's grammatical model according to your dictation language. This will enable Dragon to quickly produce proper word selections without requiring significant pauses.

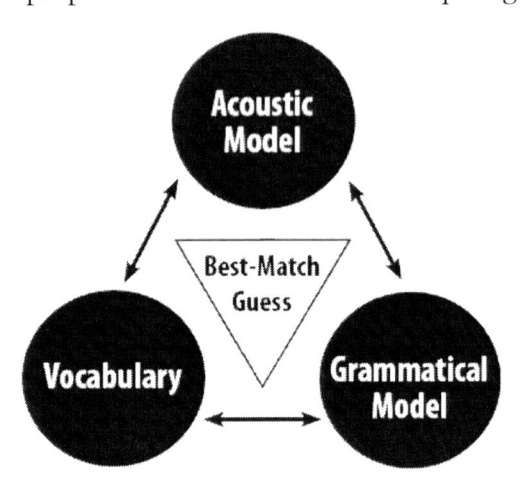

Remember that the design of Dragon's best-match guess structure is a matrix of interdependent relationships among three separate modeling algorithms: 1.) the **acoustic model**, which manages pronunciations and sound data; 2.) the **vocabulary**, which manages the word bank (or written information); and 3.) the **grammatical model** (or "language model"), which manages contextual relationships between words. Data is shared among these three when determining what the most correct word choices are and when to output them during the speech-to-text conversion process.

Dragon's grammatical model is based on proper grammar style and format and not on common speech styles of how we speak using our realtime dictation language, so we must customize it to reflect a more precise representation of words and their relationships to one another according to those patterns. The way to do this is by providing the appropriate context associated with our language by building the grammatical model with documents that are typical in style to the material we will be dictating. This means that if you use Q&A syntax for court reporting, you should build the grammatical model with that format. For CART or broadcast captioning, you can use a monologue format. For financial call reporting, you will want to use both monologue and Q&A formats.

If words and terms you will be regularly encountering in your everyday work are not a part of Dragon's base vocabulary, you already know you would need to add them if you want them to exist. But if all you do is simply add them to the vocabulary and record pronunciations for them, you end up giving to Dragon information regarding only two of the three total parts which factor into recognition: the vocabulary and acoustic model. For accurate speech recognition to take place, you must also give Dragon contextual information about any new words you add so that the third design element, the grammatical model, will enable Dragon to make well-informed best-match guesses. You will do this through a process called **vocabulary building**.

Chapter 6: Improving Accuracy

Vocabulary building involves having Dragon analyze documents that are reflective of the type of dictation work you are going to be doing. Dragon will automatically find new words from those documents and present them to you in a list so you will not have to search for and/or add them manually. The only words you want to add are specialty words that you expect to regularly encounter in your realtime work. As Dragon scans through your documents to extract new words it finds, it is also studying their context at the same time. Dragon looks for common patterns in how words occur together, how frequently they occur, and uses this information to establish sets of rules to follow for coming up with accurate conclusions when determining how to recognize the words you routinely dictate. You are, in essence, building a grammatical structure that will support the context of your dictation language. So the vocabulary-building process actually serves a two-fold purpose: 1.) to automatically find and add new words, and 2.) to increase accuracy based on context.

You begin by collecting documents (at least 20 megabytes' worth of plain text files) that are typical in style to the kind of material you are expected to produce in your realtime work. Then you reformat them to reflect the written forms of your realtime dictation words and have Dragon analyze them.

The next two sections explain how to custom-prepare your documents and run them through Dragon's vocabulary-building utility for language analysis.

Document Preparation

Understand that Dragon already contains complete data on all its base vocabulary words as far as written information, acoustic information, and grammatical information. Dragon also now contains written and acoustic information on your new realtime dictation theory entries as a result of you having added them in the previous chapter; the element of information it lacks, however, is grammar.

When you add new entries to Dragon's vocabulary, data is only being given for the spelling and pronunciation of those items. There are no grammatical relationships yet established for any of them. To make contextually-correct word selections, the grammatical model analyzes word relationships and syntax to improve its ability to distinguish between similar-sounding words and word groups. For example, two commonly occurring word sets that sound similar during high-speed dictation are "do you" and "to you." Now, suppose you are dictating Q&A material using the question-marker voice code "quesco" followed by the words "do you." You must build the grammatical model to recognize "quesco" within that syntax so Dragon will know that "do you" is the most highly-probable choice.

    Common occurrence:

        Q.   Do you. . .

    Less common occurrence:

        Q.   To you. . .

We recognize that we are not actually saying *new line, uppercase Q., tab key, cap next*, and then *do you*. Instead, we are saying *quesco do you* and allowing Dragon (or a CAT program) to produce the question marker in its proper format..

To produce this:

    Q.   Do you. . .

This is what you actually say:

quesco do you

To base the grammatical model on how we actually speak as voice writers, we must customize it to recognize the syntax of our dictation theory text in relationship to the regular English words we routinely dictate. This is accomplished by gathering text files that are representative of the type of material we expect to produce in our realtime work and then searching for all the words and other text strings within those files that we *do not say* and replacing them with the ones *we do say*.

The way you determine which word replacements need to be made is by first identifying the words and text strings you use special pronunciations for as part of your realtime dictation language. Locate those specific items in Dragon's vocabulary to note how they're spelled by their original written forms (not alternate written forms or spoken forms). Next, do search-and-replacements on all the text within your documents to make them match up to the written-form spellings of all those vocabulary items. Then run those documents through Dragon's vocabulary-building utility for contextual analysis to build the grammatical model.

Basically, for whatever special text you will be regularly producing in your dictation, the written forms which represent them need to be reflected in your documents, because that is the only way Dragon can identify which vocabulary items it should refer to for making proper grammatical decisions when you dictate them—locating them based on their written identities.

As an example, if your question-marker entry appears with a written form of "quesco" in Dragon's vocabulary (the original written-form spelling, not any alternate written-form spelling), you would do a search for all text strings associated with your question marker and replace them with "quesco" as follows:

If your document shows this:

(new line)
Q. *(tab)*

Replace the above string with this:

quesco ← This is the original written-form spelling of the question-marker entry as it appears in Dragon's vocabulary.

Note

If you are using Microsoft Word or a CAT program to handle formatting of your question marker, you may have combined a period (.) into your formatting instructions (. *(new line)* Q. *(tab)*). If so, include a period (.) in your text-string search.

While "quesco" is the original written-form spelling used in this book to represent a question marker, it is merely an example. You may have chosen to use a different written-form spelling in your own vocabulary (a different voice code, a CAT code, etc.). Whatever spelling you choose, it is important <u>not</u> to have the original written form of the question-marker entry be spelled as "Q"; otherwise, Dragon would expect you to dictate *uppercase Q*—or other pronunciations which pertain to vocabulary items possessing that same written-form identity—in every context those spellings occur throughout your documents. By using a unique written-form spelling that does not match the written-form spellings of any other vocabulary item, Dragon will not have a problem identifying specifically the vocabulary item you're referring to as it analyzes your documents.

Remember that the only way for Dragon to learn how to recognize your realtime dictation language based on context is for you to custom-prepare your documents to reflect the written forms of those special vocabulary items and run them through Dragon's vocabulary-building utility for analysis. Doing this will allow Dragon to consistently recognize them when you speak continuously without pausing.

For any new vocabulary entries you make whose written forms are not reflected in the documents you have Dragon analyze, you will need to pause before and after saying their pronunciations if you want Dragon to consistently recognize them in your dictation. Pauses separate your special dictation words from all other words and prevents them from being batch-processed together. This gives you the ability to bypass Dragon's tendency to pick common English words that would normally be chosen from its base vocabulary due to the grammatical model's influence.

To obtain accurate recognition from continuous speech without pausing, it is important to identify the written forms of all your special dictation words and do search-and-replacements for them. Pay the most attention to the commonly used parts of your realtime dictation language (see the table below).

Table 44

Main Focus of Text Search-and-Replacements
Vocabulary Item Categories for Each Career Specialty

Court Reporting	CART	Captioning	Financial Call Reporting
Punctuation, Speaker Identifiers, and Q&A Markers	Punctuation and Speaker Identifiers	Punctuation and Speaker Identifiers	Punctuation, Speaker Identifiers, and Q&A Markers

> **Note**
>
> Be sure to use the exact written-form spellings of all your special dictation entries just as they appear in your Dragon vocabulary when doing text replacements.

Now let's review a sample portion of text that needs to be prepared for Dragon to analyze. Illustration 3 shows the document screen of a regular word-processing program displaying the text from a deposition transcript. The example being used here is Q&A syntax for court reporting. To prepare that document for analysis, the only information you need is the text that appears in the body of the transcript. We do not need line numbers, headers, footers, or graphics, because that text does not represent the words you will be speaking during realtime dictation, so that information must be removed altogether.

Sample Document

Illustration 3

```
 6   Q    Doctor, we may use the term "diet drugs" loosely
 7   in your deposition today.  You understand that I'm
 8   talking about either fenfluramine or dexfenfluramine?
 9   A    Yes, sir.
10   Q    Are you aware, one way or the other, whether or
11   not Ms. Bailey has ever taken either dexfenfluramine or
12   fenfluramine?
13   A    I reviewed the records and don't recall that she
14   ever mentioned that when I visited with her.
15   Q    So there's no indication in your chart, is that
16   correct, Dr. Carter, that Ms. Bailey ever told you that
17   she took diet drugs, generally, and specifically
18   Pondimin or Redux?
19   A    I want to review real quickly just to make sure.
20   But, no, this was a surprise to me.  No, sir, she has
21   never mentioned that.
22   Q    And I mentioned brand names there, Pondimin and
23   Redux, fenfluramine, and dexfenfluramine.  No indication
24   of any of those medications in the history that she gave
25   you, correct?
```

Chapter 6: Improving Accuracy

Once you have only the text from the body of the transcript, you then need to determine the written forms of your realtime dictation words as they appear in Dragon's vocabulary so that Dragon will know which words it should expect you to dictate in order to produce the text that is present. In this portion of text, you would primarily focus on punctuation marks and Q&A markers. If the written forms of your punctuation marks and Q&A markers are different than what your document shows, you must do search-and-replacements on them so they match your vocabulary. To reduce the time it takes to reformat your documents, program macros that will automatically perform repetitive search-and-replace operations within your word-processing software.

The next three sections give examples of how the sample portion of text from Illustration 3 may appear in its finished format according to different software scenarios.

Reformatting Documents for Dragon-Alone Use

If you are using Dragon alone without CAT software and set up your vocabulary entries exactly the same way as the examples in this book, you would need to identify the written forms of your Q&A markers and punctuation marks in the sample portion of text provided in Illustration 3, as shown below. Note that your Q&A marker entries show different written forms than what appears in the document, but the punctuation marks already match up to the written-form spellings in Dragon's vocabulary. This means you will only need to do search-and-replacements for Q&A text strings and leave the punctuation marks just as they are.

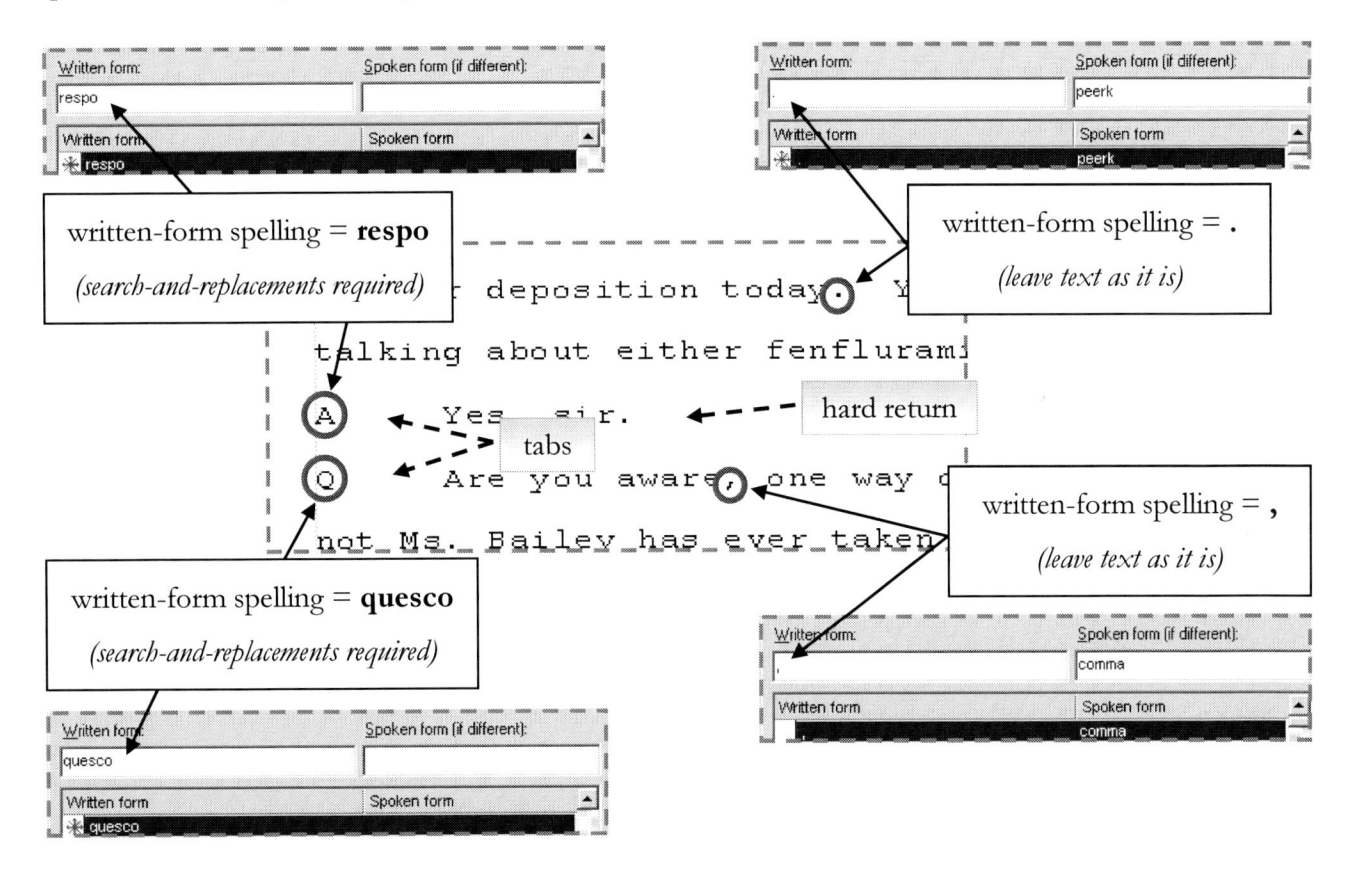

written-form spelling = **respo**
(search-and-replacements required)

written-form spelling = .
(leave text as it is)

hard return

tabs

written-form spelling = ,
(leave text as it is)

written-form spelling = **quesco**
(search-and-replacements required)

167

Of course, the sample document provided in Illustration 3 only shows Q&A text. But if you are reformatting your own documents, you may have other text present as well, such as speaker names, parentheticals, et cetera. For any text present in your own documents that relate to your special voice writing entries for your realtime dictation language, remember to make all text match up to their written forms in Dragon's vocabulary.

Below is an example of how your sample document should appear after this text has been reformatted to reflect the written forms identified in Dragon's vocabulary. Note that all formatting has been removed because you do not want Dragon to expect you to dictate formatting commands.

Sample Document Reformatted
(for Dragon-Alone Use)

quesco Doctor, we may use the term "diet drugs" loosely in your deposition today. You understand that I'm talking about either fenfluramine or dexfenfluramine? respo Yes, sir. quesco Are you aware, one way or the other, whether or not Ms. Bailey has ever taken either dexfenfluramine or fenfluramine? respo I reviewed the records and don't recall that she ever mentioned that when I visited with her. quesco So there's no indication in your chart, is that correct, Dr. Carter, that Ms. Bailey ever told you that she took diet drugs, generally, and specifically Pondimin or Redux? respo I want to review real quickly just to make sure. But, no, this was a surprise to me. No, sir, she has never mentioned that. quesco And I mentioned brand names there, Pondimin and Redux, fenfluramine, and dexfenfluramine. No indication of any of those medications in the history that she gave you, correct?

The table below outlines the basic order of steps taken to reformat the above text.

Table 45

Reformatting Steps for Dragon-Alone Use
Order of Text Search-and-Replacements

Steps	Searched for	Replaced with
1	*(hard return)* **Q** *(tab)*	*(space)* **quesco** *(space)*
2	*(hard return)* **A** *(tab)*	*(space)* **respo** *(space)*
3	*(hard return)*	*(space)*
4	*(tab)*	*(space)*

Additionally, if you had text strings for specific speaker names (e.g., *(hard return)* MR. SMITH: *(tab)* and *(hard return)* MS. BLACK *(tab)*, etc.) for which you would normally dictate voice codes such as "spee 1," "spee 2," et cetera, you would do search-and-replacements on those text strings as well. If the written forms for your voice codes are "SPEAKER-1," "SPEAKER-2," et cetera, you would replace the first speaker name text string with "SPEAKER-1," the second with "SPEAKER-2," and so on.

When you are finished cleaning up your document, save it in an ASCII or plain text (.txt) file format.

Chapter 6: Improving Accuracy

Reformatting Documents for CAT Software Use

As you already know, depending on the operational requirements of your CAT software to work effectively with Dragon, there are a variety of ways your Dragon vocabulary voice writing entries may be set up. Since it is not possible to approach this section with instructions pertaining to every conceivable CAT program, only general ideas are presented. Use the concepts provided here, but refer to your CAT software manual for specific instructions.

Let's cover two different scenarios as to how you may need to reformat your documents according to the way Dragon's vocabulary is set up to work with your particular CAT program.

CAT Software Scenario 1
(with CAT Dictionary)

This scenario assumes that your CAT program is unlimited in its formatting capabilities, so the written forms of your special voice writing entries can be anything you want because you will have full control in being able to define their output yourself by making entries for them into your CAT program's dictionary database. If this is the case, then you will likely use only voice code spellings for your Dragon vocabulary entries, as this is a simpler one-way method of inputting them without requiring you to make so many different spelling considerations. Identifying written forms are easy, because they would simply be your voice code spellings, as shown below.

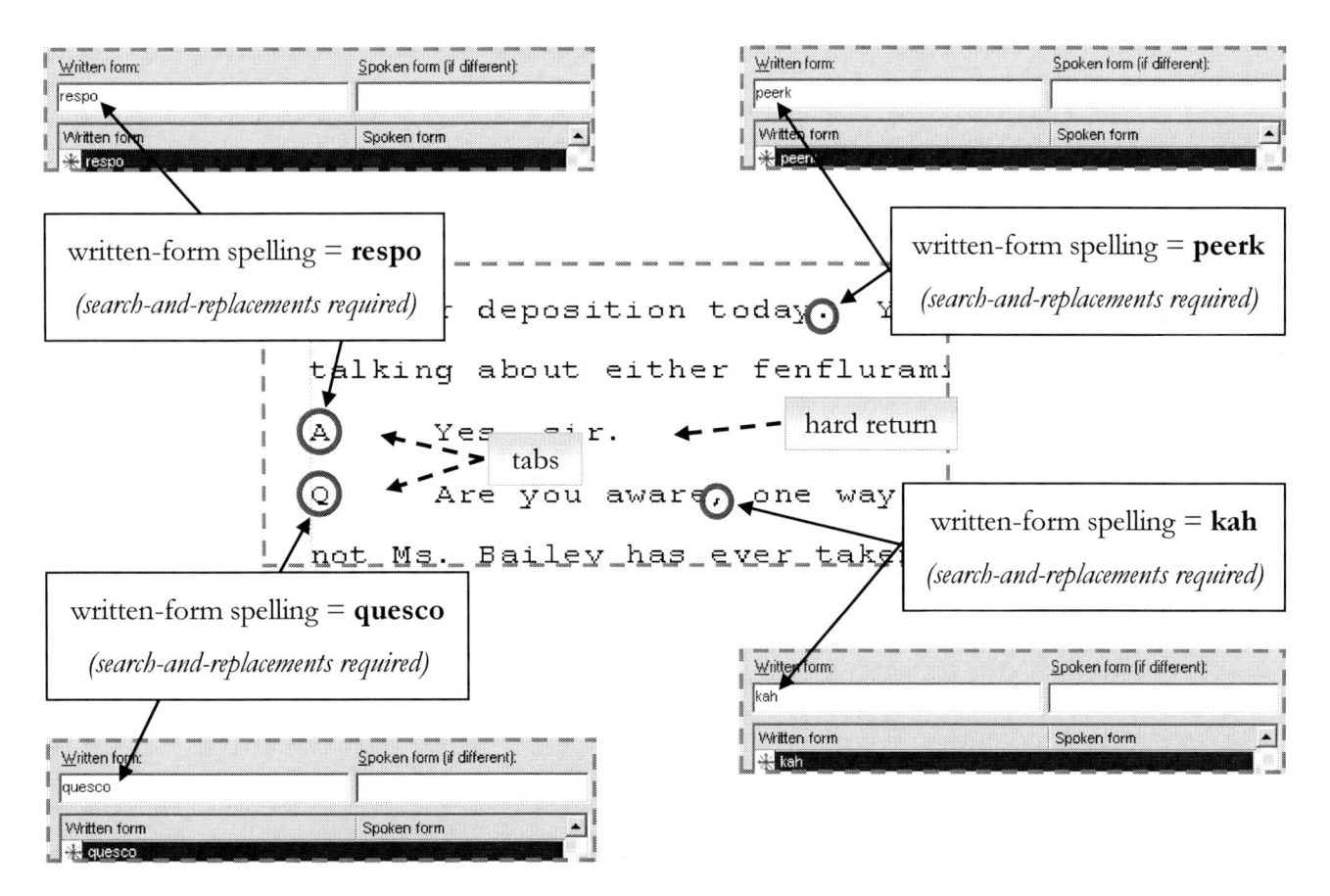

written-form spelling = **respo**
(search-and-replacements required)

written-form spelling = **peerk**
(search-and-replacements required)

written-form spelling = **kah**
(search-and-replacements required)

written-form spelling = **quesco**
(search-and-replacements required)

hard return

tabs

169

Perform search-and-replace operations for all text designated as part of your realtime dictation language— speaker identifiers, parentheticals, brief forms, et cetera—if their written forms are different than what appears in the document. Using the portion of text from the *Sample Document* in Illustration 3 as an example, the document would be reformatted as follows:

Sample Document Reformatted
(for Use with Dragon+CAT with CAT Dictionary)

quesco Doctor kah we may use the term oco diet drugs cloco loosely in your deposition today peerk You understand that I'm talking about either fenfluramine or dexfenfluramine respo Yes sir quesco Are you aware kah one way or the other kah whether or not Ms. Bailey has ever taken either dexfenfluramine or fenfluramine respo I reviewed the records and don't recall that she ever mentioned that when I visited with her quesco So there's no indication in your chart is that correct Dr. Carter kah that Ms. Bailey ever told you that she took diet drugs comma generally kah and specifically Pondimin or Redux respo I want to review real quickly just to make sure peerk But kah no kah this was a surprise to me peerk No kah sir kah she has never mentioned that quesco And I mentioned brand names there kah Pondimin and Redux kah fenfluramine kah and dexfenfluramine peerk No indication of any of those medications in the history that she gave you kah correct kwee

---

The table below outlines the basic order of steps taken to reformat the above text.

Table 46

Reformatting Steps for CAT Software Scenario 1 (with CAT Dictionary)
Order of Text Search-and-Replacements

Steps	Searched for	Replaced with
1	. *(hard return)* **Q** *(tab)*	*(space)* **quesco** *(space)*
2	*(hard return)* **Q** *(tab)*	*(space)* **quesco** *(space)*
3	? *(hard return)* **A** *(tab)*	*(space)* **respo** *(space)*
4	*(hard return)* **A** *(tab)*	*(space)* **respo** *(space)*
5	. *(space) (space)*	*(space)* **peerk** *(space)*
6	? *(space) (space)*	*(space)* **kwee** *(space)*
7	, *(space)*	*(space)* **kah** *(space)*
8	*(hard return)*	*(space)*
9	*(tab)*	*(space)*
10	*(space) (space)*	*(space)*

Chapter 6: Improving Accuracy

Let's look at the basic order of steps taken to clean up the sample document for *CAT Software Scenario 1*.

- The first replacements made were characters and formatting that are being handled by the CAT program, for example, a period (.), a hard return, a Q, and a tab. Since a CAT program always inserts a period (.) at the end of the last line before a question marker and a question mark (?) at the end of the last line before an answer marker, you will not be dictating that ending punctuation, so you should replace those text strings prior to doing replacements for all remaining punctuation marks.

- When replacing periods (.) and commas (,), you must make sure that the appropriate spaces are included in your searches because you do not want to replace punctuation symbols that should remain with certain text, such as "Dr." or "Ms." Note that if this document contained numbers (such as "$5.00" or "10,000"), you wouldn't have had to worry about the period (.) or comma (,) being replaced by a voice code in those instances because spaces were not included in your punctuation searches.

- Assuming that CAT dictionary entries exist for converting phrases such as "yes sir" and "no sir" (without commas) to "yes, sir" and "no, sir" (with commas) you would want to remove the comma (,) from those phrases within your documents since you will not be dictating them. Realize this also means you would not have had to enter pre-punctuated phrases in your Dragon vocabulary since CAT dictionary entries take care of inserting all punctuation.

- You'll also notice that ", is that correct," was replaced with the voice code "ITC." Assuming this was entered as a brief form into the CAT program's dictionary and the CAT program has user-programmable artificial-intelligence capabilities, you would only need to dictate *ITC* for every occurrence of "is that correct"; depending on the context in which that phrase occurs—whether it be the beginning, middle, or end of a sentence—this phrase would appear as just the words by themselves or be automatically punctuated with either commas (,) or a semicolon (;).

- If speaker names, parentheticals, or other punctuation symbols were present, you would have needed to do search-and-replacements on those as well so their written forms would be reflected in the document.

- Lastly, you would remove all leftover formatting because you do not want Dragon to expect you to dictate formatting commands. You would search for tabs, hard returns, indents, et cetera, and replace those with single spaces. To remove multiple spaces, repeat the operation of replacing double spaces with single spaces until only a single space remains between all words of text.

When you are finished cleaning up your document, save it in an ASCII or plain text (.txt) file format.

CAT Software Scenario 2
(without CAT Dictionary)

In this scenario, your CAT program may be limited to formatting only speaker names and Q&A markers, so you may need to use CAT codes as the written-form spellings for only those particular voice writing entries. The written forms of all other voice writing entries you make would be regular English text spellings, because your CAT program will not be able to convert custom text spellings into other words or formats. This also means you will have to rely on Dragon's formatting capabilities for punctuation and all remaining text requiring special formatting needs.

For purposes of this example, let's begin by assuming your CAT program requires you to use **{Q}** as the CAT code for a question marker and **{A}** as the CAT code for an answer marker. Next, using the sample portion of text provided in Illustration 3, you would need to identify the written forms of your Q&A markers and punctuation marks to determine which search-and-replacements need to be made. Note that your Q&A marker entries show different written forms than what appears in the document, but the punctuation marks already match up to the written-form spellings in Dragon's vocabulary. This means you will only need to do search-and-replacements for Q&A text strings and leave the punctuation marks just as they are.

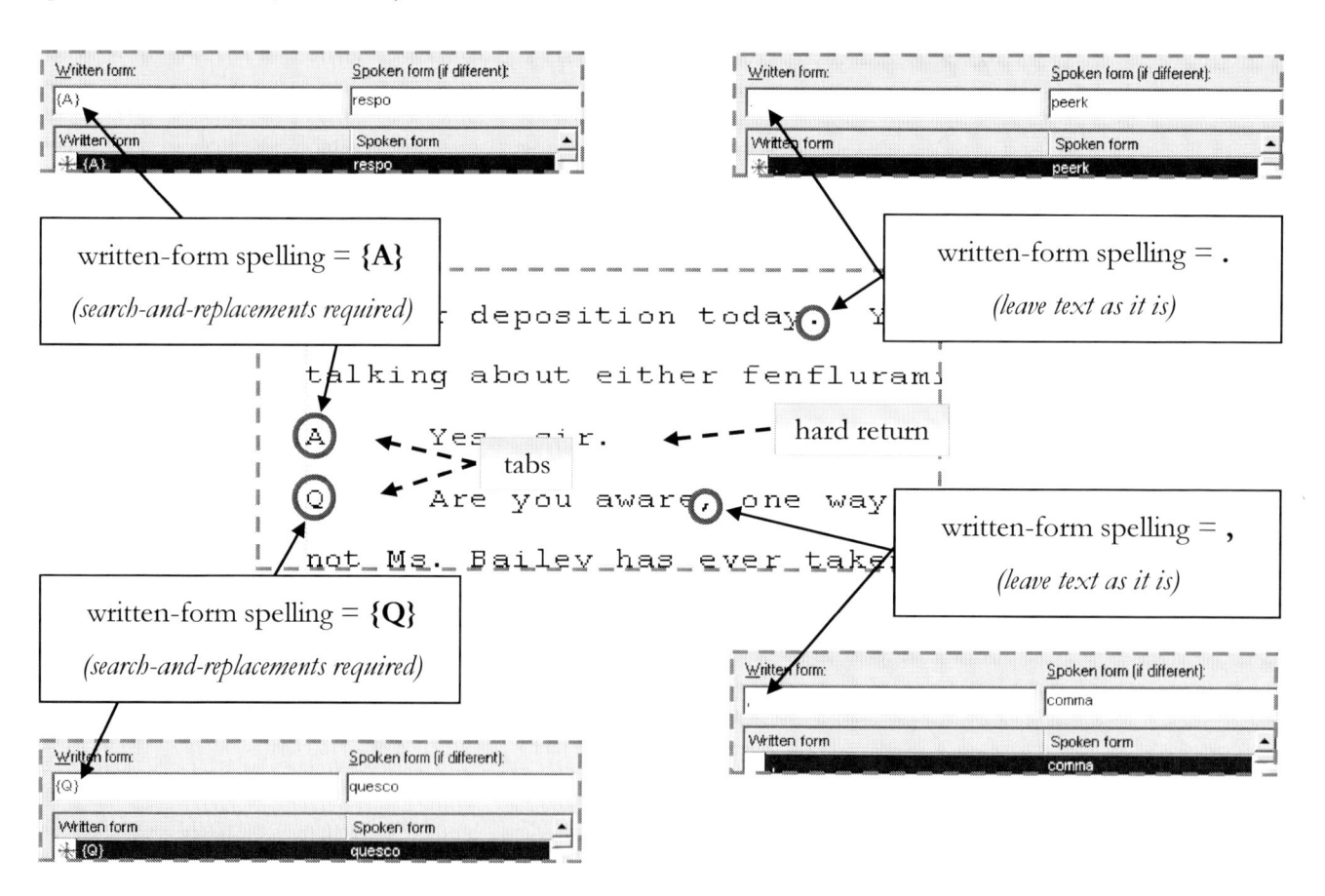

written-form spelling = **{A}**
(search-and-replacements required)

written-form spelling = .
(leave text as it is)

written-form spelling = ,
(leave text as it is)

written-form spelling = **{Q}**
(search-and-replacements required)

172

Below is an example of how your sample document would appear after this text has been reformatted to reflect the written forms identified in Dragon's vocabulary.

Sample Document Reformatted
(for Use with Dragon+CAT without CAT Dictionary)

{Q} Doctor, we may use the term "diet drugs" loosely in your deposition today. You understand that I'm talking about either fenfluramine or dexfenfluramine {A} Yes, sir {Q} Are you aware, one way or the other, whether or not Ms. Bailey has ever taken either dexfenfluramine or fenfluramine {A} I reviewed the records and don't recall that she ever mentioned that when I visited with her {Q} So there's no indication in your chart, is that correct, Dr. Carter, that Ms. Bailey ever told you that she took diet drugs, generally, and specifically Pondimin or Redux {A} I want to review real quickly just to make sure. But, no, this was a surprise to me. No, sir, she has never mentioned that {Q} And I mentioned brand names there, Pondimin and Redux, fenfluramine, and dexfenfluramine. No indication of any of those medications in the history that she gave you, correct?

The table below outlines the basic order of steps taken to reformat the above text.

Table 47

Reformatting Steps for CAT Software Scenario 2 (without CAT Dictionary)
Order of Text Search-and-Replacements

Steps	Searched for	Replaced with
1	. *(hard return)* Q *(tab)*	*(space)* {Q} *(space)*
2	*(hard return)* Q *(tab)*	*(space)* {Q} *(space)*
3	? *(hard return)* A *(tab)*	*(space)* {A} *(space)*
4	*(hard return)* A *(tab)*	*(space)* {A} *(space)*
5	*(hard return)*	*(space)*
6	*(tab)*	*(space)*
7	*(space) (space)*	*(space)*

Additionally, if you had text strings for specific speaker names (e.g., *(hard return)* MR. SMITH: *(tab)* and *(hard return)* MS. BLACK *(tab)*, etc.) for which you would normally dictate voice codes such as "spee 1," "spee 2," et cetera, you would do search-and-replacements on those text strings as well. The written forms for your speaker identifiers would be CAT codes, so that's what you would replace your speaker text strings with.

Let's look at the basic order of steps taken to clean up the sample document for *CAT Software Scenario 2*.

- The first replacements made were characters and formatting that are being handled by the CAT program, for example, a period (.), a hard return, a Q, and a tab. Since a CAT program always inserts a period (.) at the end of the last line before a question marker and a question mark (?) at the end of the last line before an answer marker, you will not be dictating that ending punctuation. These text strings need to be replaced or else Dragon will expect you to dictate that ending punctuation before dictating Q&A markers.

- All leftover formatting is removed so that Dragon does not expect you to dictate formatting commands. You would search for tabs, hard returns, indents, et cetera, and replace those with single spaces. To remove multiple spaces, repeat the operation of replacing double spaces with single spaces until only a single space remains between all words of text.

When you are finished cleaning up your document, save it in an ASCII or plain text (.txt) file format.

Programming Macros to Automate Search-and-Replace Functions

Any word-processing program may be used to clean up texts. To make the text-cleanup process faster when reformatting multiple documents, you can program macros to record your key strokes for performing search-and-replace functions. You can then program a master macro to run the entire list of macros you created for individual search-and-replacement steps and instantly be able to clean up a document using a single key stroke. Refer to the Help menu of your word-processing software to learn how to record macros.

Saving and Storing Documents

All files should be saved in an ASCII or plain text (.txt) format since other file formats may contain special formatting that can prevent text from being read properly when those files are analyzed by Dragon. The recommended amount of data to gather for building your grammatical model is a minimum of 20 megabytes' worth of plain text (.txt) files. Create a special *Vocabulary Building Documents* folder on your computer's hard drive for storing all your vocabulary-building files, and also make backup copies onto external storage media to protect them from loss should you experience computer malfunctions.

Preparing Documents Before a Realtime Job

Be aware that as part of your future preparatory work before taking realtime jobs, you will need to clean up new documents for Dragon to analyze. Do this by obtaining files of transcripts or other text material previously completed in the subject matter you will be reporting. If you do not have personal access to such files, ask your employer or the client ordering your services to furnish you with as many of those files as they can, explaining that it is a requirement for you to be able to build your vocabulary properly. If you are provided with hard-copy materials, use a scanner with optical character recognition (OCR) capabilities from which to create a text file.

Chapter 6: Improving Accuracy

Vocabulary Customization

Once you have prepared a recommended minimum of 20 megabytes' worth of plain text (.txt) files for building the grammatical model, make sure that you separate the monologue material from the Q&A. You want to build a vocabulary that is specific to a certain type of syntax. If your realtime work involves different dictation formats—for example, Q&A style material for court reporting versus monologue style material for CART—you should create separate vocabularies for each, using your current user's vocabulary as their foundational bases. As your realtime work evolves to include a wider array of subject matter, you can use those vocabulary bases for creating unique vocabularies in those specialized topics. See Diagram 3 below for illustration of this concept.

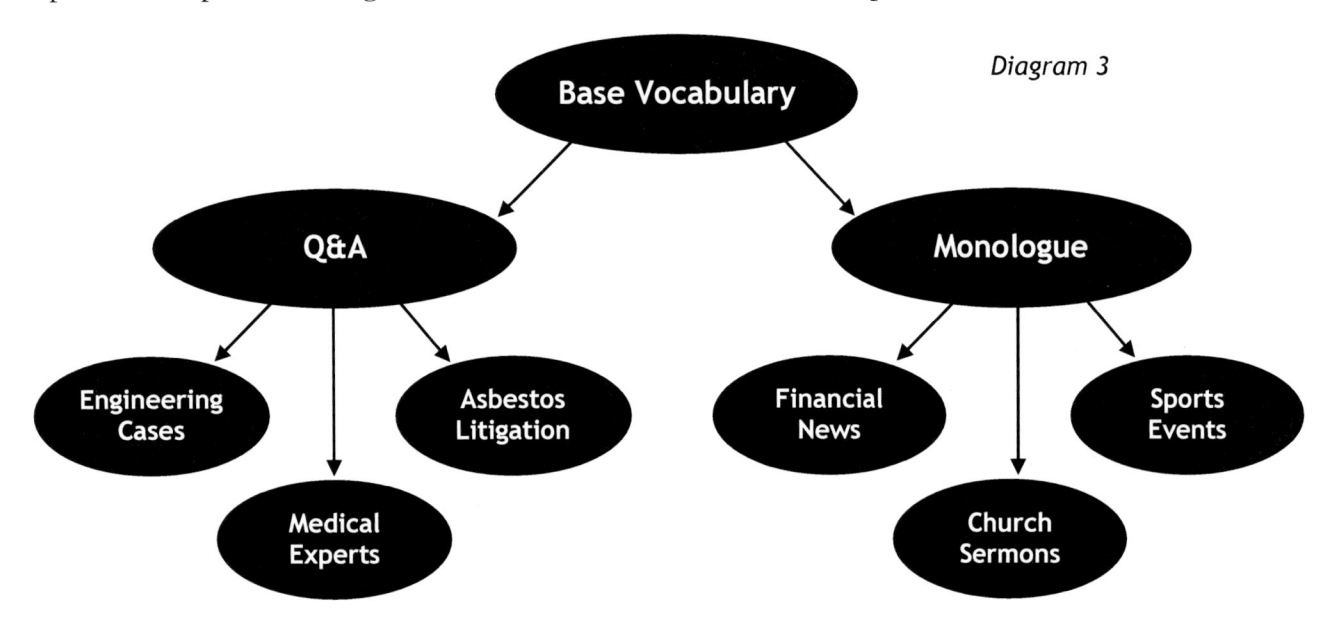

Diagram 3

```
                    Base Vocabulary
                   /              \
                Q&A              Monologue
              / | \              / | \
    Engineering Medical Asbestos  Financial Church Sports
       Cases   Experts Litigation   News   Sermons Events
```

If you will be constructing two grammatical formats, one for monologue and the other for Q&A, you can do so in your current voice model by creating separate vocabularies with those specific styles ("Monologue" and "Q&A") based on your existing vocabulary. This will allow you to safely retain generic vocabularies on which to base other, more specific vocabularies in the future. For example, if you are doing court reporting work in asbestos litigation, you can create a vocabulary called "Asbestos Litigation" based on your Q&A vocabulary; for doing CART work of church sermons, you can base a "Church Sermons" vocabulary on your Monologue vocabulary; for captioning financial news, you can base a "Financial News" vocabulary on your Monologue vocabulary. However, in financial call reporting, where you will often be producing text in both monologue and Q&A format during the same financial call, you will be building your vocabulary with a mixture of these styles, so you should have a separate vocabulary base for "Financial Calls."

Be aware that if you only do one type of work as a realtime voice writer, you only need one vocabulary.

The method used to create separate vocabularies for your work depends on your edition of Dragon. If you have the Dragon NaturallySpeaking® Professional edition, you can easily create a new vocabulary by going to the *Vocabulary* menu > *Manage Vocabularies*. (See the *Creating a New Vocabulary* subsection under the *Vocabulary Management* section of Chapter 8 for instructions.)

Voice Writing Method • FIFTH EDITION
Dragon NaturallySpeaking® 12

If you have the Dragon NaturallySpeaking® Premium edition, you do not have the ability to create multiple vocabularies for a single Dragon user. To create a new vocabulary in Dragon NaturallySpeaking® Premium, you would have to maintain separate Dragon users. But instead of having to recreate a new Dragon user from scratch, you would simply export your existing user, then import it under a new user name, and then modify the vocabulary of the newly imported user.
(See the *Exporting/Importing User Files* subsection under the *User File Management* section of Chapter 8.)

Document Analysis

This section guides you through the process of analyzing documents to build and customize the grammatical model for your particular Dragon vocabulary. Use your custom-prepared files when following the steps below.

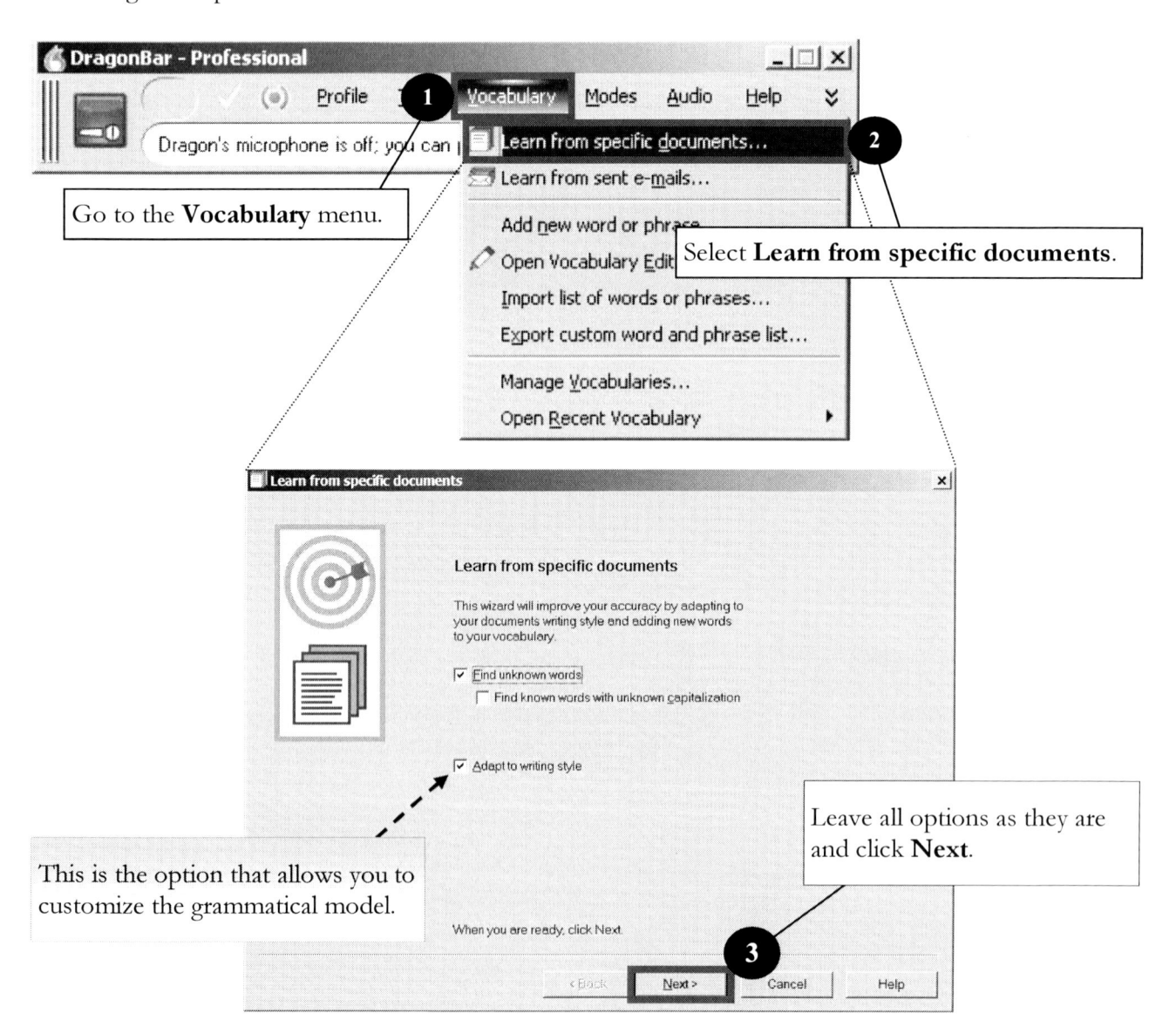

1 Go to the **Vocabulary** menu.

2 Select **Learn from specific documents**.

This is the option that allows you to customize the grammatical model.

3 Leave all options as they are and click **Next**.

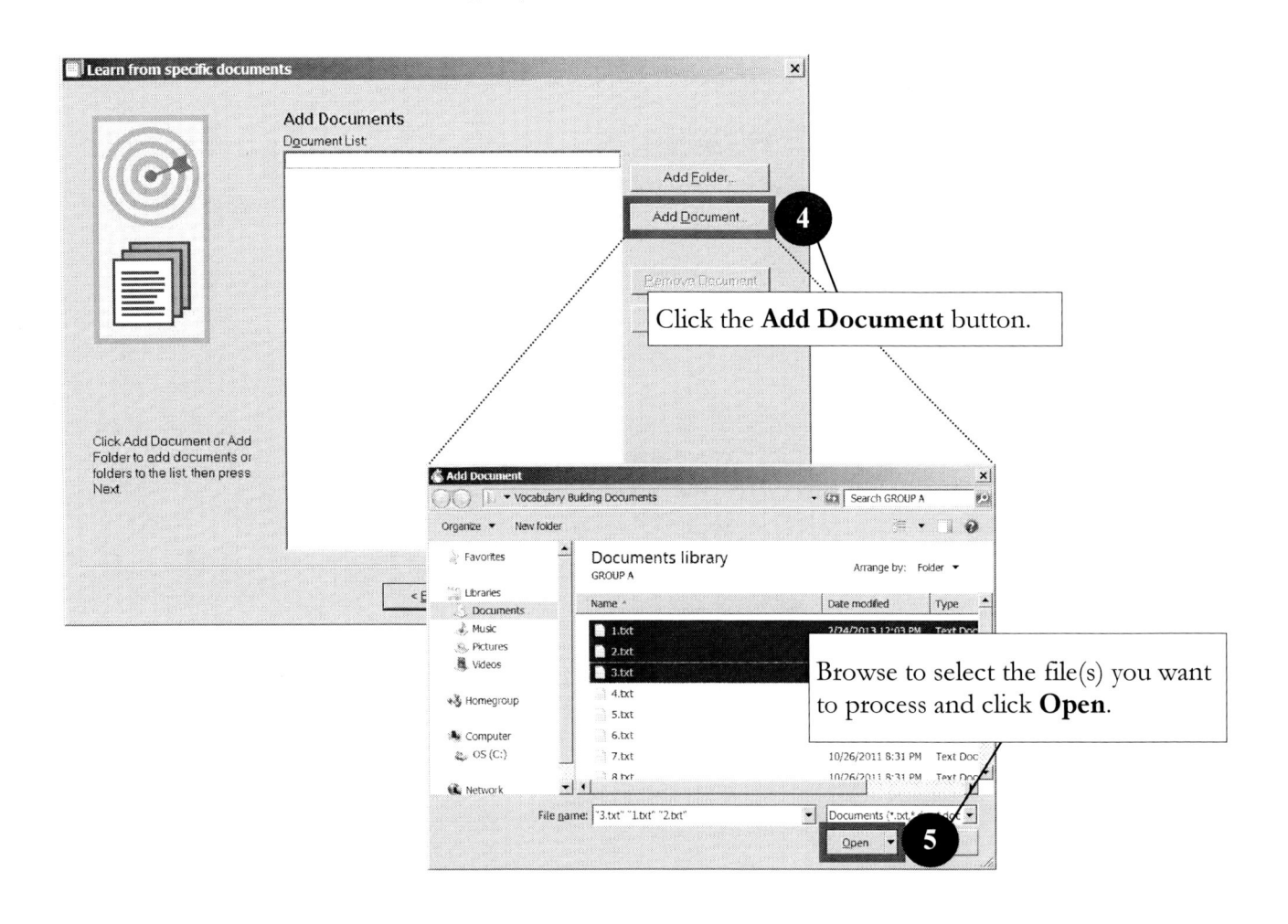

Chapter 6: Improving Accuracy

Click the **Add Document** button.

Browse to select the file(s) you want to process and click **Open**.

Note

It is recommended that you limit the number of files selected to under 2 megabytes per each document-analysis session for best processing quality.

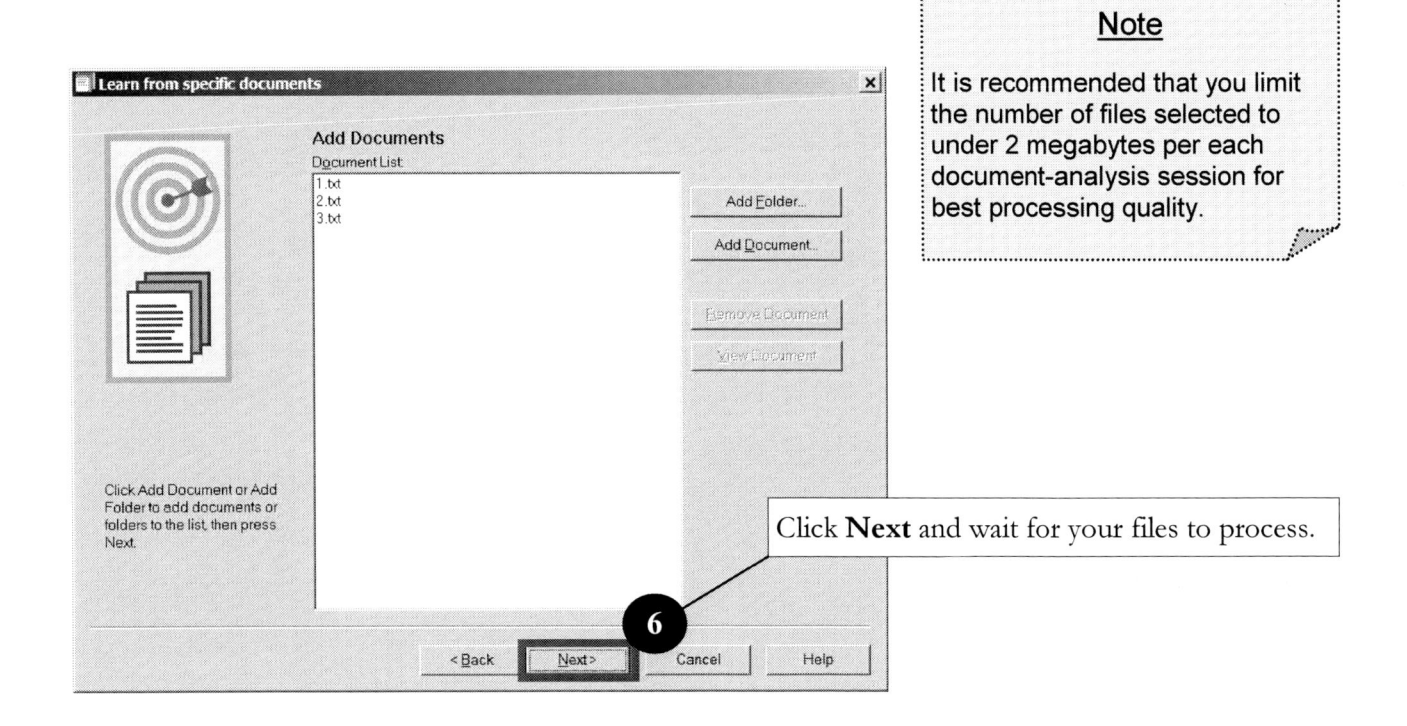

Click **Next** and wait for your files to process.

177

Dragon has processed all files to extract any new words from your documents.

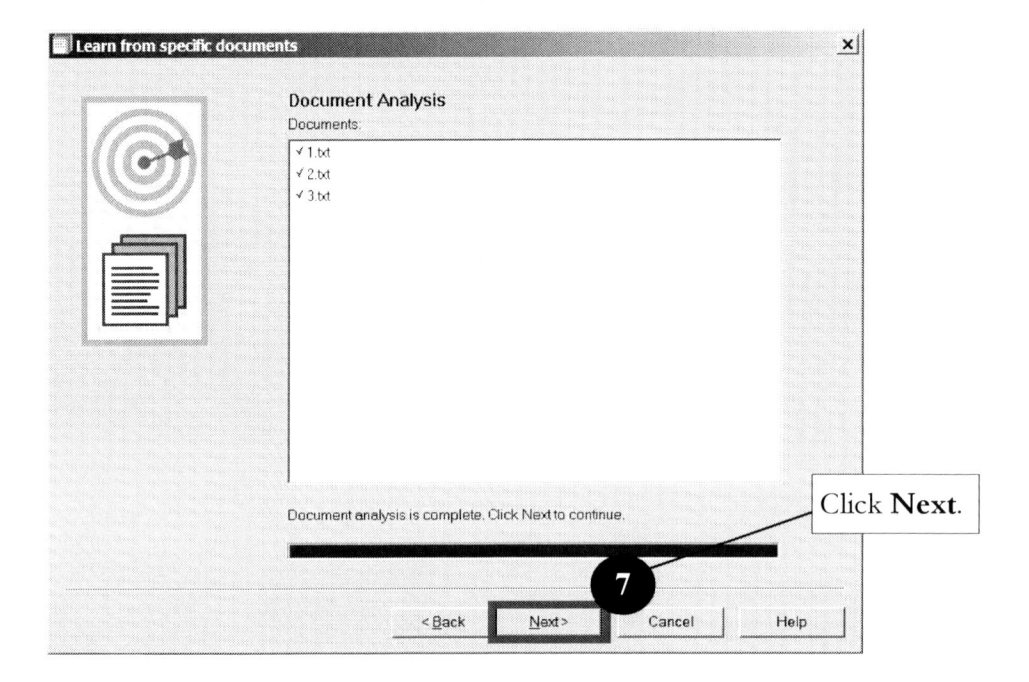

Click **Next**.

The next screen displays all the new words found. You should never automatically add all of them without reviewing the entire list first.

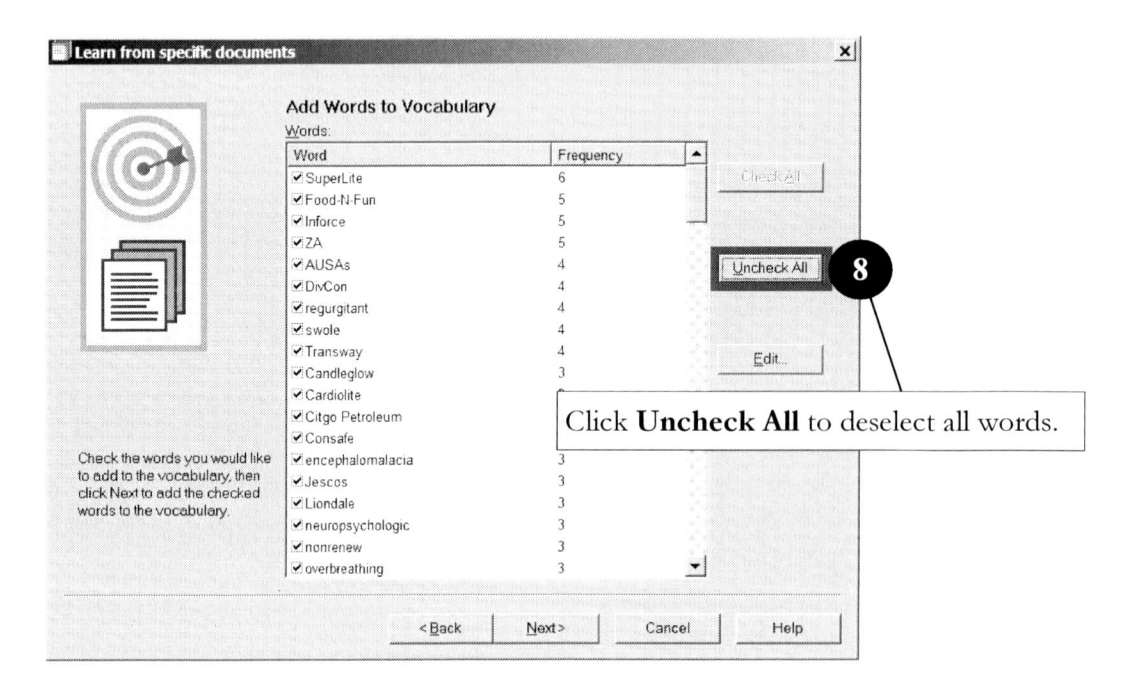

Click **Uncheck All** to deselect all words.

Chapter 6: Improving Accuracy

Now you will scroll through the list and select any words you would like to add to your vocabulary, choosing only the ones you think you will encounter during dictation. This step is crucial when processing transcripts for specific subject matter as part of your preparatory work to provide realtime services.

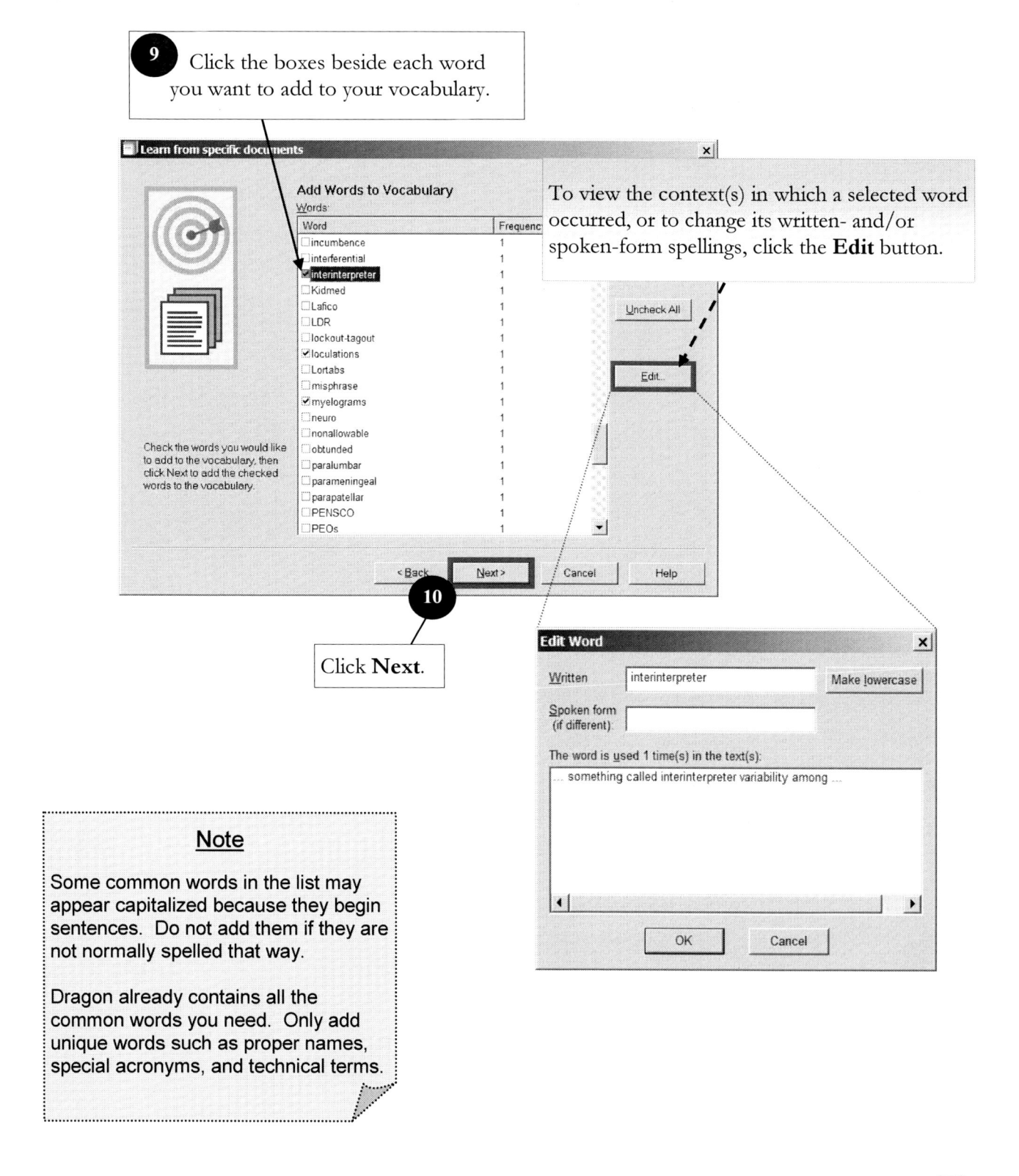

9 Click the boxes beside each word you want to add to your vocabulary.

To view the context(s) in which a selected word occurred, or to change its written- and/or spoken-form spellings, click the **Edit** button.

10 Click **Next**.

> **Note**
>
> Some common words in the list may appear capitalized because they begin sentences. Do not add them if they are not normally spelled that way.
>
> Dragon already contains all the common words you need. Only add unique words such as proper names, special acronyms, and technical terms.

179

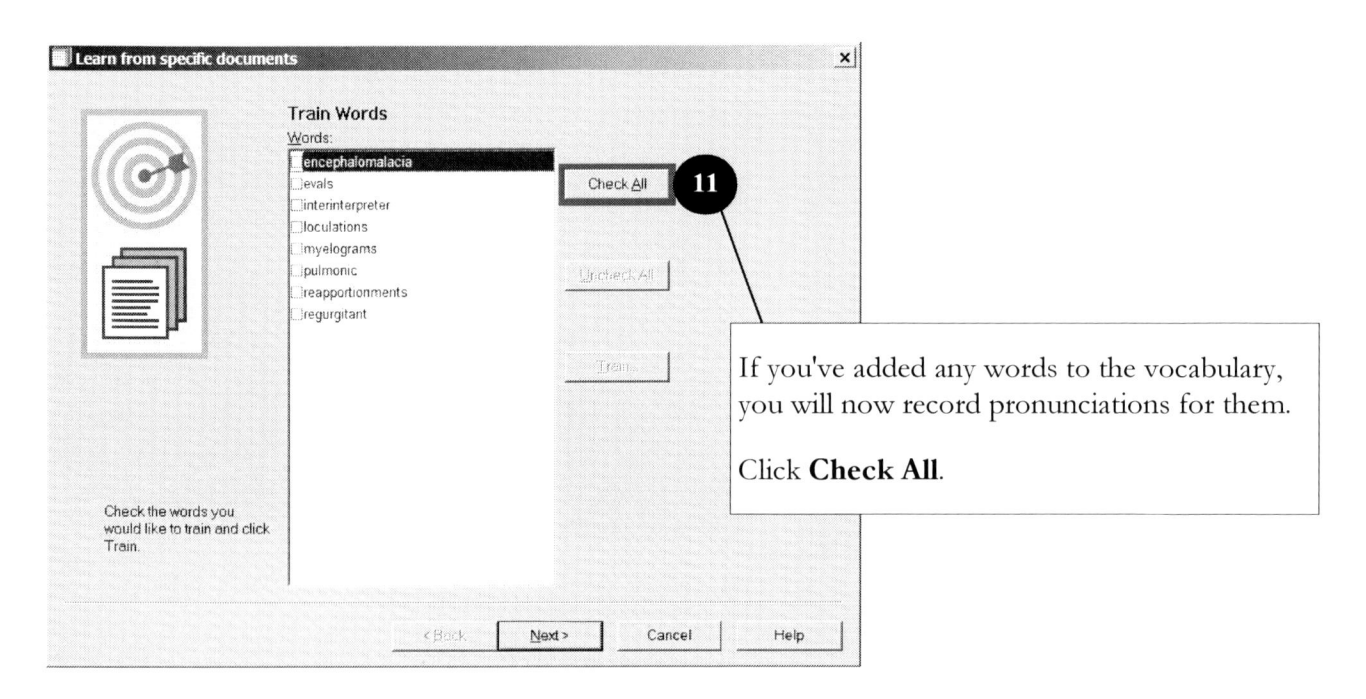

Voice Writing Method • FIFTH EDITION
Dragon NaturallySpeaking® 12

If you've added any words to the vocabulary, you will now record pronunciations for them.

Click **Check All**.

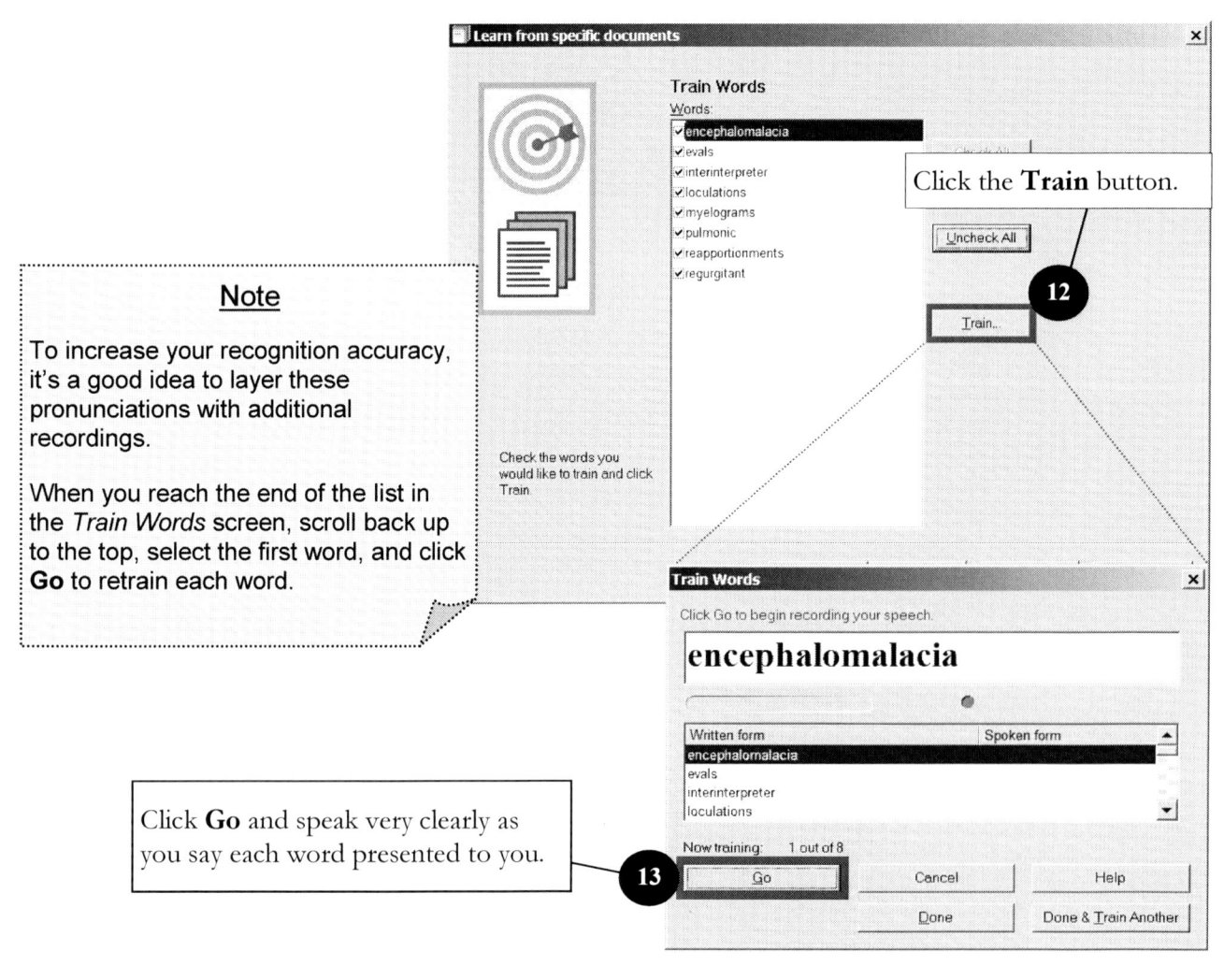

Click the **Train** button.

Note

To increase your recognition accuracy, it's a good idea to layer these pronunciations with additional recordings.

When you reach the end of the list in the *Train Words* screen, scroll back up to the top, select the first word, and click **Go** to retrain each word.

Click **Go** and speak very clearly as you say each word presented to you.

180

Chapter 6: Improving Accuracy

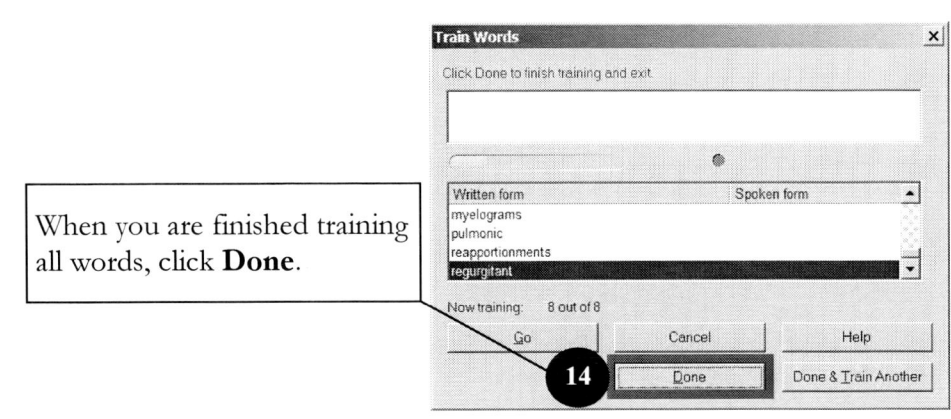

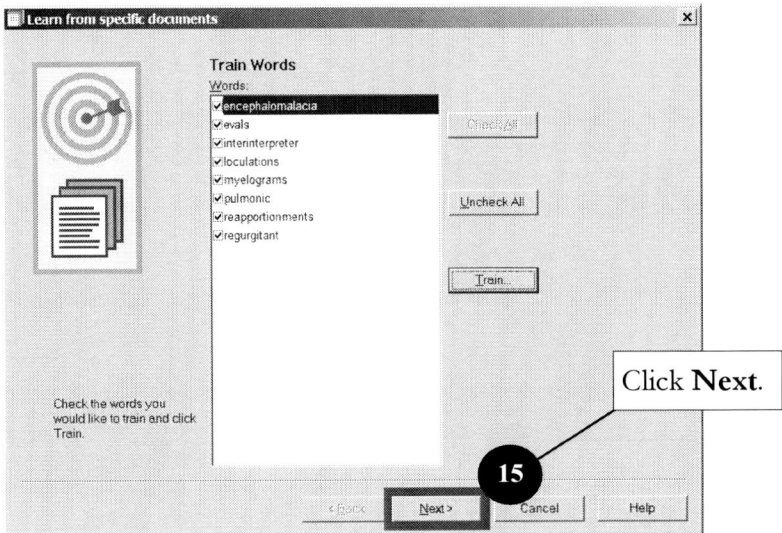

When you are finished training all words, click **Done**.

Click **Next**.

Wait as the grammatical model is adapted. This may take up to a minute or longer, depending on document file sizes. When complete, Dragon will be able to recognize the grammatical context of your dictation language from the file(s) just analyzed.

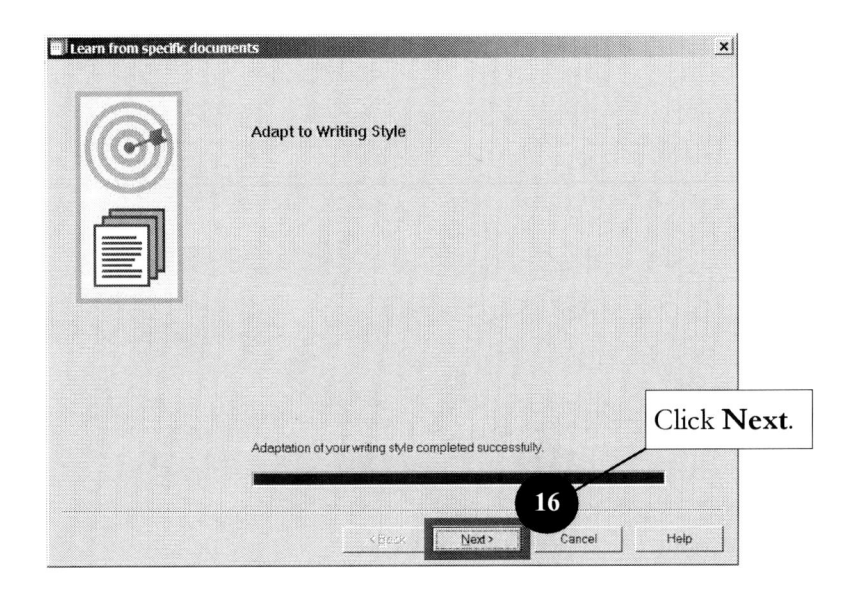

Click **Next**.

Click **Finish**.

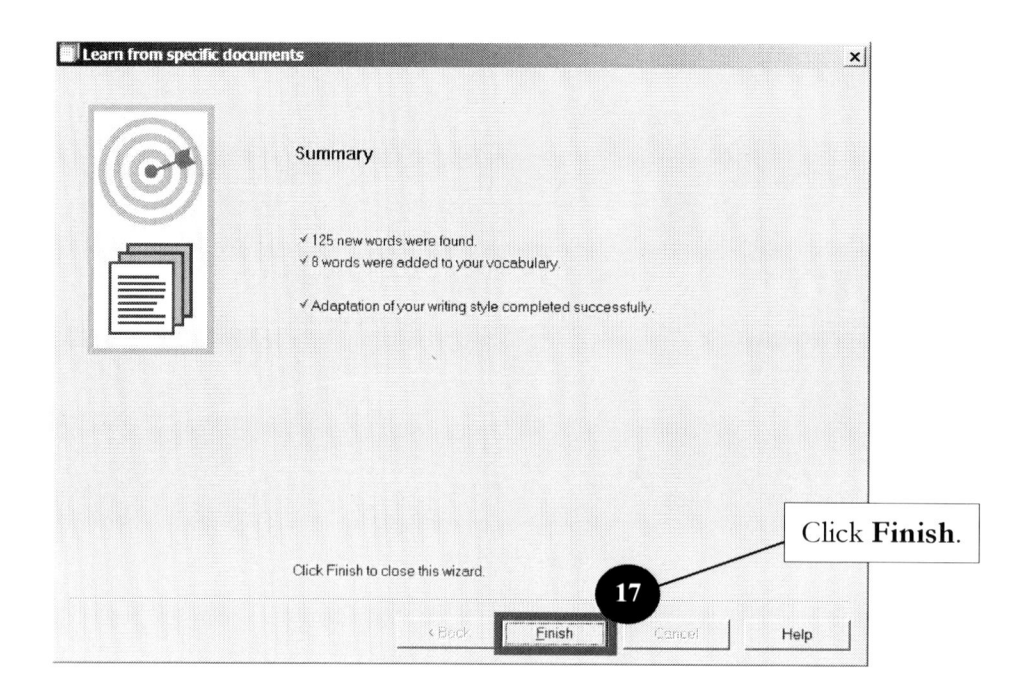

17

18 If you have additional documents to analyze, repeat Steps 1 though 17.

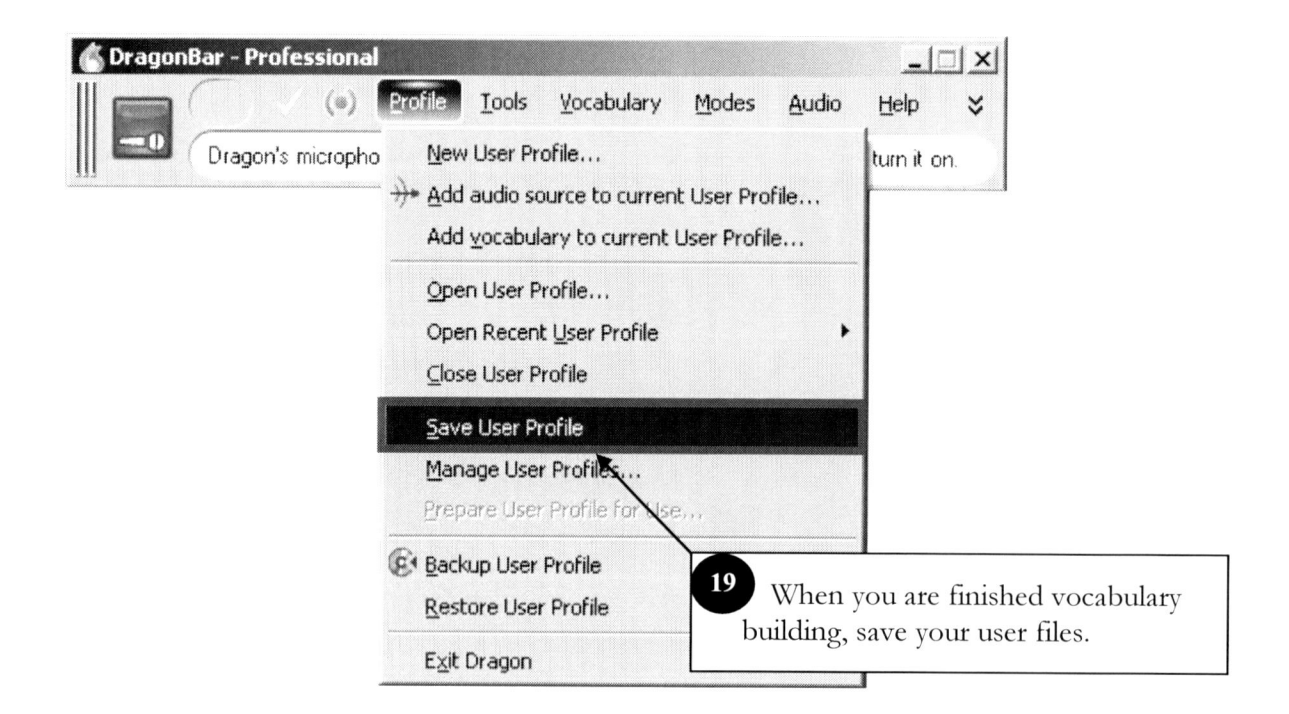

19 When you are finished vocabulary building, save your user files.

Chapter 6: Improving Accuracy

Resolving Recognition Problems

Dragon will likely have trouble recognizing newly added voice writing entries. Since they are a primary part of your basic dictation language, it is important to focus on improving your recognition accuracy of them first before increasing your accuracy for common English words. The point at which all your special dictation items are successfully recognized is the point at which you should begin focusing on correcting all other misrecognitions.

The majority of recognition improvement will come from the traditional method of using the correction tool that was previously explained in the *Correcting Misrecognitions* section of this chapter. If correct translations are not being produced with consistency after having made at least five corrections to the same type of error, you may have to take other corrective actions. This section explores those alternatives.

Removing Unnecessary Words That Conflict with Necessary Words

If Dragon continuously misrecognizes any words you regularly dictate for other words you never expect to encounter in your dictation, remove those words from your vocabulary altogether so they never pose conflicts again. (See the *Deleting an Entry* subsection under the *Working with the Vocabulary* section of Chapter 5.)

Making Additional Vocal Recordings to Distinguish Between Words with Similar Pronunciations

If certain base vocabulary words are necessary to retain, though they pose conflicts with other necessary words, utilizing the correction tool will usually resolve recognition problems. But if you still do not see significant improvement after continuous and repeated attempts to correct those errors, you should record your pronunciations for them. Vocal recordings should be clearly enunciated so Dragon can make distinctions between the pronunciations of the words in question. You may train pronunciations directly from within the correction window (*Train* button) as you are typing in your corrections, or you may locate the words in the vocabulary and train them individually.

To resolve conflicts between words that sound identical, a pronunciation distinction will have to be made using a unique spoken form for the less common word. Refer to the *Homophones* subsection under the *Conflict Resolution* section in Chapter 3 for more details.

Making Multiple Vocabulary Entries to Obtain Correct Recognition for a Single Result

Oftentimes, recognition conflicts may be overcome simply by creating multiple entries for a single word using various spoken-form spellings. For example, if your pronunciation of "would you" sounds more like *woojew* during high-speed dictation and Dragon regularly produces "which you," you may enter the phrase "would you" as two separate entries: 1.) using a spoken-form spelling of "woojew" and 2.) using a spoken-form spelling of "woochew." Once those vocal pronunciations have been recorded, it is a lot less likely for you to experience a recurrence of these conflicts.

Using Misrecognitions to Produce Recognitions

Another way to improve accuracy is to use Dragon's actual errors to invoke proper recognitions. This manipulation technique works best when the erroneous word selections sound identical to intended word selections during high-speed dictation. As an example, if your pronunciation of "how do you" sounds more like *howdy you* and Dragon continuously recognizes that phrase just like it sounds, "howdy you," then instead of working against that misrecognition, you can use it to your benefit by making a new phrase entry for "how do you" and using the spoken-form spelling of "howdy you."

Recording a vocal pronunciation using such a spoken-form spelling will result in significant accuracy improvement, since Dragon already has a tendency to choose the words you're now using as your spoken form. Looking at Dragon's results and analyzing the reasoning behind what it thought you said (*howdy you*), as opposed to what you slurringly said (*how do you*), will guide you in drawing the right conclusions of how to properly resolve similar recognition conflicts in the future.

Accuracy Reinforcement

As you know, Dragon works from a best-match guess structure, which means that even the words it does accurately recognize it was not a hundred percent sure about. So the words it gets correct right now, it may not necessarily get correct in the future. Those words were just the most likely choices at that very moment in time, and of course this will always be the case. Many factors can be weighed into the outcome of future recognition results, such as vocal fluctuations, measurements during periods of adjustment, and available computer resources at the time. But we can significantly improve Dragon's ability to consistently recognize our speech by improving its ability to correctly "guess."

A special technique you may use to reinforce Dragon's ability to correctly "guess" again in the future is called **accuracy reinforcement**. This means using the correction tool not merely for correcting misrecognitions, but to reinforce correct recognitions. To explain how the accuracy reinforcement process works, let's assume Dragon properly recognized all the words in the following sentence:

Would you please state your name for the record?

Chapter 6: Improving Accuracy

Since we cannot physically see Dragon's guessing routines behind the scenes, we never have a way of knowing whether any words make it just barely, or by a landslide, into the finals category of best-match guesses. These specific words were obviously the most probable choices this time around, but we want to be sure they will also be the most probable choices in the future. Whether those same words will be chosen in the future or not is dependent on priority selections. We don't know, for certain, the beginning priority order placed on any words within Dragon's ranking system, but we can certainly have influence over priority-order outcomes in the future. By identifying the common word groups that are likely to occur again in the future and reinforcing to Dragon that its current selections are, in fact, correct, best-match guesses involve less guess work, because the accuracy-reinforcement process has specifically instructed Dragon to assign these words even higher probability values within its ranking system.

Reinforcing proper word selections is a simple matter of opening the correction window (Alt C) with those specific words appearing in it and pressing the *Enter* key (or click OK).

(Note: Most CAT programs allow you to perform accuracy reinforcements in this way. If you have CAT software, refer to either your CAT program's manual or Help menu to see how or if this can be done using the correction tool within that program. Also note that the selected portion of text appearing within the correction window may not be the exact output originally generated by Dragon when using certain CAT programs, in which case you should change the text in the correction window to reflect the actual words you dictated before making the reinforcement.)

Using the sentence from the previous page as an example, the common phrases to target first are "would you," "state your," and "for the." Highlight each phrase independently, then open the correction window and press *Enter*.

The next time this sentence correctly occurs in your dictation, select other phrase combinations such as "would you please," "please state your," "state your name," and "for the record," and repeat the accuracy reinforcement steps for each.

Remember to make sure the text you've selected matches up to the audio of your speech. If not, listen to a playback of the words occurring before and after to determine if more words need to be included with that selection, then make the reinforcement.

Accuracy reinforcement lets Dragon collect more specific acoustic information than it previously had about the words you speak. Dragon will be able to use this more specific data in the future and retrieve it, if necessary, to better determine whether or not it should select those same words again. The result will be that your desired word choices are more likely to consistently make it into the finals category of subsequent best-match guesses.

Chapter 7:
Speed Building

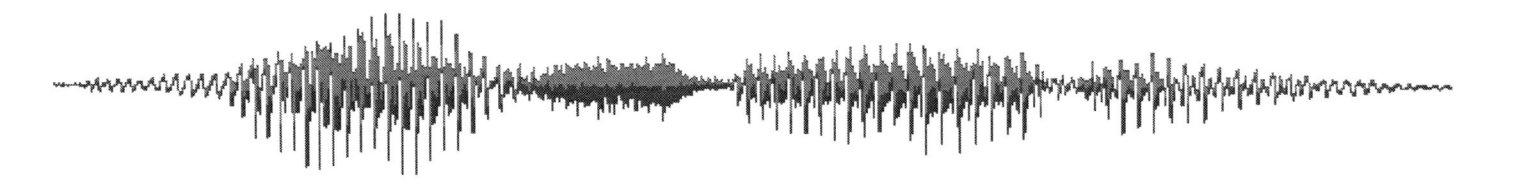

This chapter offers a simple speed-building approach for quickly achieving realtime competency with speech recognition to meet the reporting industry's generally accepted realtime speed and accuracy performance standards (a minimum of 96% accuracy at variable speeds ranging between 180 and 200 words per minute). This is accomplished with the assistance of SpeedMaster™, which is provided on CD along with this book; also included are general voice training texts and sample practice dictation files.

SpeedMaster™ is a speed-building software program designed to help you rapidly build dictation speed as you work to increase speech recognition accuracy, so you can take the guesswork out of measuring speed-building progress. All you do is load a document file, click the green start button, and follow the red arrow as you dictate the words that appear on the SpeedMaster™ screen. You can adjust the speed setting to any number between 100 and 350 words per minute and SpeedMaster™ will automatically scroll the text at your desired rate. Any document may be used, either your own custom material or text you've obtained elsewhere, such as the Internet—the resources are unlimited—as long as it is saved in plain text (.txt) file format. This is the most effective tool you will ever find for mastering your speed-building goals inexpensively.

While SpeedMaster™'s primary purpose is for speed-building practice, it may also be used to determine how fast you should dictate when creating your voice model. Prior to SpeedMaster™'s release, there was no assurable way to know how fast you were dictating when creating a voice model; one could only guesstimate at realtime speed ranges while reading general voice training texts. A text file of each general voice training story is provided on the SpeedMaster™ CD for your convenience.

The majority of speed-building practice has traditionally been performed using professionally pre-recorded audio, which is an excellent source of material to use for training once you have mastered proper dictation skills for speech recognition. But audio material is limited in scope and availability, making it difficult to determine reporting competency levels in highly specialized subjects. The array of specialty subjects you are likely to encounter in the real world is so vast, it would be impossible to make such a variety of pre-recorded material commercially available.

SpeedMaster™ gives you the ability to practice realtime dictation with your own custom material, targeting the subjects you need to focus on most for the types of words you expect to encounter in your realtime career. Gathering your own custom material may be a simple matter of copying and pasting text from the Internet into your regular word-processing software screen and saving the file in a plain text (.txt) file format. You may also be interested in joining NVRA (www.nvra.org) to benefit from the member forum, where you can openly ask experienced professionals to share with you their own transcript text files to use as speed-building practice material.

Speed-Building Plan

Acquiring proper dictation skills for speech recognition requires dedicated focus to personal linguistic programming—before working on additional skill sets—and there is no better way to do this than by reading a text display in SpeedMaster™. As you incorporate the dictation techniques described in Chapter 2, you will find it most practical to begin speaking at an average rate, around 140 words per minute, since you are also learning to dictate punctuation at the same time. You will be able to gradually increase your dictation rate at your own comfortable pace. The goal is to dictate out loud and get comfortable with a given speed until you reach 180 words per minute, then dictate into Dragon at that rate and begin improving your voice model through the correction process. When your SR accuracy reaches 96%, increase the rate in five-word-per-minute increments, continuously dictating and correcting at each speed interval until your SR accuracy reaches a sustained 96% minimum accuracy level at 200 words per minute.

After mastering your linguistic skills by reading and your voice model is perfected based on proper dictation, you can then begin listening to audio recordings and repeating what you hear so that you may merge both skill sets. A good way to practice listening and repeating is by dictating what you hear from television news and programs. You may also contact NVRA about purchasing audio practice material.

SpeedMaster™ Software

Using SpeedMaster™ is simple. Just open a file, set your speed, then dictate or read along following the red arrow as it scrolls through the text. All you need is a collection of document files saved in plain text (.txt) format, or you may use the sample practice dictation files from the CD to get started.

How to Install

Your computer must have Windows XP Home edition or above to meet program system requirements. (Please note that program performance may be affected by your computer's configuration and setting selections.)

Insert the CD in the CD-ROM drive for the installation process to begin and follow the on-screen instructions to complete setup.

Chapter 7: Speed Building

Upon installation, SpeedMaster™ automatically creates a folder for you on your computer's hard drive at the following location:

C:\Program Files\SpeedMaster

Obtaining Files From the SpeedMaster™ CD

Before you eject the SpeedMaster™ CD, be sure to copy and paste all the files from the CD into your *SpeedMaster* folder so you may access them at any time while the SpeedMaster™ CD is no longer inserted. These files include Dragon's general voice training texts as well as other practice files for your dictation. As you collect your own files to use as practice dictation material, remember to save them as plain text (.txt) files so they can be read by SpeedMaster™. Store all dictation files in your SpeedMaster™ folder to keep them together in one place where you can easily find them.

How to Operate

Select SpeedMaster™ from your computer's *Programs* list or double-click on the SpeedMaster™ icon from your desktop screen.

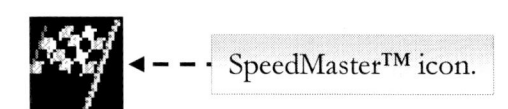

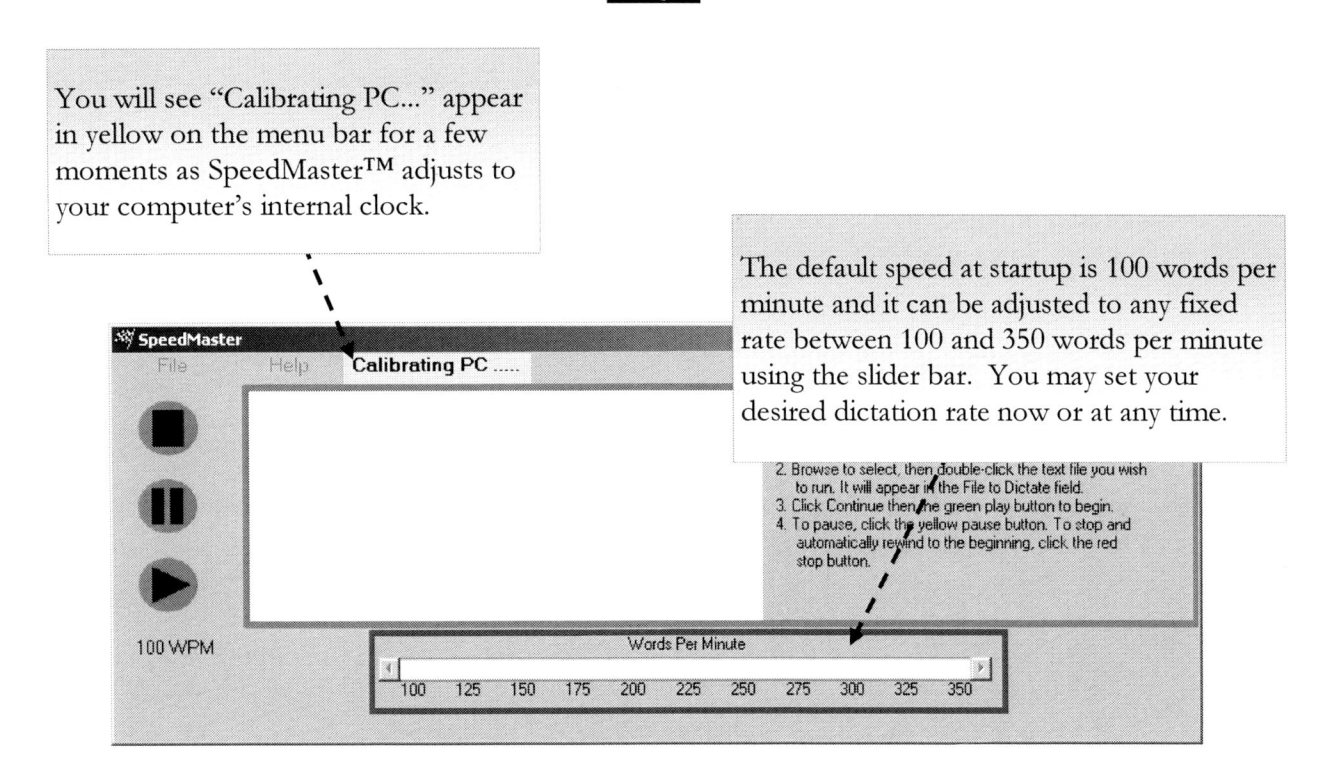

◄- - - SpeedMaster™ icon.

You will see "Calibrating PC..." appear in yellow on the menu bar for a few moments as SpeedMaster™ adjusts to your computer's internal clock.

The default speed at startup is 100 words per minute and it can be adjusted to any fixed rate between 100 and 350 words per minute using the slider bar. You may set your desired dictation rate now or at any time.

2. Browse to select, then double-click the text file you wish to run. It will appear in the File to Dictate field.
3. Click Continue then the green play button to begin.
4. To pause, click the yellow pause button. To stop and automatically rewind to the beginning, click the red stop button.

189

Voice Writing Method • FIFTH EDITION
Dragon NaturallySpeaking® 12

Click **File** and browse to the location of the file you want to open.

Double-click on the name of the file you want to open.

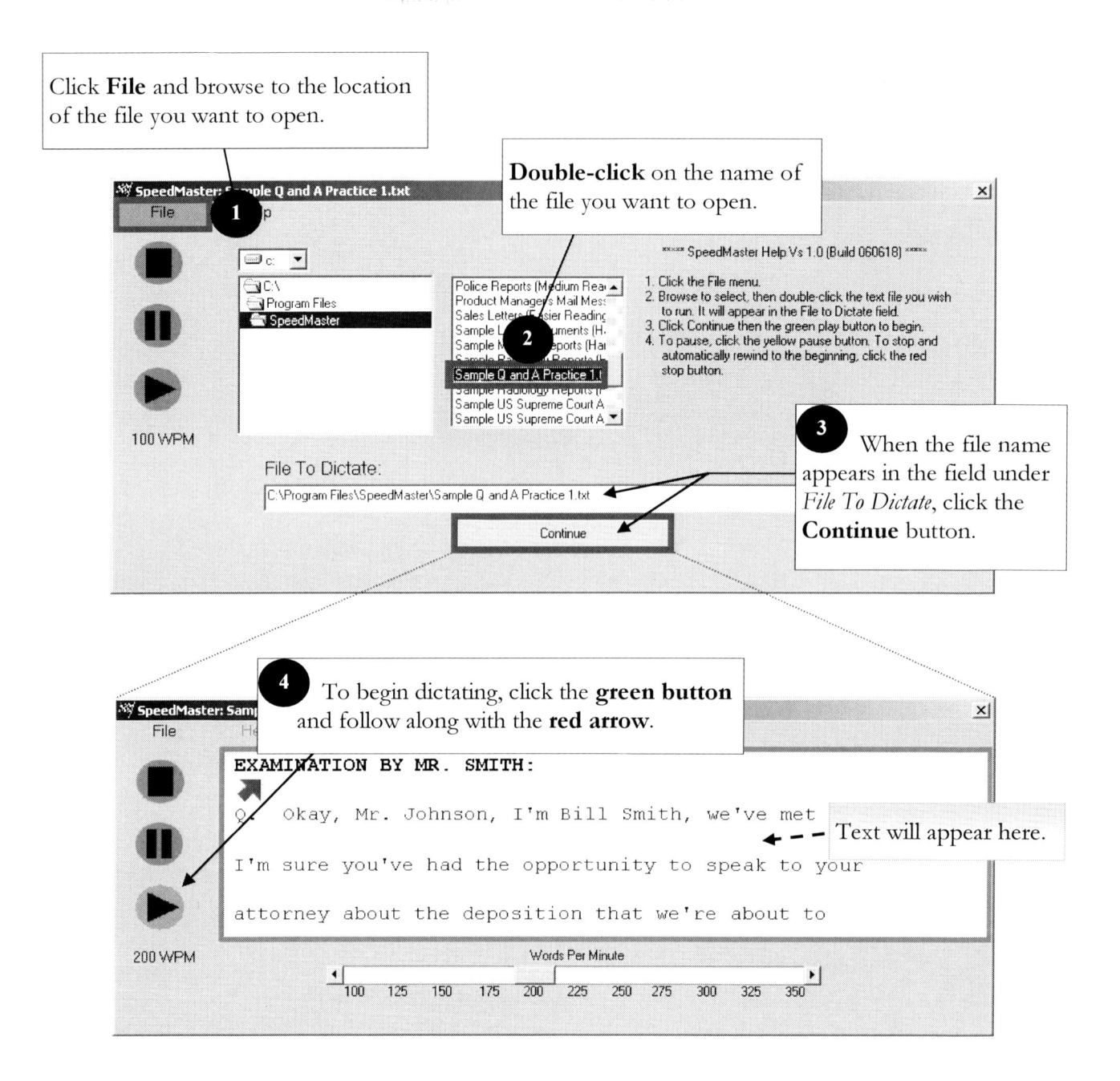

1. Click the File menu.
2. Browse to select, then double-click the text file you wish to run. It will appear in the File to Dictate field.
3. Click Continue then the green play button to begin.
4. To pause, click the yellow pause button. To stop and automatically rewind to the beginning, click the red stop button.

When the file name appears in the field under *File To Dictate*, click the **Continue** button.

To begin dictating, click the **green button** and follow along with the **red arrow**.

Text will appear here.

Click to stop and go to the beginning of the file.

Click to pause.

Click to begin.

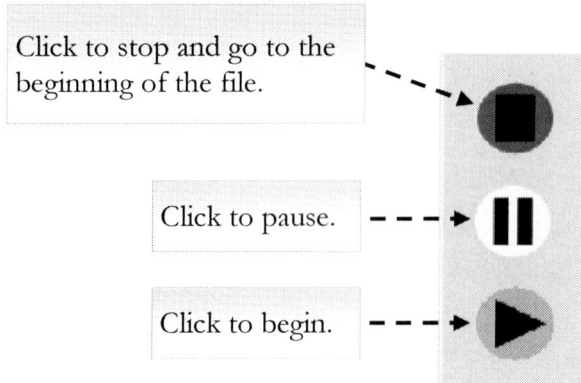

To change the speed during dictation, readjust the word-per-minute setting using the slider bar, then click the green button again.

To adjust in 25-word-per-minute increments, click in the white space on either side of the slider bar.

Click the arrows to adjust in one-word-per-minute increments.

190

Dictation Material

You can download practice dictation material for just about any topic in the world, from anywhere in the world, through the Internet.

For court reporting transcripts, the United States Supreme Court hosts a permanent public archive of oral arguments, found at the following Web address:

>   http://www.supremecourtus.gov/oral_arguments/argument_transcripts.html

Popular sites where captioning and CART material can be found are:

>   http://www.scriptcrawler.com • http://www.script-o-rama.com
>   http://www.simplyscripts.com • http://www.twiztv.com

Here is a good site to visit for financial calls:

>   http://biz.yahoo.com/cc

SpeedMaster™ works best with single-spaced document files. To ensure that your documents are single-spaced, using a Supreme Court transcript as an example, you would simply click on a transcript to open it in Adobe Reader, go to the *File* menu and select *Save as Text*, then save the file to a specified location on your computer's hard drive. Once the file is opened in your regular word-processing program, you can then use the search-and-replace feature to reduce line spacing within the document. In Microsoft Word, you would replace two consecutive paragraph marks (^p) with one. In Corel WordPerfect, you would replace two consecutive hard returns (HRt) with one.

Determining Accuracy Performance

To calculate your accuracy percentage, you must first determine the total number of given errors within a practice dictation session. You can do this as you work to improve your accuracy by marking your errors on a sheet of paper each time you encounter a wrong, missing, or added word, then tallying the total when you're finished making corrections. Next, find the total number of words within the practice dictation file by opening the original plain text (.txt) document in your regular word-processing program and use its word-count utility. Finally, divide your total number of errors by the total word count; then subtract that number from one whole, as shown in the table below.

Table 48

Accuracy Percentage Calculation

Total Number of Errors: 33
Total Number of Original Words: 950

>   Errors (33) / Total Words (950) = Error Rate (0.0347)
>   One (1) - Error Rate (0.0347) = Accuracy Rate (0.9653)
>
>   **Accuracy Percentage = 96.53%**

Chapter 8:
Computer Maintenance and File Management

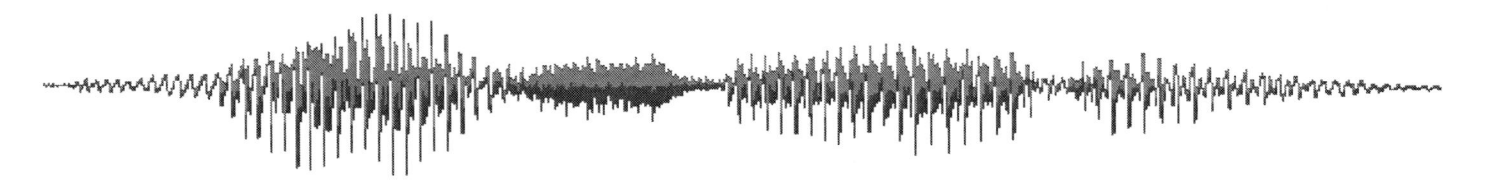

This chapter covers basic information about maintaining your computer and managing your Dragon user and vocabulary files. Included are steps on how to create a new Dragon user and vocabulary and also how to back up/restore and export/import your Dragon user files, vocabularies, and word lists.

Since Dragon NaturallySpeaking® Premium does not allow users the capability to create specialized vocabularies, the *Vocabulary Management* section and all its related subsections apply to only Dragon NaturallySpeaking® Professional users. All other sections apply to users of both Dragon NaturallySpeaking® Professional and Dragon NaturallySpeaking® Premium editions.

If you have CAT software, remember that the instructions provided herein may not apply, so refer to your CAT software manual for related details.

Computer Maintenance

This section lists a few general tasks involved in the proper care of your computer, including the removal and archival of data from your hard drive to make room for new data, defragmenting your hard drive to speed up the file reading and writing processes, and running anti-virus scans to detect and remove computer infections. Regular schedules should be set to perform these operations.

Archiving Data

File sizes vary according to the type of data they hold. Audio recordings in .wav format take up much more hard drive space than document files containing only text. So if you create a lot of audio files, you should archive that data at least once a week. Since all computers have a limited amount of data they can store, old files and infrequently accessed information you want to permanently keep should be saved onto an external storage source, such as a CD or external hard drive, and removed from your computer's hard drive. To help determine your archiving schedule, never let data accumulate past 50 percent of your hard drive's storage capacity.

Disk Defragmentation

Defragmenting your hard drive will result in higher efficiency when performing general computing tasks, and this relates to SR performance as well. As your computer reads and writes data on your hard drive, it stores that information in any available space, resulting in the distribution of data in noncontiguous segments. As more data is compiled and updated over time, the array of storage points that must be accessed for retrieval and dissemination slows the machine's processing of every operational task.

Having data build up in noncontiguous sectors increases the computational overhead involved in input/output processing. Defragmenting your hard drive will consolidate fragmented data and place it in contiguous locations, freeing up blocks of hard drive space for more efficient access and storage of new data. Directly relating to SR, this ensures that the computer will read and write your voice files and recordings in the shortest time possible.

Depending on the amount of data that is stored, how often you have defragmented your hard drive in the past, your computer's speed, and other factors, this process could take hours the first time it is run. Thereafter, regularly scheduled defragmentation consumes much less time.

It is not unusual for the program to report that some files could not be defragmented. You may wish to repeat this process until the Disk Defragmenter's results window is free of fragmented files, which are indicated by red stripes. You can defragment your hard drive as often as you would like, but make sure it is done at least once a month.

If you wait more than a month to defragment your hard drive, it is best to export your Dragon user files onto external storage media (an external hard drive, CD, flash drive, etc.) first before defragging. Otherwise, you may notice deterioration in your accuracy performance when you work in your Dragon user after the defrag; intensive reorganization of your Dragon user files can result in corruption to sensitive bits of data. Do not work in your Dragon user until you have done a fresh import of your Dragon user files after your disk defragmentation is complete.

Anti-Virus Scans

Anti-virus software is important to keep on your computer even if you never intend to access the Internet. Your realtime work computer is always vulnerable to the risk of receiving viruses from other computers as a result of data transfers through external storage devices and other media. Be sure not to uninstall this software; you should occasionally perform virus scans for detecting and removing viruses in the event your computer is infected.

Run your virus-scanning software at least once a week if you transfer data between other computers and your realtime work computer. To initiate the virus scan, double-click the anti-virus program's icon located in the *System Tray* of your computer's desktop screen and follow the instructions.

Chapter 8: Computer Maintenance and File Management

User File Management

The next five sections address Dragon user file management. You'll learn when and how to create a new user, open and close users to make them active or inactive, rename and delete users, and preserve your user files to avoid having to unnecessarily create a new user from scratch if your user files somehow get corrupted.

As with all important files, it's always a good idea to store additional copies of them on your computer's hard drive as well as external media, such as a backup hard drive or CD, to prevent permanent loss of data in the event you experience file corruptions or a major computer malfunction. When storing a copy of your user files so you'll be able to easily restore them directly from your computer's hard drive, follow the steps in the *Backing Up/Restoring User Files* subsection. Should you ever need to transfer your user files from one computer to another, follow the steps in the *Exporting/Importing User Files* subsection.

Creating a New User

If you decide to use a different dictation microphone or USB sound card, you need to create a new Dragon user so Dragon will be able to recognize your speech utilizing those devices. This is especially important to be aware of if your career is cross-functional between different voice writing professions where you may be dictating into a speech silencer for court reporting one day, an open-mic headset for captioning the next, et cetera. In that case, you will be working with multiple Dragon users; you should have a separate Dragon user dedicated for your work with each set of dictation input devices.

All speech silencers, open-mic headsets, and USB sound cards—whether they be from different or the same manufacturers or even the same exact models of the same exact brands—have certain unique mechanical characteristics that will cause your speech recognition to be processed differently. Once you have determined which dictation input devices to use, you must continue to use those same devices when working in the voice model you created based on that component configuration. Whether you switch out your dictation microphone, USB sound card, or both, with different devices, remember that you need to create a new voice model based on that specific configuration of components. This will ensure you obtain the best accuracy results with those newer devices. If you do not create a new user in these instances, you will experience degraded accuracy results.

Another time you may want to create a new Dragon user is when switching from one computer to another with significantly lower- or higher-end components. Only do this if you find that the speech recognition performance you're experiencing on your new computer is not as accurate as the results you are used to experiencing with your other computer. Be sure to test the accuracy results of your existing Dragon user on your new computer first before making the determination to create a new user all over again from scratch. See the *Exporting/Importing User Files* section for instructions on transferring your existing Dragon user files from one computer to another.

Be sure that your dictation input devices are properly connected before following the steps on the next page to create a new Dragon user.

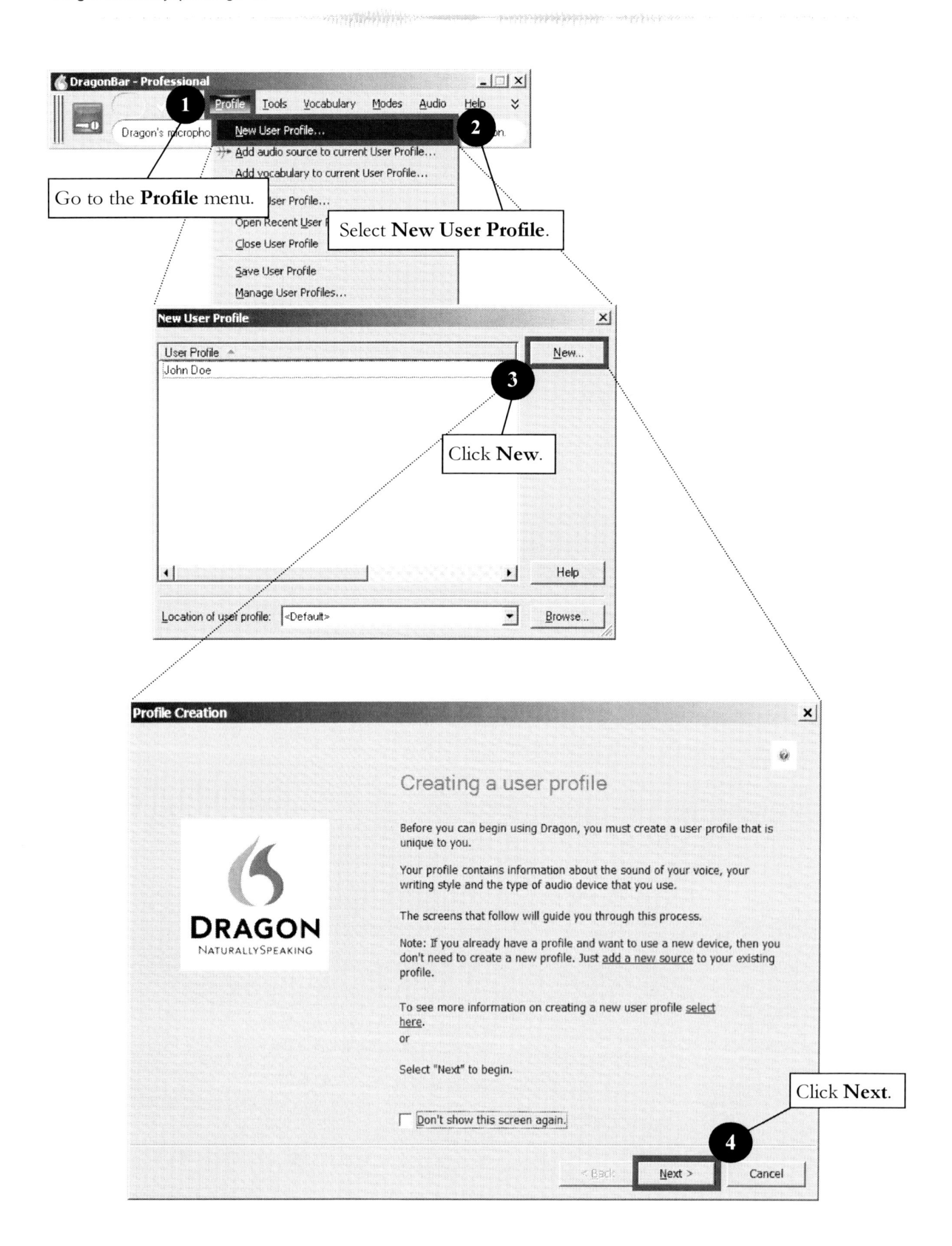

Voice Writing Method • FIFTH EDITION
Dragon NaturallySpeaking® 12

1 Go to the **Profile** menu.

2 Select **New User Profile**.

3 Click **New**.

4 Click **Next**.

196

Chapter 8: Computer Maintenance and File Management

[Screenshot of "Profile Creation" dialog titled "Name your user profile" with instruction "Please give your profile a unique name. Most people use their first and last name." and an "Enter a name:" field containing "John Doe - Captioning (Sennheiser mic and Andrea USB)", marked with callout **5**. Buttons: < Back, Next >, Cancel, with Next marked **6**.]

5 You may want to give a name that includes a description of the dictation devices you will be working with in this new user.

6 Click **Next** and follow the on-screen instructions.

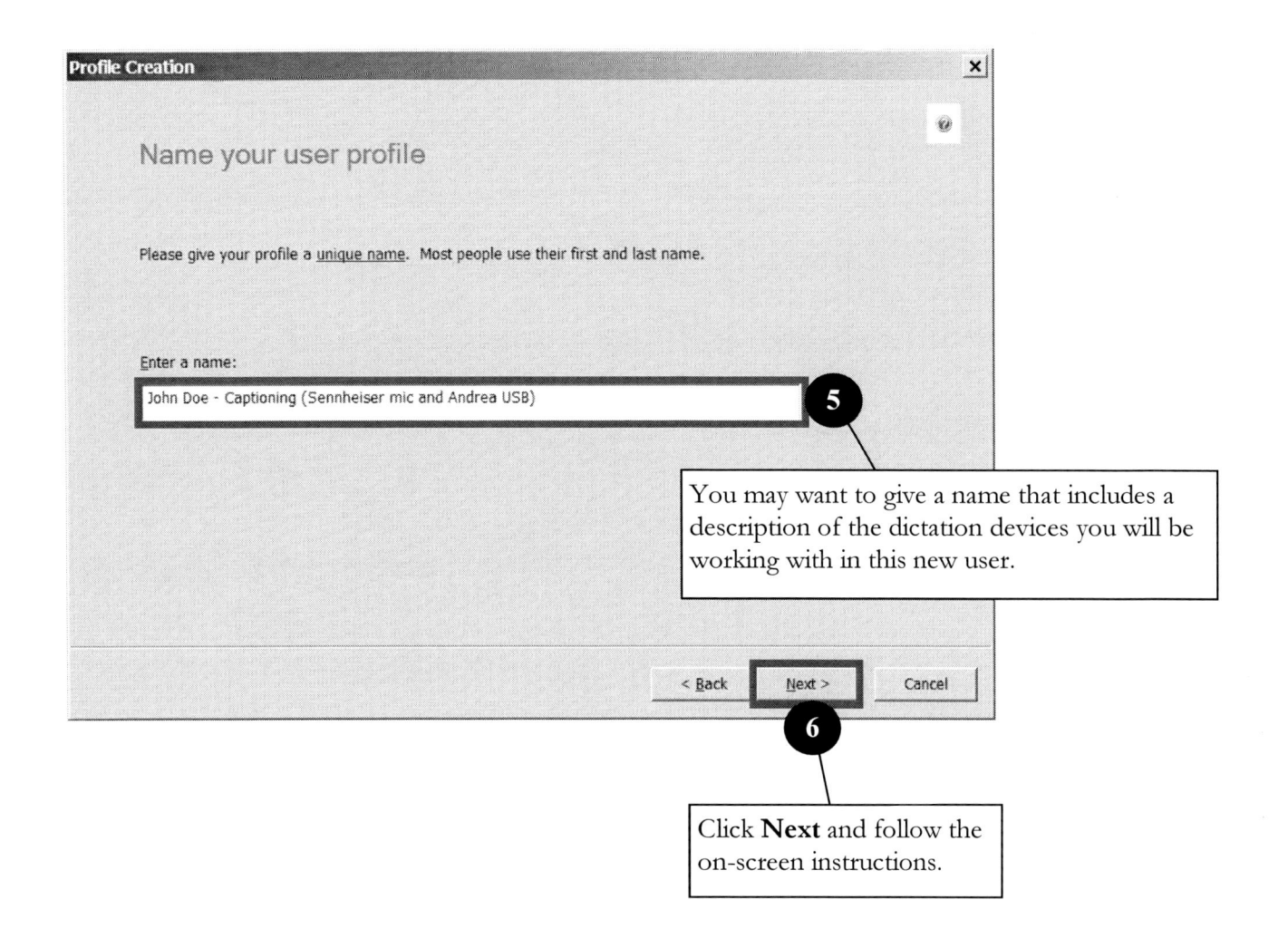

Note

Under the *Profile* menu, the *Add audio source to current User Profile* option gives you a way to adapt your current user to recognize other forms of audio input when you switch audio devices. However, the audio source options are limited to general device types and connection categories. You cannot add multiple audio sources that are similar, such as one USB sound card versus another USB sound card. Since you will likely always use a USB sound card, this option wouldn't be used in lieu of having to create a new user when switching your dictation input devices; regardless of the type of your dictation microphone, USB will always be your microphone source selection, and Dragon only allows you to have one USB audio source per Dragon user.

Opening/Closing Users

You can open and close Dragon users from the *Profile* menu as shown below.

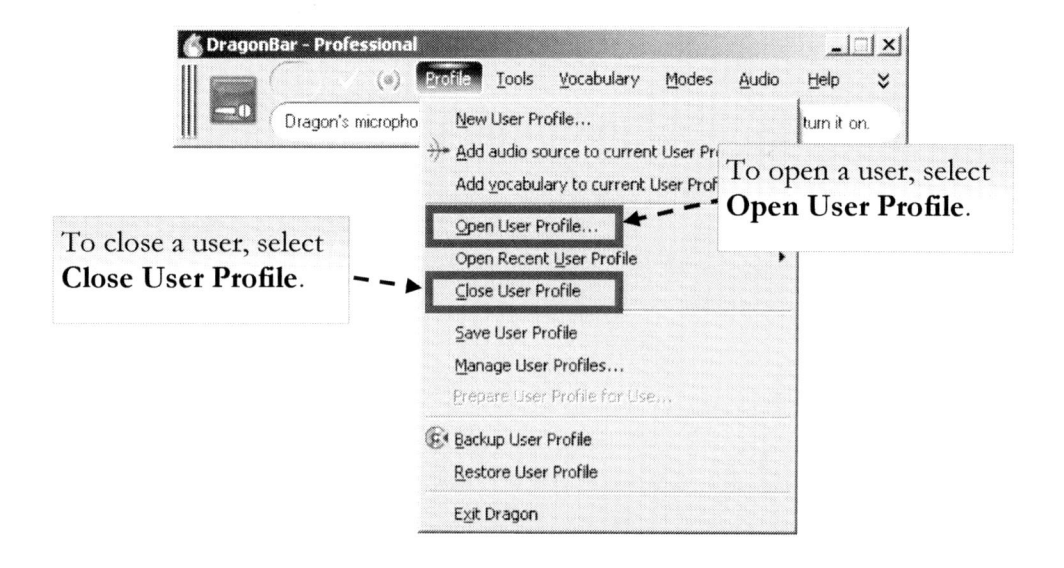

To close a user, select **Close User Profile**.

To open a user, select **Open User Profile**.

Renaming/Deleting Users

If working with multiple Dragon users, you will probably end up renaming some of them to reflect more specifically what type of work or dictation devices those users apply to, or you may want to delete old users no longer applicable to your work in order to free up space on your computer's hard drive. This can all be done from the *Manage User Profiles* screen as follows:

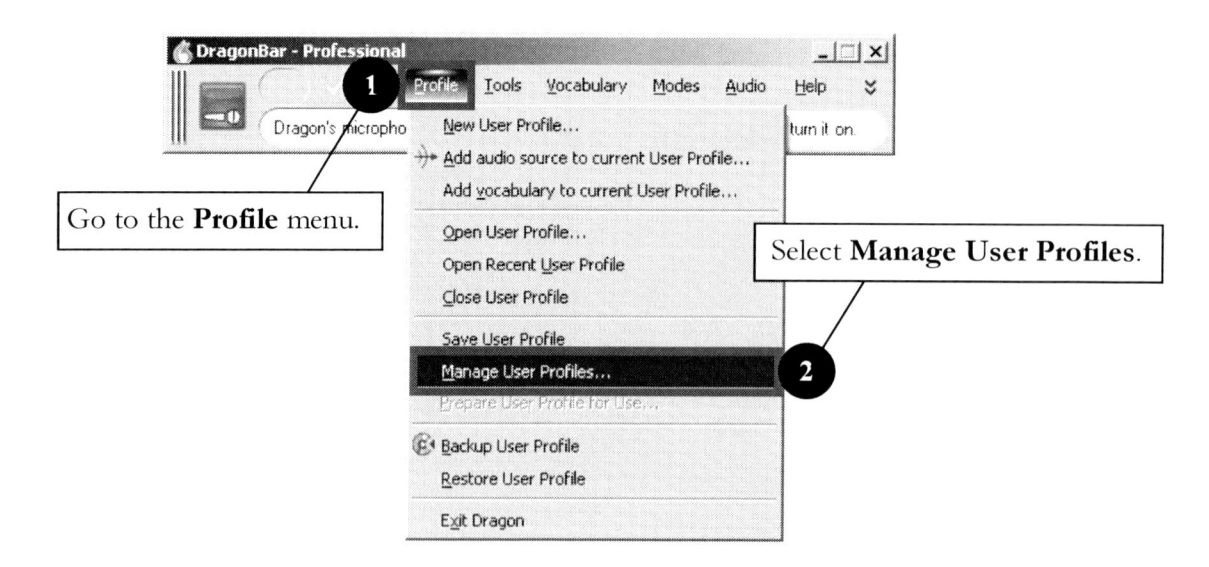

1. Go to the **Profile** menu.
2. Select **Manage User Profiles**.

Chapter 8: Computer Maintenance and File Management

3 — Select the user you want to rename or delete.

4 — Select either **Rename** or **Delete**.

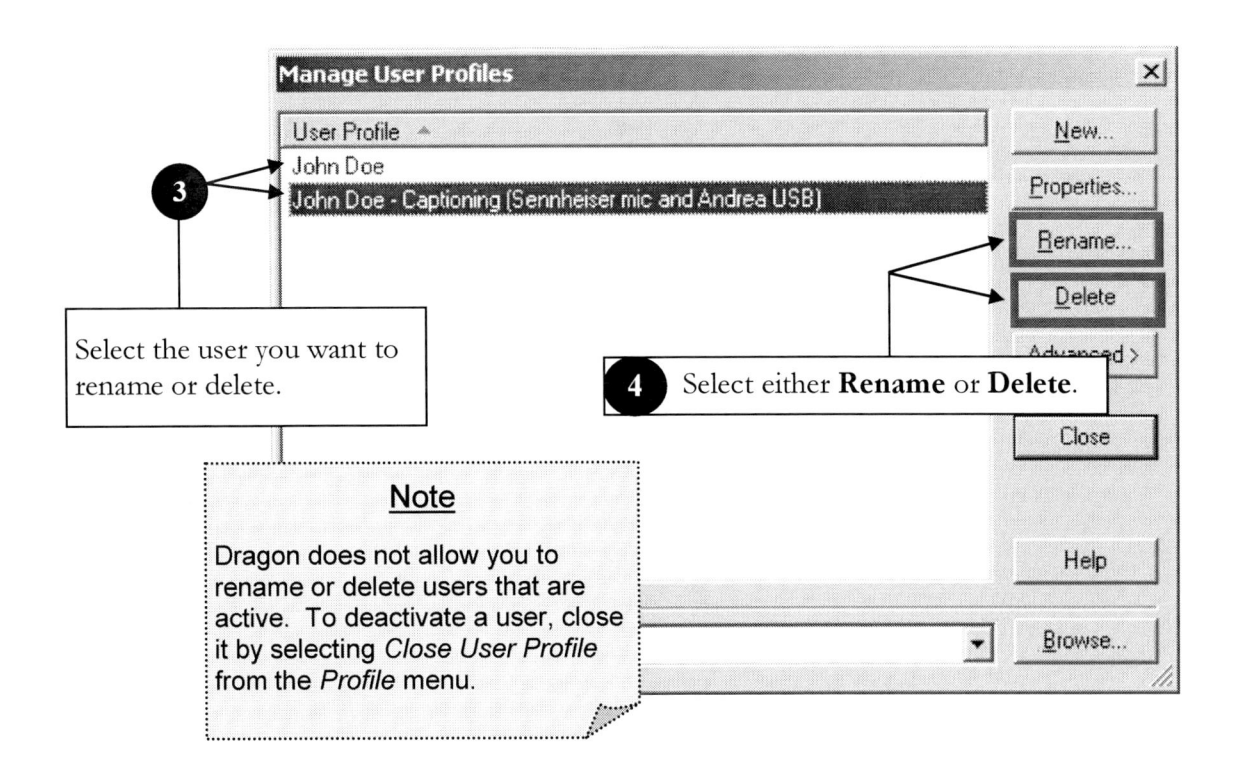

Note

Dragon does not allow you to rename or delete users that are active. To deactivate a user, close it by selecting *Close User Profile* from the *Profile* menu.

Backing Up/Restoring User Files

You should make a backup copy of your Dragon user files on your computer's hard drive for easy retrieval in case your files somehow get corrupted and you need to restore them. Follow the steps in this section to either back up or restore your Dragon user files.

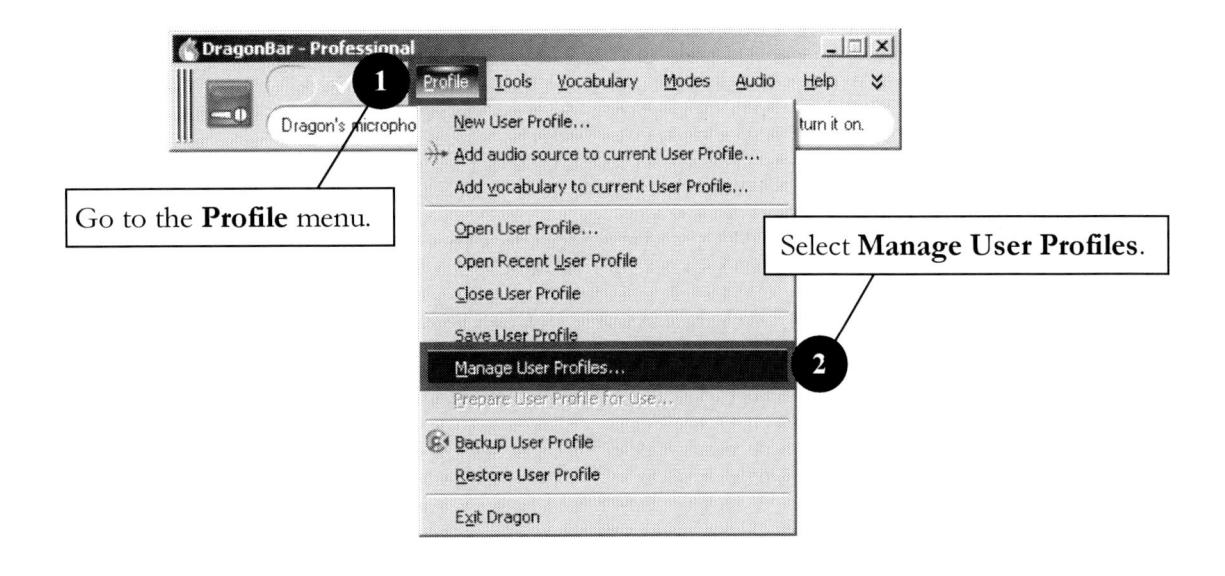

1 — Go to the **Profile** menu.

2 — Select **Manage User Profiles**.

199

Voice Writing Method • FIFTH EDITION
Dragon NaturallySpeaking® 12

3 Select the user you want to back up or restore.

4 Click the **Advanced** button.

5 Select either **Backup** or **Restore**.

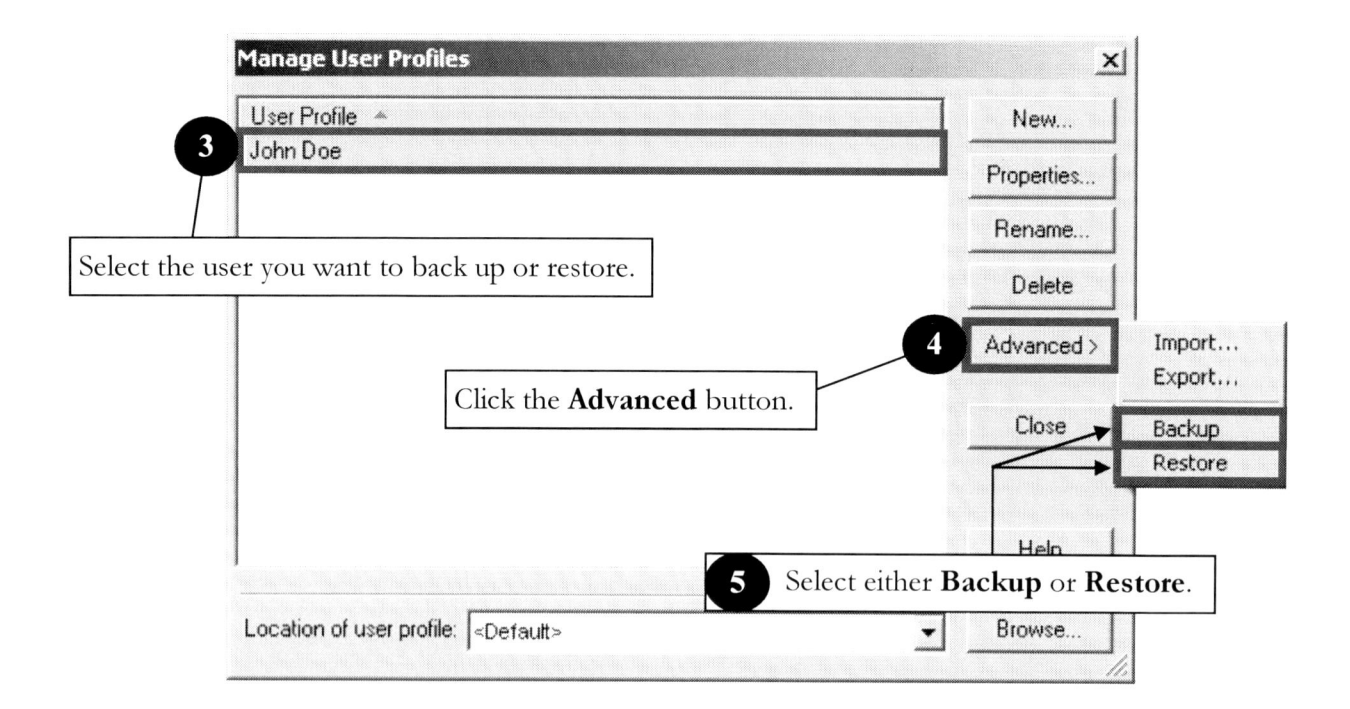

If the user you want to back up or restore is the one you already have open, you can do a backup or restore faster by going through the *Profile* menu and selecting *Backup User Profile* or *Restore User Profile*.

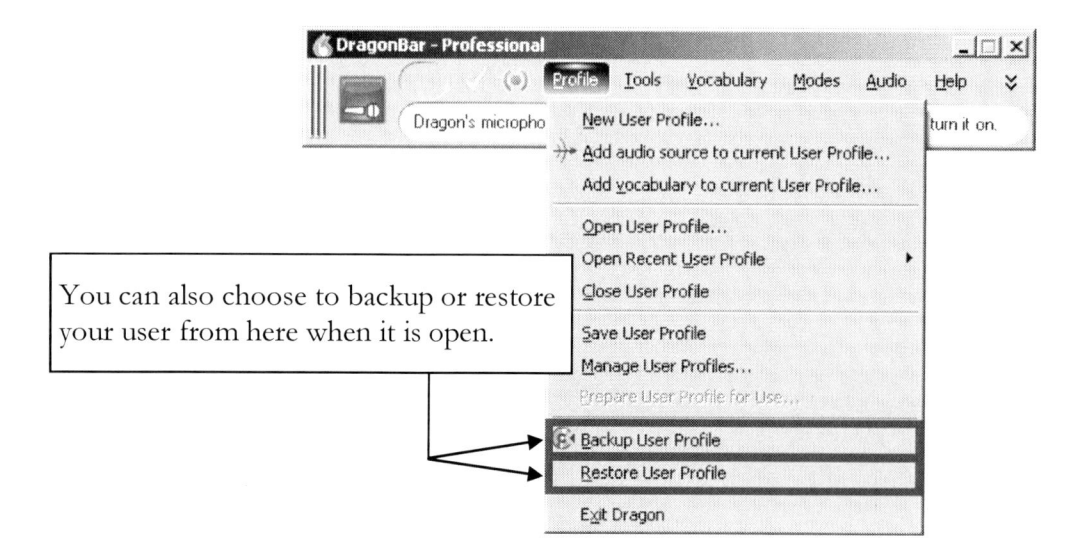

You can also choose to backup or restore your user from here when it is open.

Chapter 8: Computer Maintenance and File Management

Exporting/Importing User Files

Should you ever need to move your Dragon user files from one computer to another, you can export and save them onto an external storage device (i.e., flash drive, backup hard drive, or CD), then import them on a different computer. The steps on how to do this are below.

Exporting User Files

Follow these steps to export your Dragon user files.

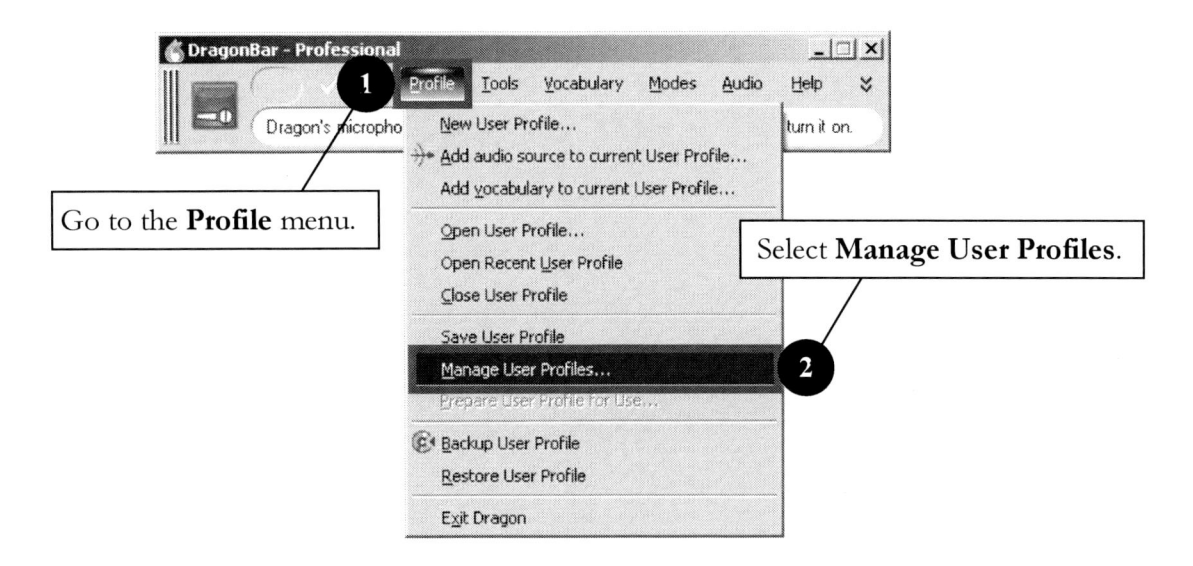

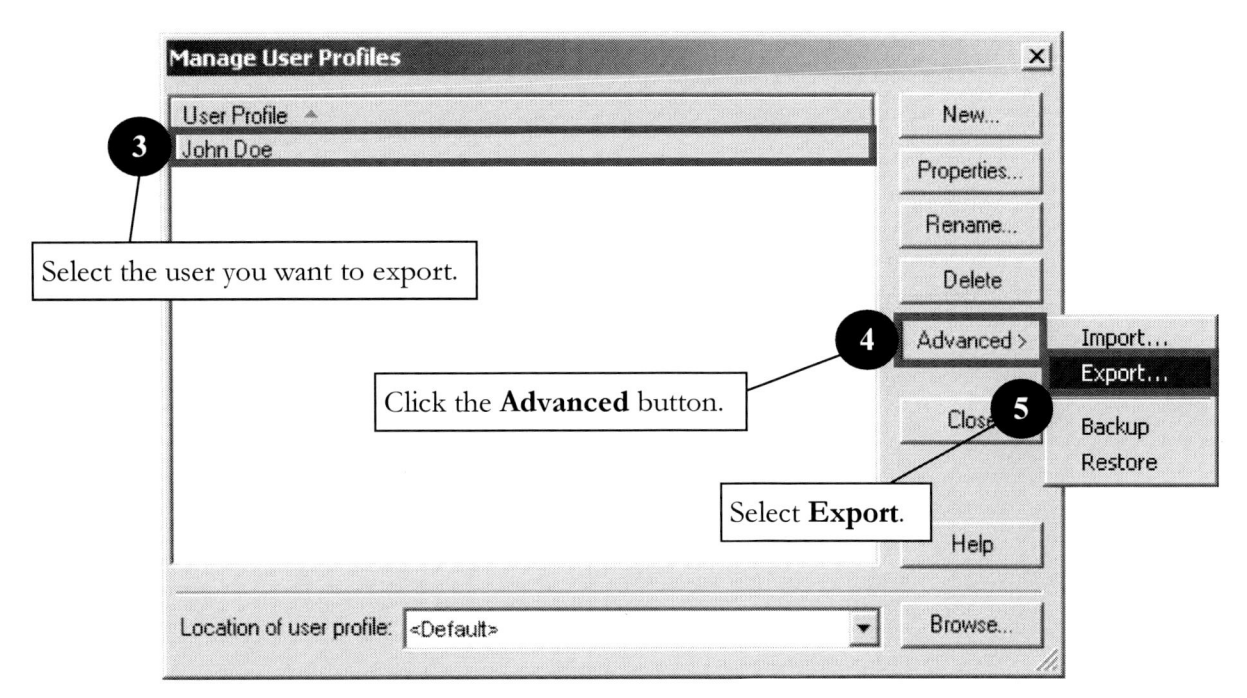

1. Go to the **Profile** menu.
2. Select **Manage User Profiles**.
3. Select the user you want to export.
4. Click the **Advanced** button.
5. Select **Export**.

201

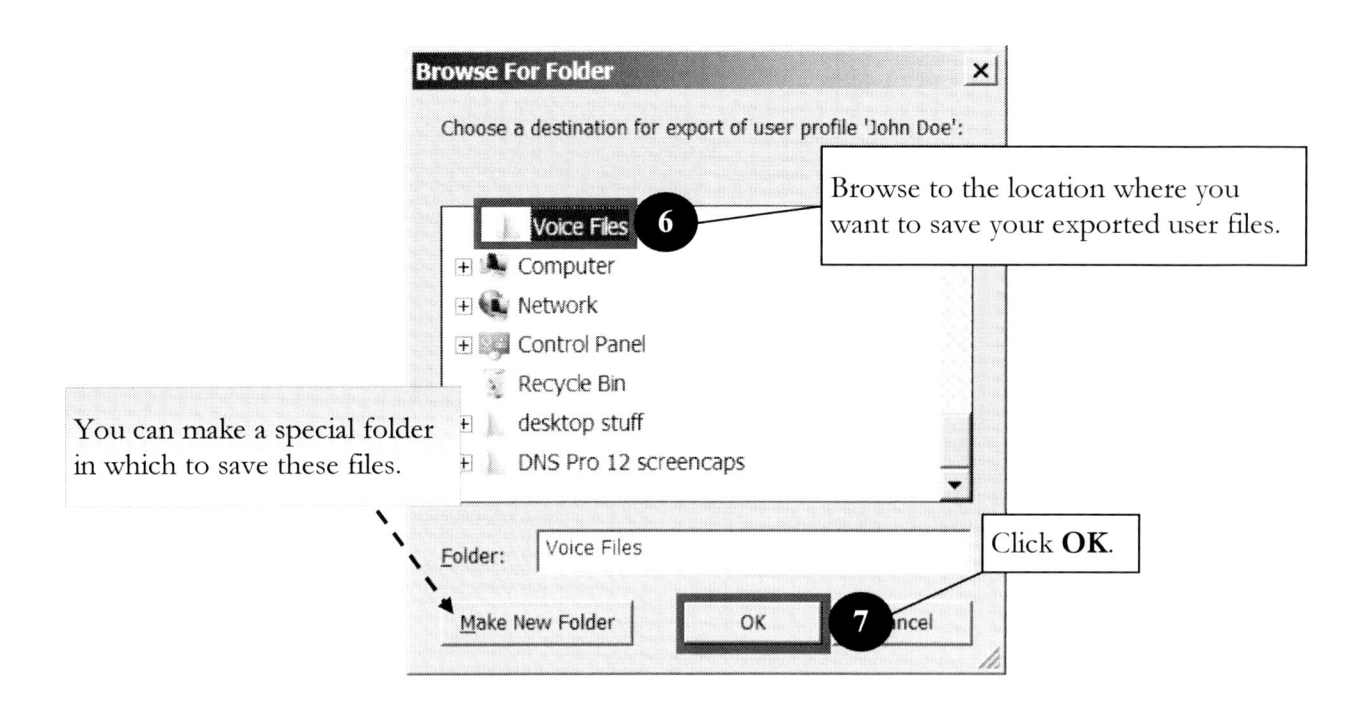

Voice Writing Method • FIFTH EDITION
Dragon NaturallySpeaking® 12

Browse to the location where you want to save your exported user files. — 6

You can make a special folder in which to save these files.

Click **OK**. — 7

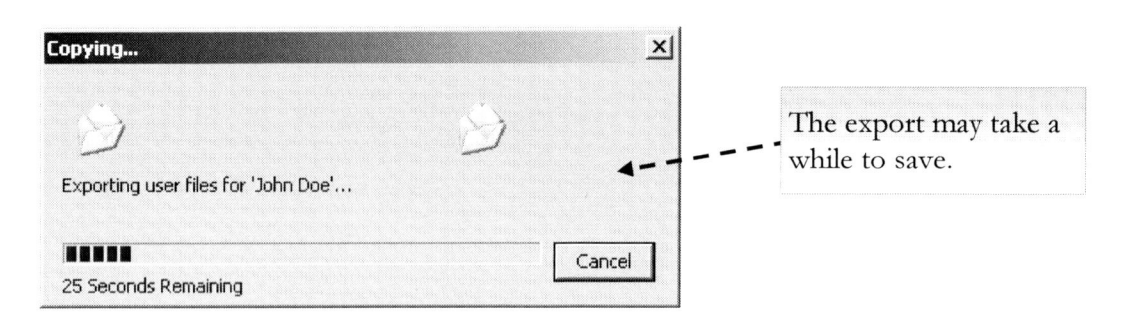

The export may take a while to save.

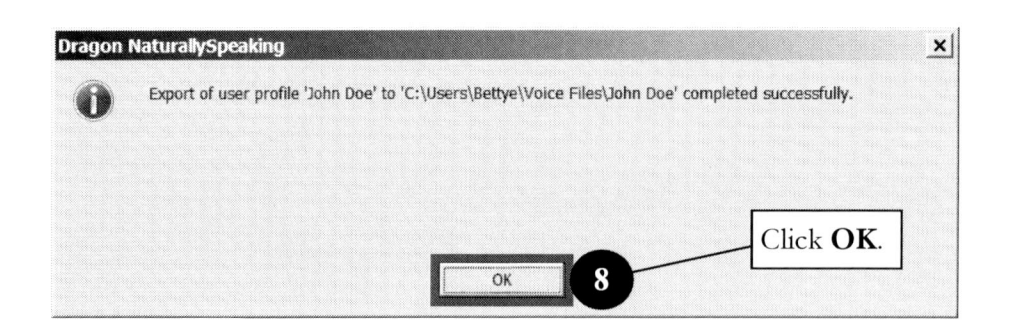

Click **OK**. — 8

202

Chapter 8: Computer Maintenance and File Management

Importing User Files

Import your Dragon user files from the same screen you performed your export: *Profile > Manage User Profiles*.

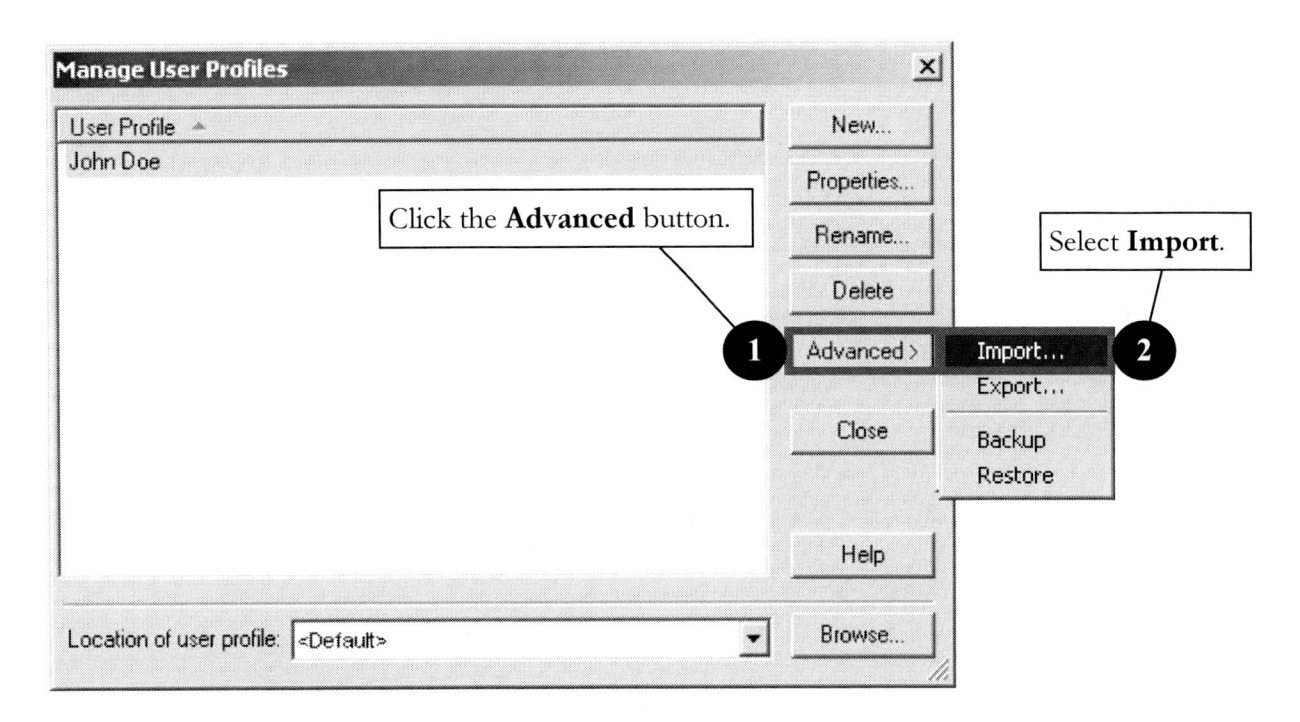

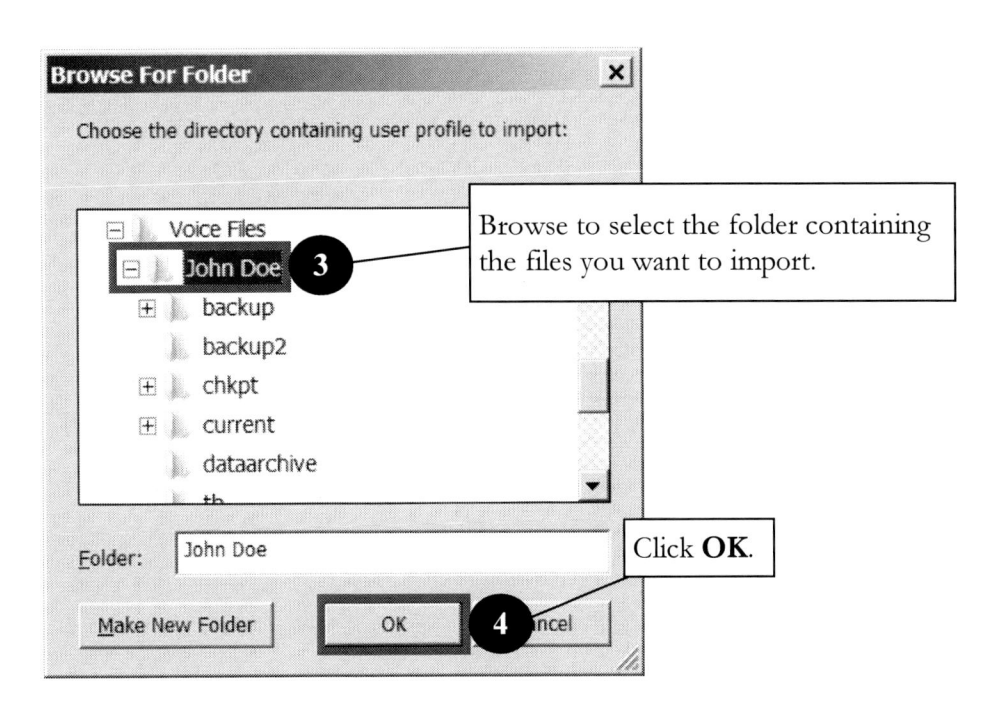

1 Click the **Advanced** button.

2 Select **Import**.

3 Browse to select the folder containing the files you want to import.

4 Click **OK**.

203

Voice Writing Method • FIFTH EDITION
Dragon NaturallySpeaking® 12

The import may take a while to process.

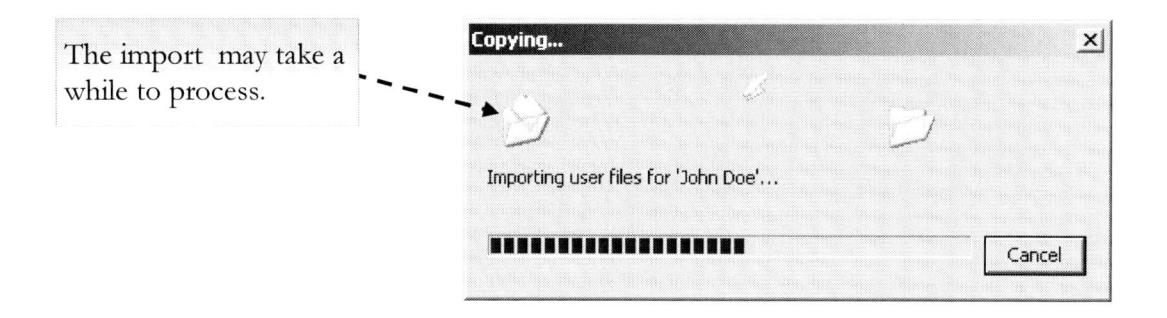

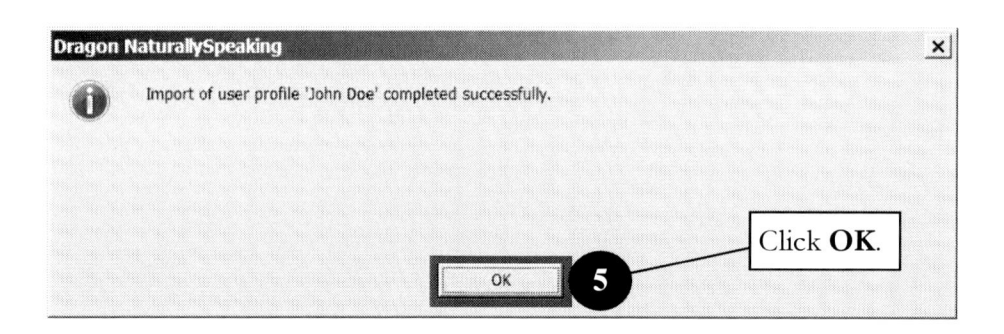

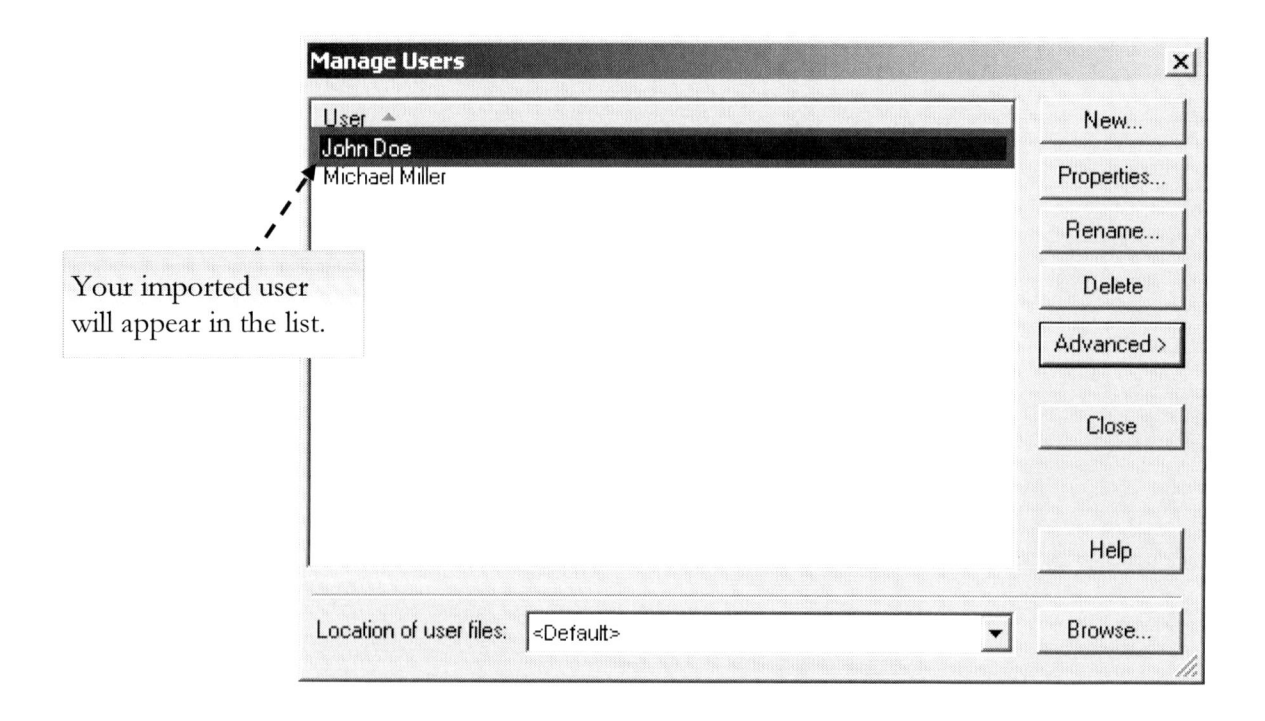

Importing user files for 'John Doe'...

Import of user profile 'John Doe' completed successfully.

Click **OK**.

5

Manage Users

John Doe
Michael Miller

Your imported user will appear in the list.

Location of user files: <Default>

204

Chapter 8: Computer Maintenance and File Management

Vocabulary Management

The next five sections cover details related to Dragon vocabulary management, where you'll learn how to create, open, rename, delete, and export/import vocabularies if your edition of Dragon supports those capabilities. You'll also learn how to export and import a custom word list to save yourself the trouble of having to manually type in your custom words all over again in the event you need to create a new vocabulary from scratch.

(Note: All information presented in the following sections applies to the Professional edition of Dragon NaturallySpeaking®. But if you have the Dragon NaturallySpeaking® Premium edition, the only section that will apply to you is *Exporting/Importing a Word List*. Dragon Professional edition users have the ability to create multiple vocabularies based on a single Dragon user; whereas, the Dragon Premium edition does not allow users to create more than one vocabulary per Dragon user.)

Should your work involve realtime reporting in multiple subject matters, you will want to maintain separate vocabularies catering to the different types of work you do. For example, if you will be doing captioning work which requires one Dragon vocabulary to be customized to reflect the type of grammar that will be used during your dictation of the Larry King Live show and you will also be doing court reporting work requiring you to dictate Q&A material, you will need to have a different vocabulary for each type of syntax . For a general idea on how to structure multiple vocabularies, see *Diagram 3* on page 175 under the *Vocabulary Customization* subsection of the *Vocabulary Building* section in Chapter 6.)

If you have the Professional edition of Dragon, simply follow the steps under *Creating a New Vocabulary* for each new vocabulary you'd like to build. Also, you may share your customized vocabularies with others by following the instructions provided in *Exporting/Importing a Vocabulary*; this will allow you to transfer your vocabulary's custom words and all the grammatical information it contains so someone else with Dragon Professional can benefit from all your vocabulary-building work.

Even though Premium users must create multiple Dragon users if they wish to work with multiple Dragon vocabularies, this doesn't mean that Premium users are stuck with having to create a brand-new user from scratch each time they want to create a new vocabulary. The work-around solution is found through Dragon's user backup and restore functions; simply restore a backup of your existing user, then rename it with some type of description indicating the vocabulary type you will develop under the new user, and proceed with your vocabulary-building tasks to customize it accordingly. (See the *Backing Up/Restoring User Files* subsection under the *User File Management* section earlier in this chapter for instructions.)

Remember that you will need to process your custom-prepared documents through Dragon for analysis in order to customize the grammatical model when building a new vocabulary (see the *Vocabulary Building* section of Chapter 6).

To simply share custom words from your vocabulary, follow the steps in the *Exporting/Importing a Word List* section. Keep in mind that when you export and import word lists, grammatical data is not transferred.

Voice Writing Method • FIFTH EDITION
Dragon NaturallySpeaking® 12

Creating a New Vocabulary

(Note: This section applies only to Dragon NaturallySpeaking® Professional edition users.)

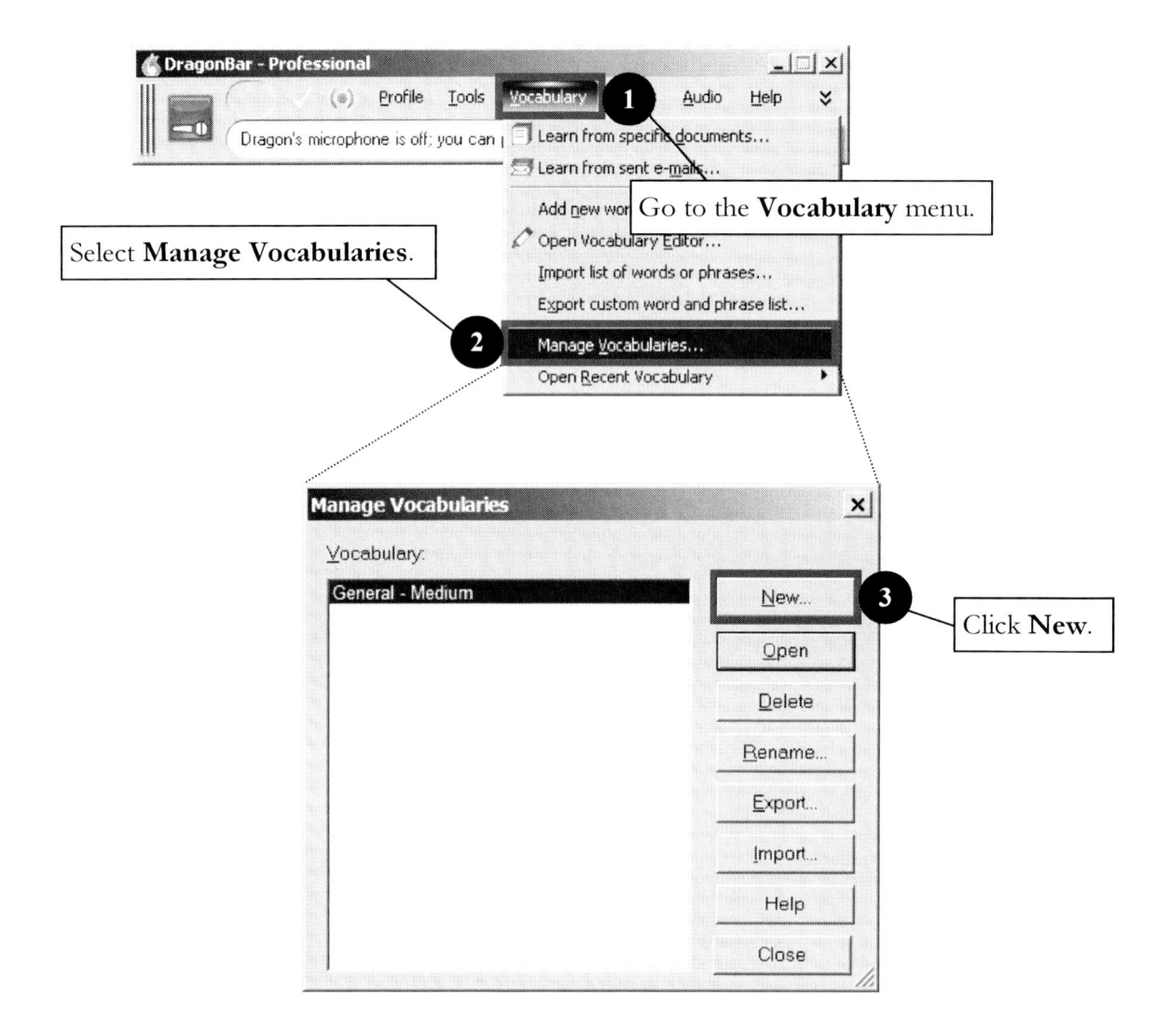

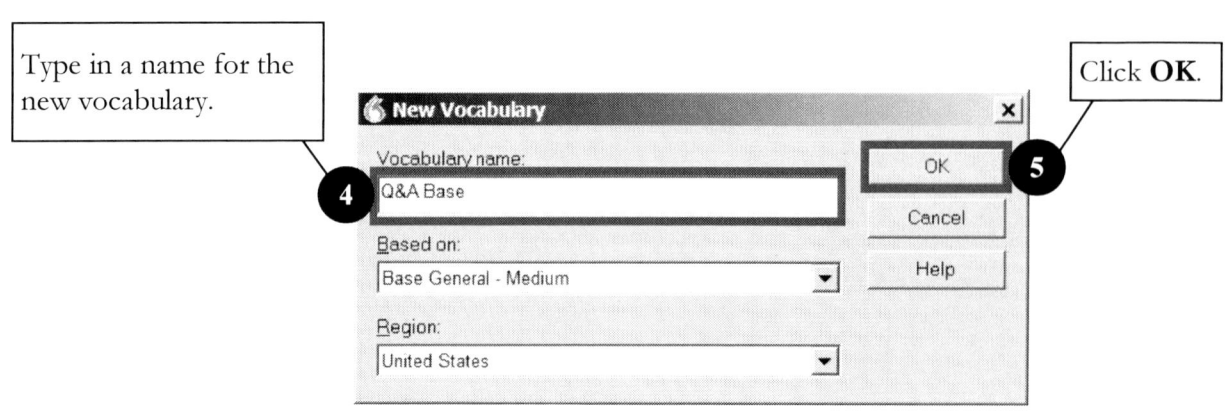

1 Go to the **Vocabulary** menu.

2 Select **Manage Vocabularies**.

3 Click **New**.

4 Type in a name for the new vocabulary.

5 Click **OK**.

Chapter 8: Computer Maintenance and File Management

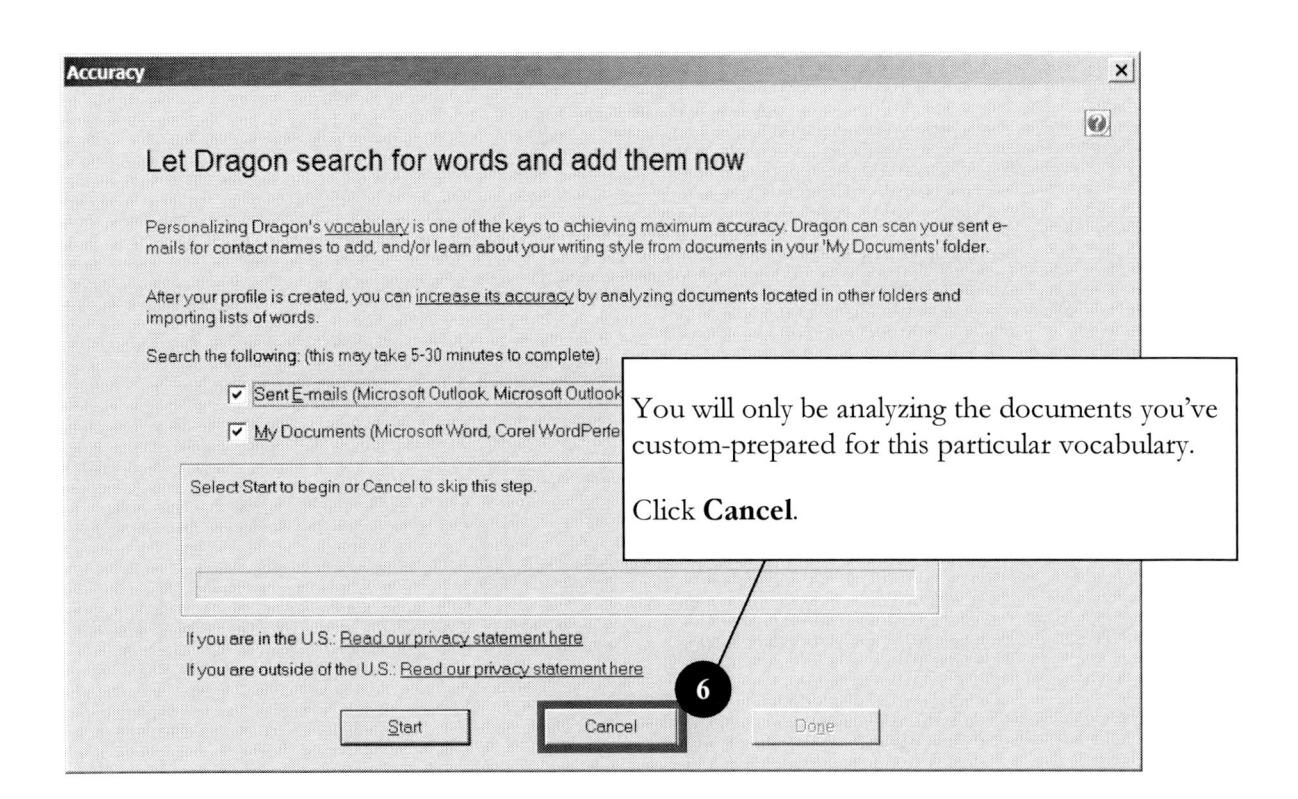

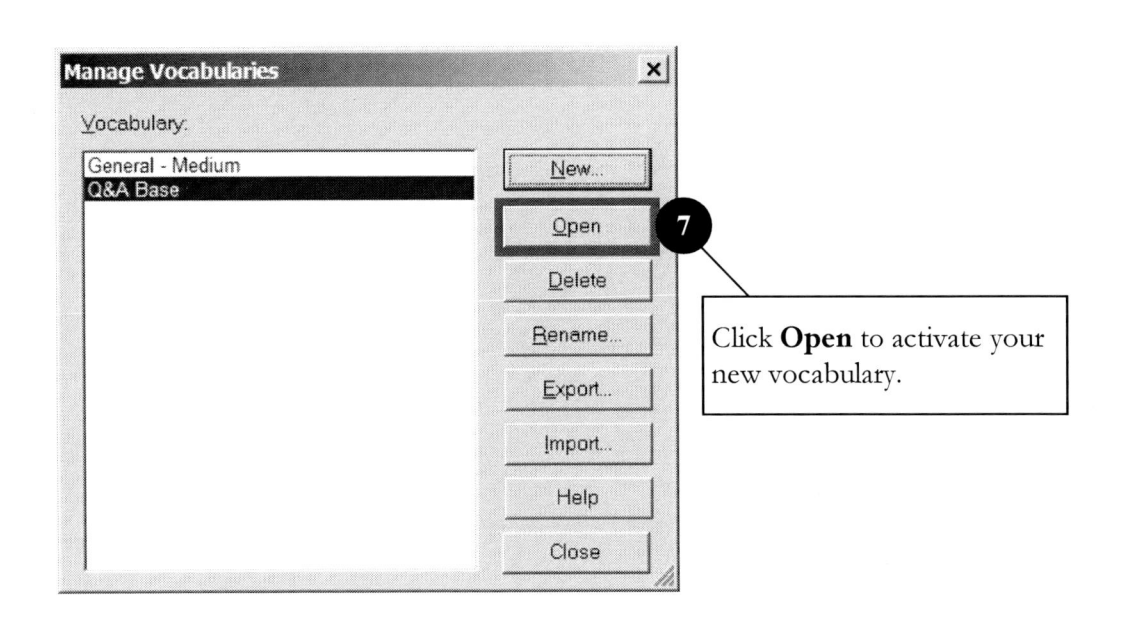

You will only be analyzing the documents you've custom-prepared for this particular vocabulary.

Click **Cancel**.

6

Click **Open** to activate your new vocabulary.

7

8 Follow the steps in the *Document Analysis* subsection under *Vocabulary Building* in Chapter 6 to customize this vocabulary using your custom-prepared documents.

207

Opening, Renaming, and Deleting Vocabularies

(Note: This section applies only to Dragon NaturallySpeaking® Professional edition users.)

When working with multiple Dragon vocabularies, you will need to know how to switch from using one vocabulary to another. You also probably end up renaming some of them to reflect their custom language style or subject matter. In order to free up space on your computer's hard drive, you will want to eventually delete any old vocabularies as well when they are no longer applicable to your work. This can all be done from the *Manage Vocabularies* screen as follows:

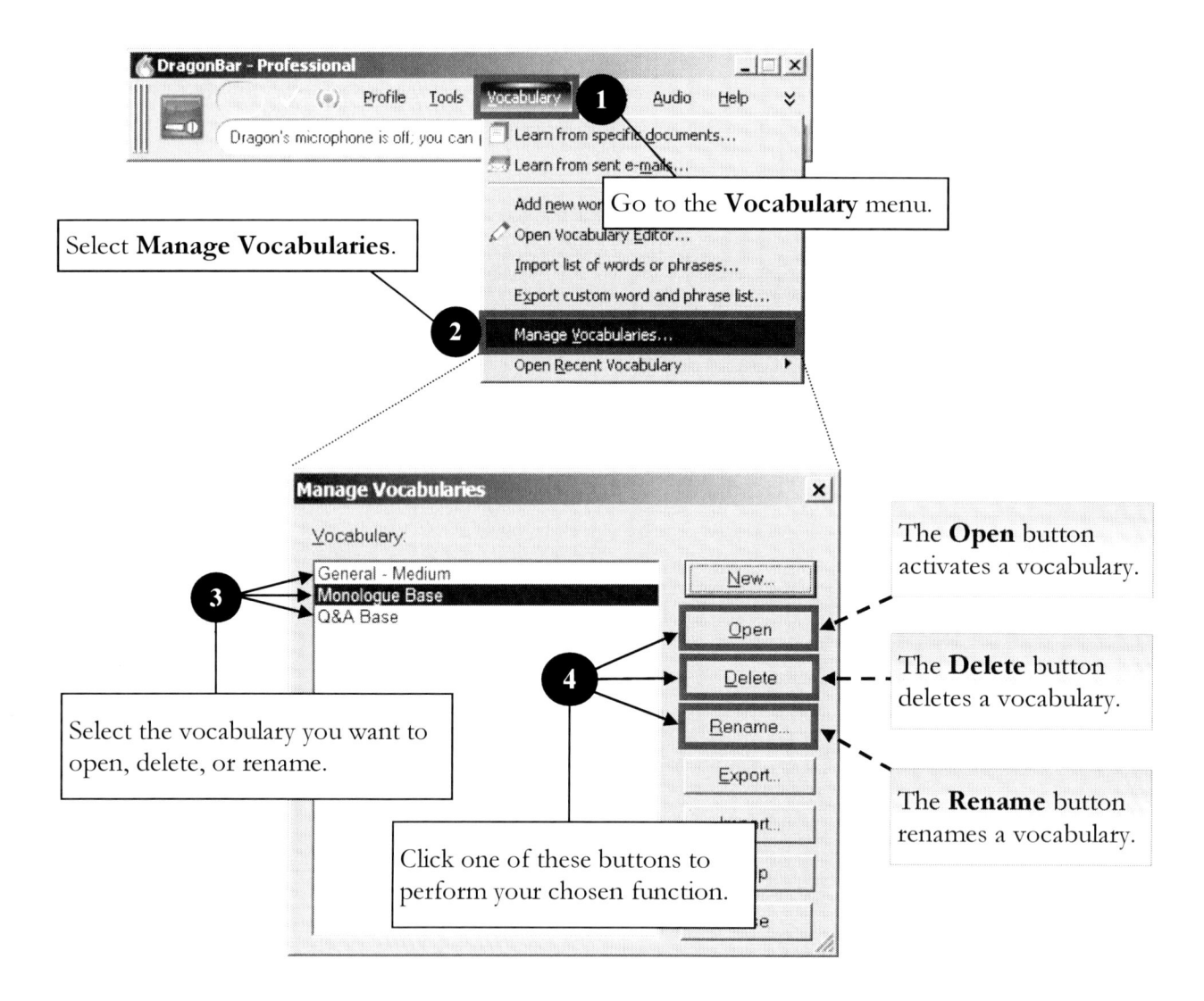

1. Go to the **Vocabulary** menu.
2. Select **Manage Vocabularies**.
3. Select the vocabulary you want to open, delete, or rename.
4. Click one of these buttons to perform your chosen function.

The **Open** button activates a vocabulary.

The **Delete** button deletes a vocabulary.

The **Rename** button renames a vocabulary.

Chapter 8: Computer Maintenance and File Management

Exporting/Importing a Vocabulary

(Note: This section applies only to Dragon NaturallySpeaking® Professional edition users.)

The sections below show you how to export and import entire vocabularies—including all custom words and grammatical data—so you will be able to share them with others.

Exporting a Vocabulary

Follow the steps below to export a vocabulary.

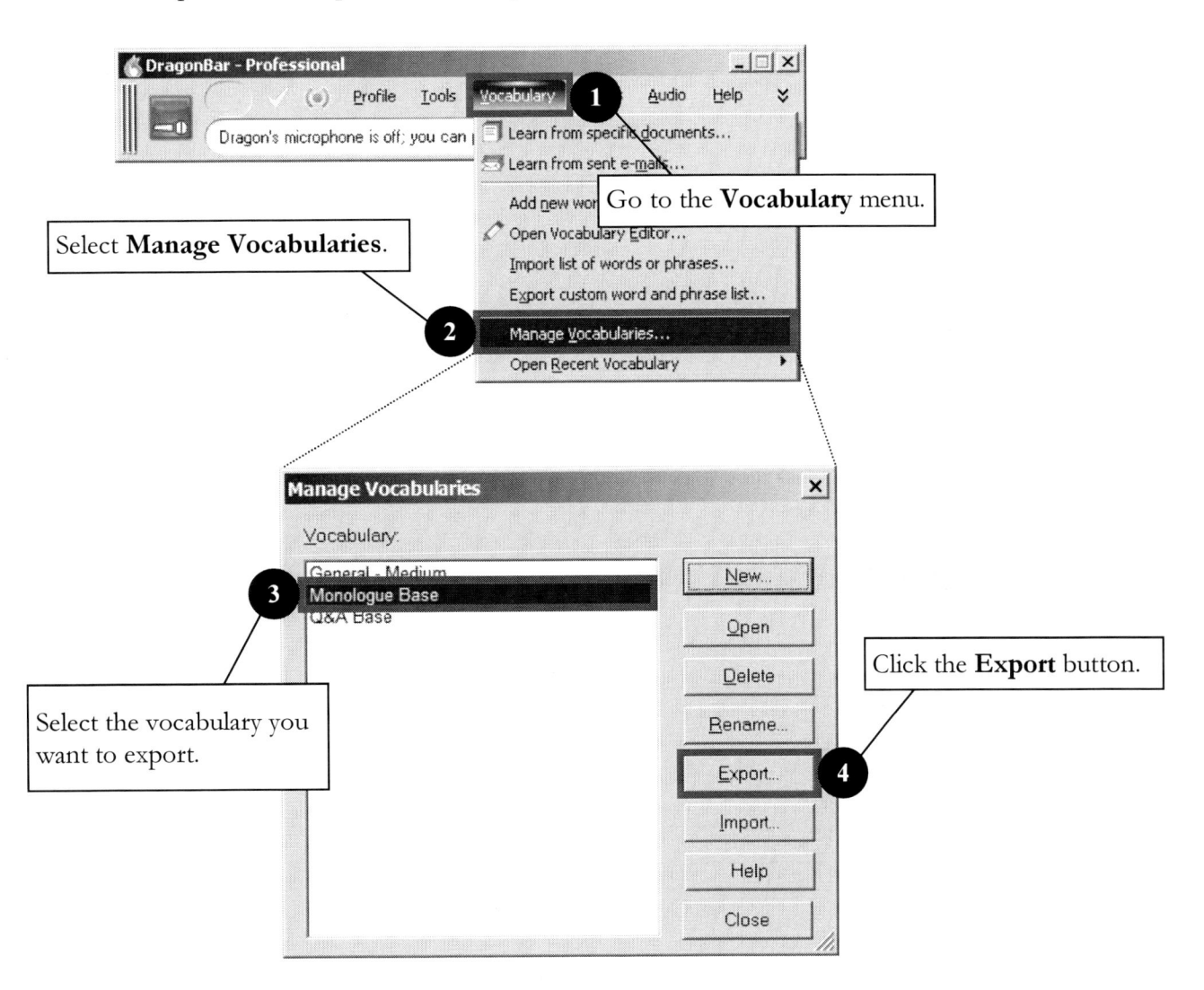

1. Go to the **Vocabulary** menu.
2. Select **Manage Vocabularies**.
3. Select the vocabulary you want to export.
4. Click the **Export** button.

209

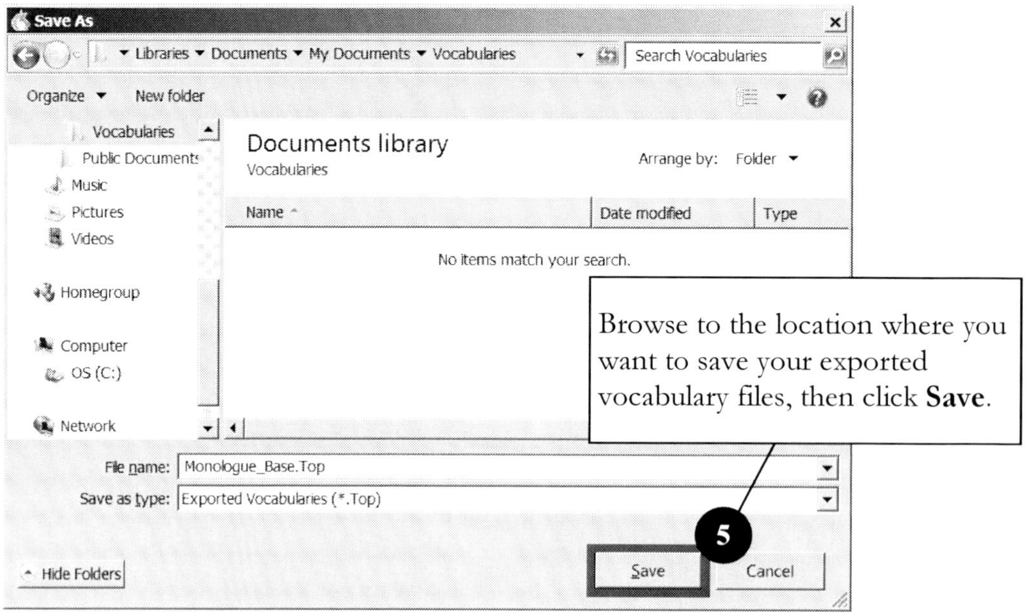
Save As
Libraries ▸ Documents ▸ My Documents ▸ Vocabularies
Search Vocabularies
Organize
New folder
Vocabularies
Public Documents
Music
Pictures
Videos
Homegroup
Computer
OS (C:)
Network
Documents library
Vocabularies
Arrange by: Folder
Name
Date modified
Type
No items match your search.
File name:
Monologue_Base.Top
Save as type:
Exported Vocabularies (*.Top)
Hide Folders
Save
Cancel
5
Browse to the location where you want to save your exported vocabulary files, then click **Save**.

Importing a Vocabulary

To import a vocabulary, go to the *Vocabulary* menu and select *Manage Vocabularies*, then follow the steps below.

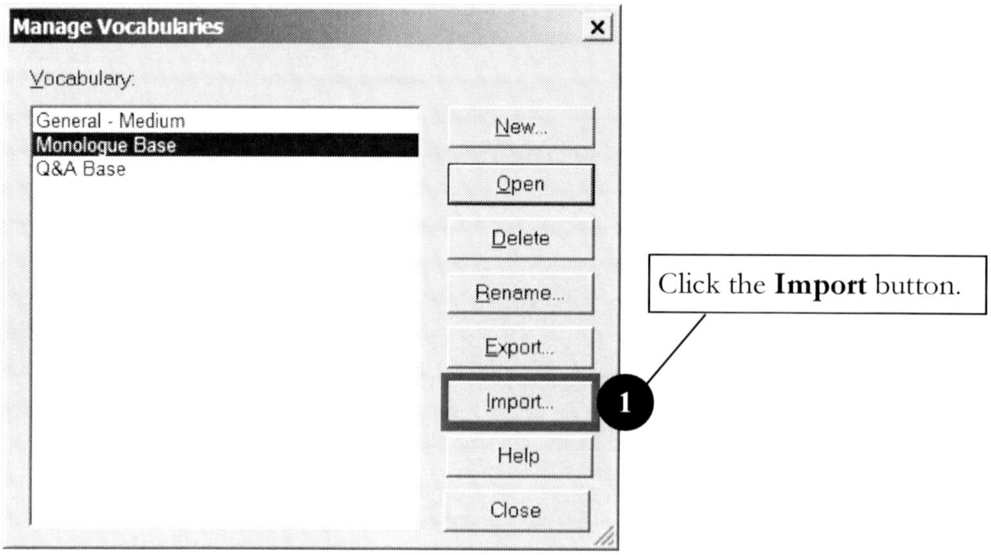
Manage Vocabularies
Vocabulary:
General - Medium
Monologue Base
Q&A Base
New...
Open
Delete
Rename...
Export...
Import...
Help
Close
1
Click the **Import** button.

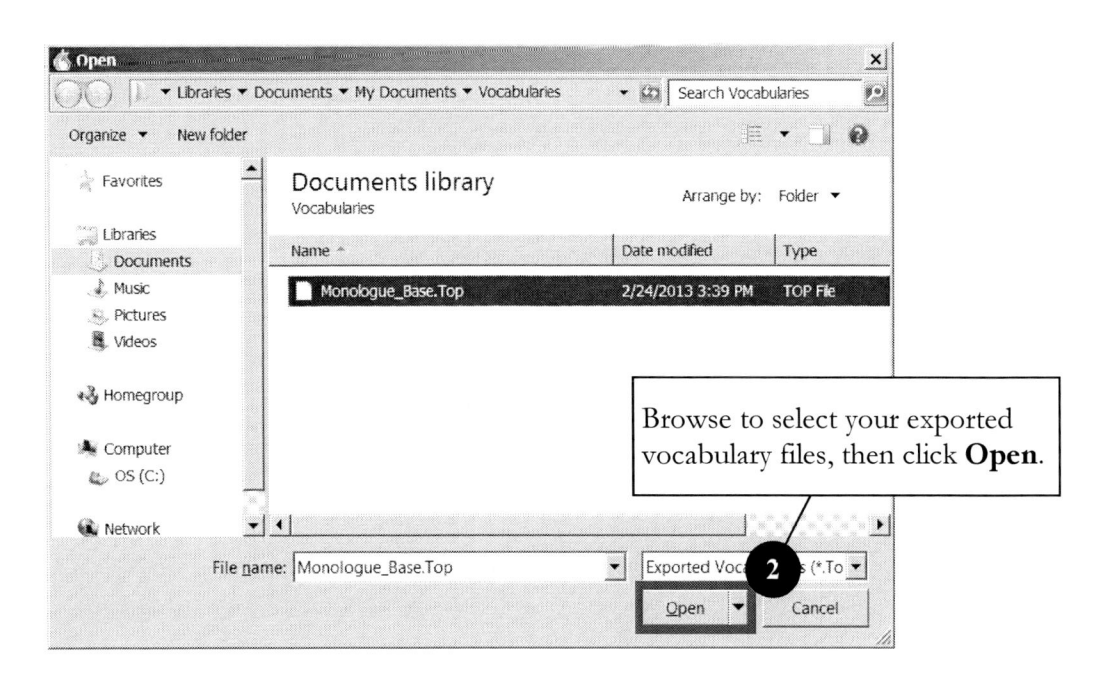

Chapter 8: Computer Maintenance and File Management

Browse to select your exported vocabulary files, then click **Open**.

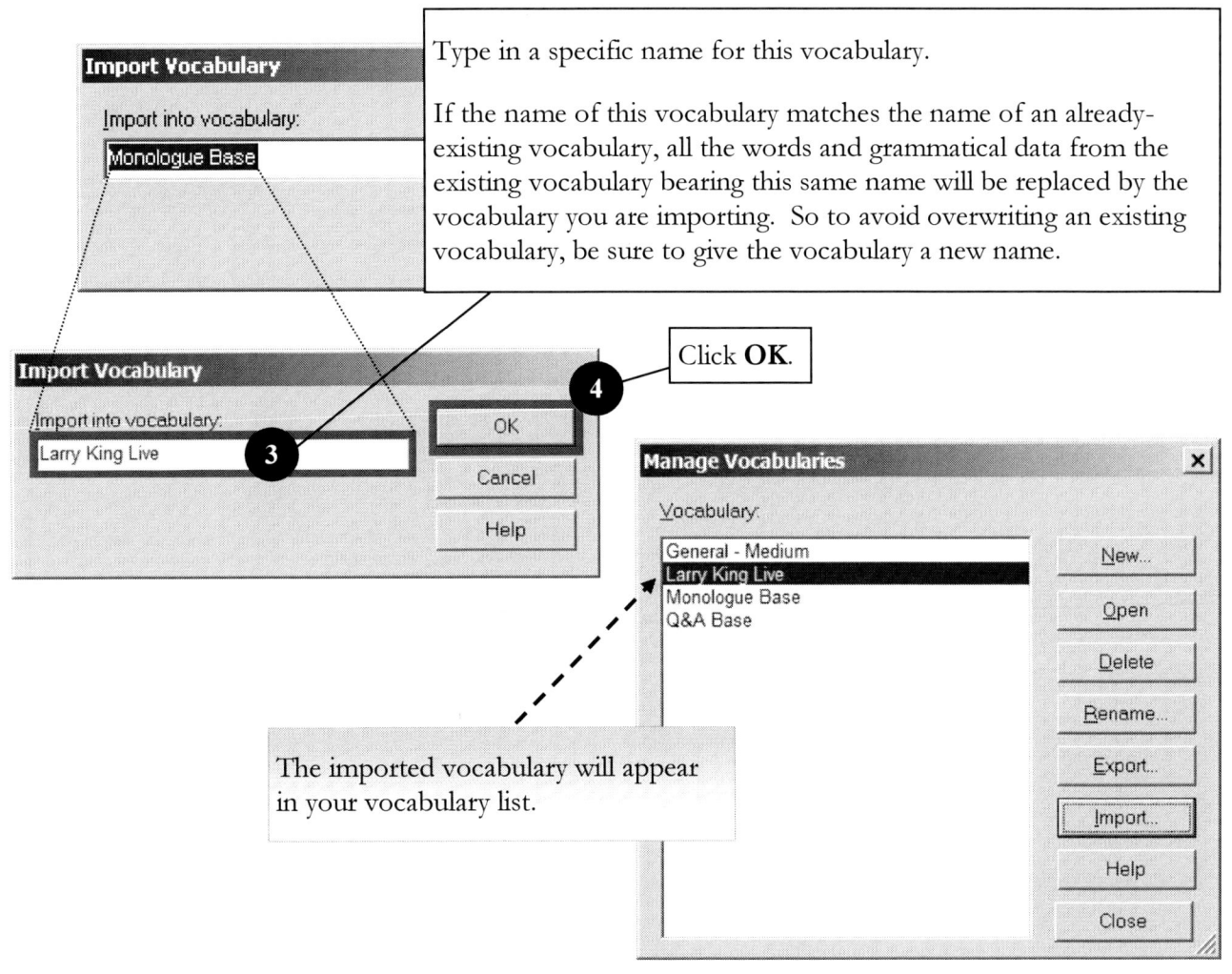

Type in a specific name for this vocabulary.

If the name of this vocabulary matches the name of an already-existing vocabulary, all the words and grammatical data from the existing vocabulary bearing this same name will be replaced by the vocabulary you are importing. So to avoid overwriting an existing vocabulary, be sure to give the vocabulary a new name.

Click **OK**.

The imported vocabulary will appear in your vocabulary list.

211

Exporting/Importing a Word List

Dragon gives you the ability to share lists of your custom vocabulary words with other Dragon users. This will be especially advantageous to you should you ever need to create a new Dragon user and don't want to have to manually re-input your special dictation theory entries into your vocabulary all over again from scratch. You can add the whole list automatically, choosing to include written- and spoken-form information and/or any formatting properties associated with those vocabulary entries. Be aware, however, that word lists are limited to exporting and importing custom words only, not your entire Dragon vocabulary list, and no grammatical data is transferred when simply exporting and importing word lists. If you wish to export or import an entire Dragon vocabulary replete with grammatical data, follow the steps in the *Exporting/Importing a Vocabulary* section of this chapter.

(Note: Dragon NaturallySpeaking® Professional edition users have export/import capabilities for both written- and spoken-form information and formatting properties of custom vocabulary words, whereas Dragon NaturallySpeaking® Premium edition users are limited to exporting/importing only written- and spoken-form information. Since word lists exported from the Premium edition do not contain formatting properties, formatting properties must be modified manually for the custom words that require them after the word list has been imported.)

Exporting a Word List

To export a list of your custom words, make sure the user and vocabulary from which you want to do the export is active, then follow the steps below. (Note: If you are using the Dragon NaturallySpeaking Premium® edition, you will not need to worry about which vocabulary is active since your Dragon user only has one vocabulary.)

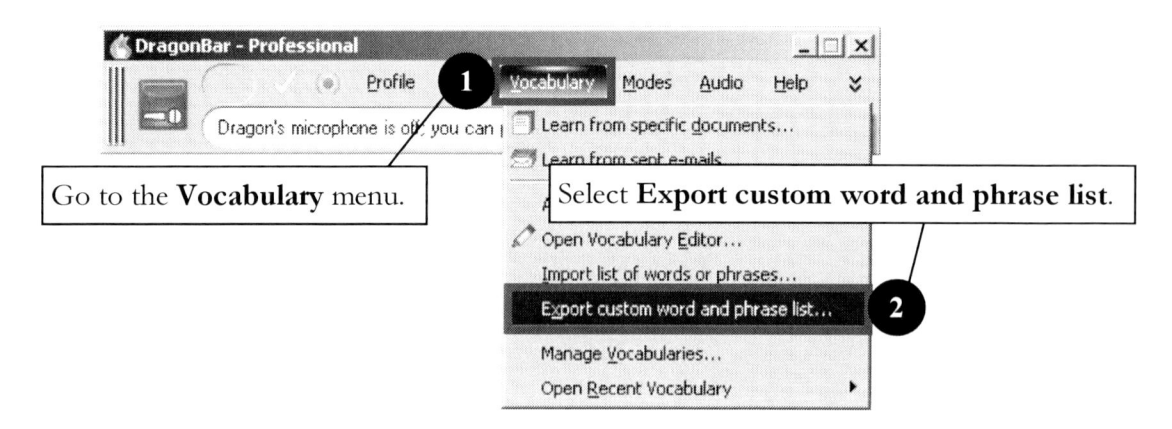

Go to the **Vocabulary** menu.

Select **Export custom word and phrase list**.

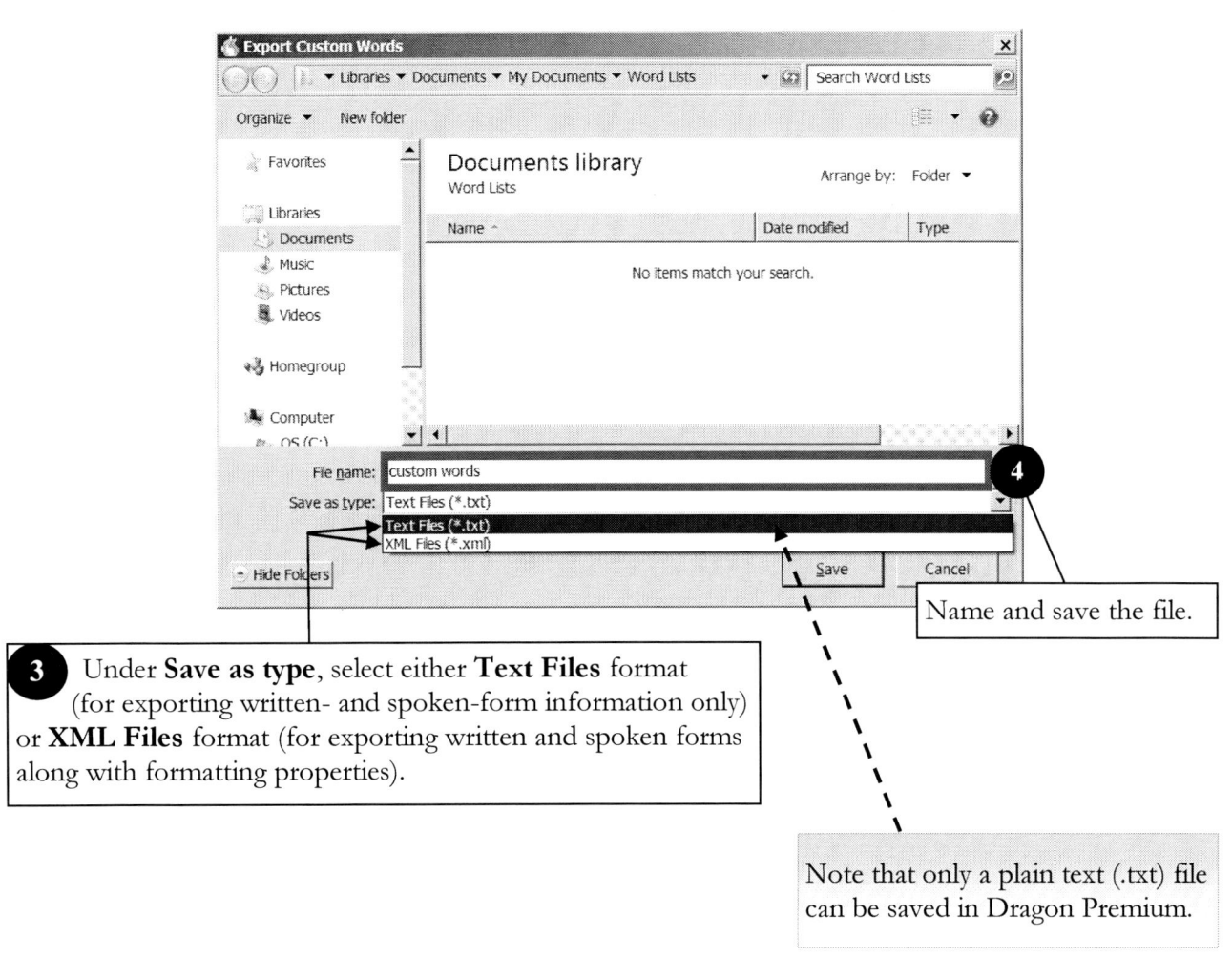

Chapter 8: Computer Maintenance and File Management

3 Under **Save as type**, select either **Text Files** format (for exporting written- and spoken-form information only) or **XML Files** format (for exporting written and spoken forms along with formatting properties).

Name and save the file.

Note that only a plain text (.txt) file can be saved in Dragon Premium.

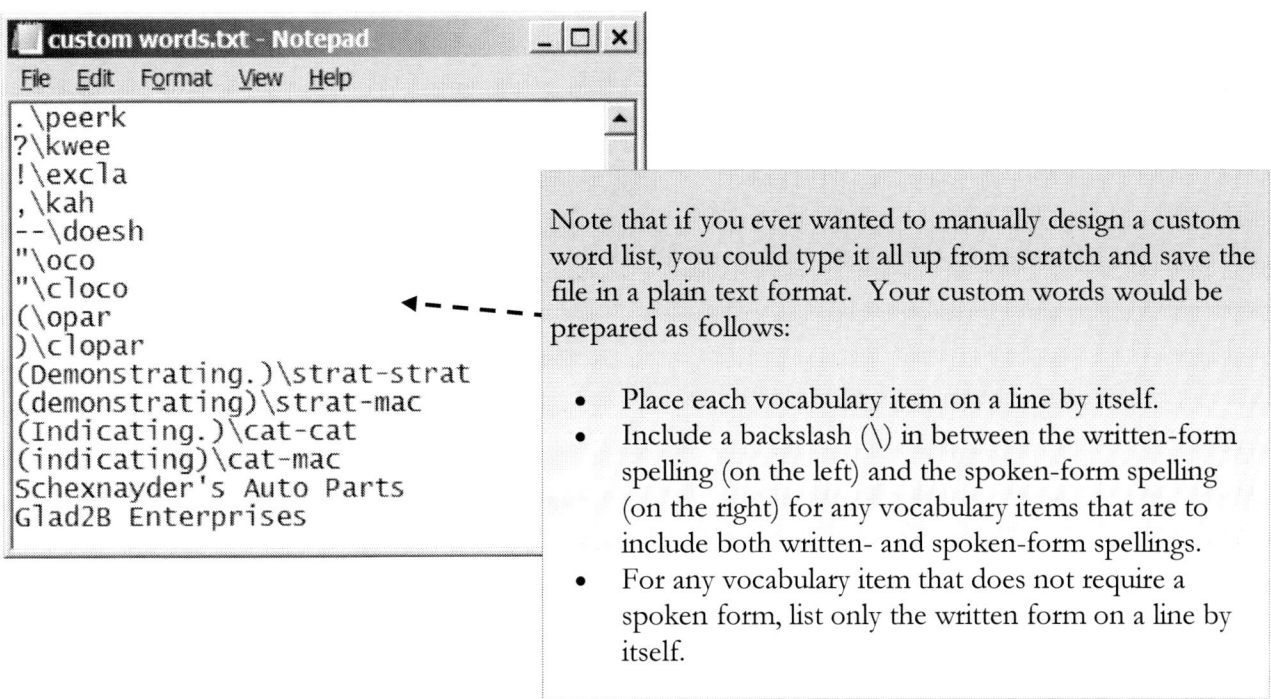

```
.\peerk
?\kwee
!\excla
,\kah
--\doesh
"\oco
"\cloco
(\opar
)\clopar
(Demonstrating.)\strat-strat
(demonstrating)\strat-mac
(Indicating.)\cat-cat
(indicating)\cat-mac
Schexnayder's Auto Parts
Glad2B Enterprises
```

Note that if you ever wanted to manually design a custom word list, you could type it all up from scratch and save the file in a plain text format. Your custom words would be prepared as follows:

- Place each vocabulary item on a line by itself.
- Include a backslash (\) in between the written-form spelling (on the left) and the spoken-form spelling (on the right) for any vocabulary items that are to include both written- and spoken-form spellings.
- For any vocabulary item that does not require a spoken form, list only the written form on a line by itself.

Importing a Word List

To import your custom words, make sure the user and vocabulary to which you want to add the list is active, then follow the steps below. (Note: If you are using the Dragon NaturallySpeaking Premium® edition, you will not need to worry about which vocabulary is active since your Dragon user only has one vocabulary.)

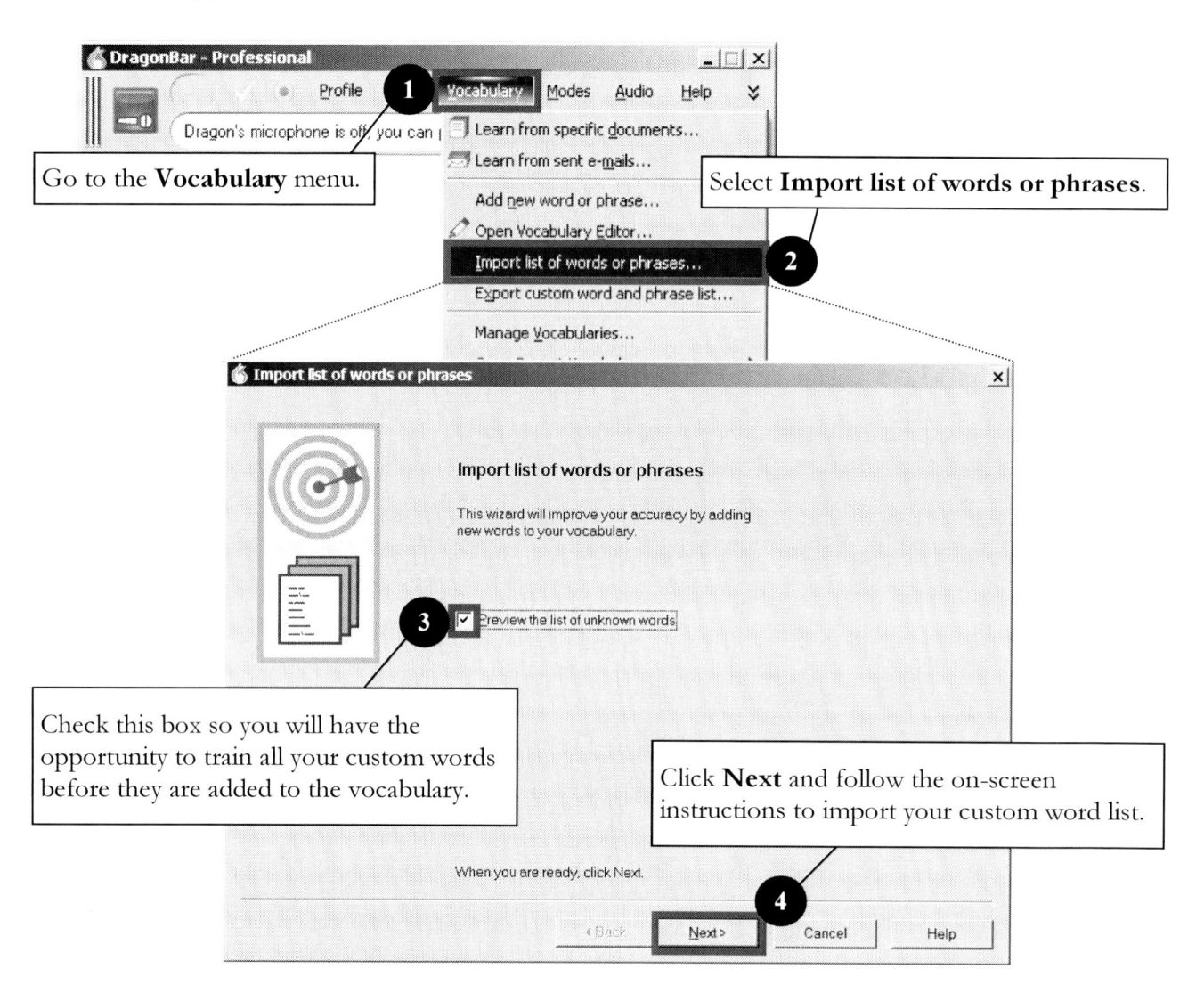

Go to the **Vocabulary** menu.

Select **Import list of words or phrases**.

Check this box so you will have the opportunity to train all your custom words before they are added to the vocabulary.

Click **Next** and follow the on-screen instructions to import your custom word list.

Comments About CAT Software

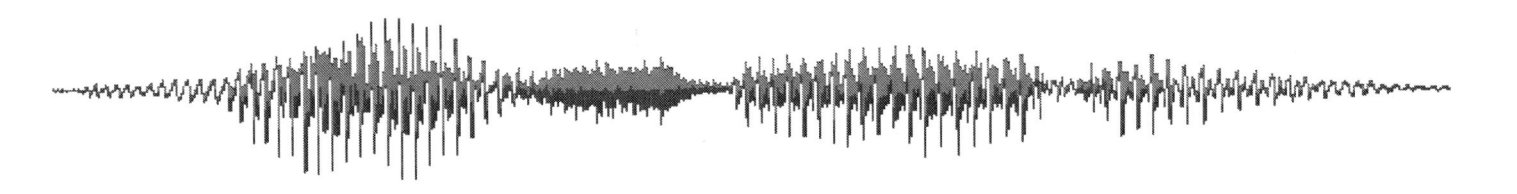

You will not have a need for a CAT program if you are using SR software simply for the purpose of doing transcription. But in voice writing professions requiring that you send realtime text feeds, it is a requirement that you purchase professional CAT software.

There are many wonderful features and tools available in CAT programs that are not mentioned in this book but are very useful. CAT software will allow you to perform globaling functions, instantly resolve conflicts during the middle of your dictation, change the letter casings of certain words when they appear in special contexts, such as always capitalizing the "S" in the word "street" when it occurs after a noun, or producing a lowercase letter to make your translations easier to read in captions that would ordinarily appear in uppercase (e.g., '70s versus '70S). Other features may include a word-swap utility for easy editing, digital recording with audio-synced text, automatic indexing, condensing, and electronic file conversions to include index hyperlinks.

These are just a few of the capabilities of CAT software. If you do not already have a CAT program and are ready to provide realtime services in the reporting industry, visit the www.nvra.org website for professional certification requirements and a list of vendors to contact. If you already have CAT software, refer to the Help menu or user's manual of your CAT program for more information on its available features.

GLOSSARY

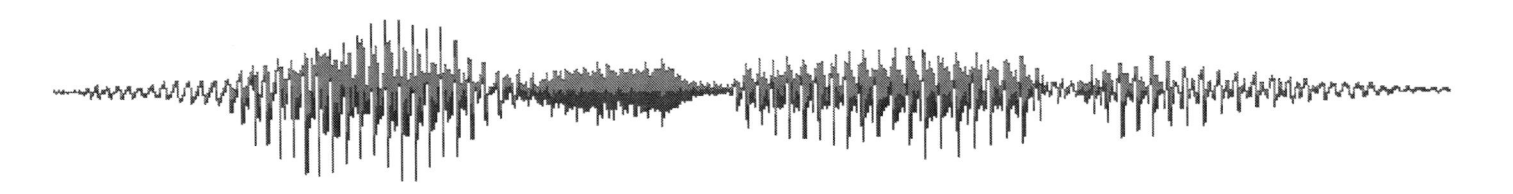

A

accuracy reinforcement – The process of using the correction tool to improve recognition performance by giving specific data to the SRE so that correct best-match guesses will become less of a "guess" in the future.

acoustic model – An algorithm-based framework within a speech recognition program which stores and manages sound data.

ADC – Analog-to-digital converter. A component found within a USB sound card device which converts continuous electrical signals into digital number sequences.

ad-hoc combining – The method of piecing together word parts to represent word groups in the creation of voice codes.

AI – Artificial intelligence. A branch of computer science dealing with intelligent behavior, learning, and adaptation in machines.

ASCII – American standard code for information interchange. A universally readable computer file used to exchange text among dissimilar computers and computer programs.

audio setup – An adjustment made for speech input levels within a speech recognition program for enhancing its ability to measure speech signals and filter out extraneous sounds.

audio-syncing – Synchronization of text with an audio recording.

B

base vocabulary – The default word bank provided in a speech recognition program.

best-match guess – The process by which a speech recognition program uses algorithms to determine the most statistically-probable word selection.

bigram – Pairs of words which are commonly used by a speech recognition program as the basis for statistical analysis of text.

blurb – Miscellaneous block of text or other data that may be inserted during realtime translations.

bottom-up processing – A computer's means of distinguishing words based upon analysis of basic sound structures.

brief form – A dictation shortcut used to represent common word groups and phrases.

C

captioning – Displaying the audio portion of a television program, film, or motion picture as text on a video screen to communicate what is said and by whom, and indicating other relevant sounds. (Also referred to as "closed-captioning.")

CAT – Computer-aided transcription.

CAT code – Combinations of characters and/or symbols interpreted by a CAT program as instructions on how to produce text and/or perform formatting functions.

CAT dictionary – A database within a CAT program which stores text, voice codes, CAT codes, and other data.

CAT software – A computer-aided transcription program that converts speech recognition output into a final realtime result, used for sending realtime feeds, editing, digital recording, and other functions.

CART – Communication access realtime translation.

CART provider – One whose job is to produce realtime translations for any purpose besides broadcast captioning and legal proceedings. (Also referred to as "CART reporter.")

CM – Certificate of Merit. NVRA's highest-standard verbatim reporting certification. A CVR designation must first be attained before a candidate is allowed to sit for the CM examination. A recording of the dictation is used to manually transcribe text, and silence requirements must be maintained.

condensed transcript – A miniaturized version of a transcript which prints out in multiple pages on a single sheet of paper. (Also referred to as "compressed transcript.")

conflict – A common speech recognition error produced for words with similar sounds.

conflict resolution – The process of eliminating recognition conflicts by any means other than using the correction tool to improve accuracy.

correction tool – A speech recognition program's utility for correcting misrecognitions.

court reporting – A profession specializing in the transcription and preservation of records for legal proceedings.

CPU – Central processing unit. The brain of a computer where calculations are processed and data is manipulated.

CVR – Certified Verbatim Reporter. NVRA's entry-level reporting certification. A recording of the dictation is used to manually transcribe text, and silence requirements must be maintained. A written knowledge test is also given.

Glossary

D

dictation – The vocal repetition of what other persons say, either directed into a speech silencer or into a headset microphone for the purpose of producing a realtime and/or verbatim record.

doubling-up method – A voice code design technique where a word or word part is doubled to create a distinctive sound while maintaining a close resemblance with the root word from which it was derived.

drop – Purposely refraining from repeating certain words during the dictation process. Mostly used for insubstantive words in paraphrasing when summarized reports are allowed.

F

flash drive – A compact, portable memory card that plugs into a computer's USB port and functions as a portable hard drive, usually small enough to be carried in a pocket or on a keychain. (Also referred to as "thumb drive.")

force-produce – Use of special voice codes to ensure that a desired word will be recognized by a speech recognition program, regardless of grammar or context.

G

general training – (See "voice training.")

grammatical model – An algorithm-based framework within a speech recognition program which stores and manages language context data. (Also referred to as "language model.")

H

headset microphone – A dictation input device that is worn as a headset and has an open noise-canceling microphone.

hold-dictation technique – In very quiet environments, refraining from dictating during periods of silence and dictating only when speakers talk.

L

language model – (See "grammatical model.")

large speech silencer – A dictation device containing a built-in microphone which covers the nose, mouth, and a portion of the chin and cheek areas.

layering technique – Making various vocal recordings at different rates of speech and using different qualities of pronunciations to increase the likelihood of correct word selection during live dictation.

lettering technique – A voice code design method of combining letters of the alphabet to represent whole words in phrases.

M

malware – Malicious software which can adversely affect speech recognition performance by slowing down a computer.

marker – A character, word, or word group used as a standardized abbreviated form of indication in a transcript (i.e., "Q" and "A" to represent questions and answers).

mask – Another term for "speech silencer."

N

n-gram – Any number of word sequences. Used in speech recognition algorithms for processing language based on statistical measurements.

NVRA – National Verbatim Reporters Association. A professional association and certifying body of verbatim reporters.

O

on-the-fly translation – Producing a word or term in a realtime translation when it does not exist within a speech recognition program's vocabulary.

P

parenthetical – A form of text appearing within parentheses to convey meaning to the reader.

phoneme – The smallest phonetic unit in a language that is capable of conveying a distinctive meaning, such as the "b" of "bat" and the "p" of "pat" in English.

phrase-dictation technique – The method of dictating words in phrase groupings to enhance a speech recognition program's ability to select correct words based on context.

pre-punctuated phrase – A group of words containing punctuation, designed to be output by a speech recognition program in response to the dictation of words without punctuation.

punctuation technique – The part of voice theory language that involves resolution of punctuation-versus-word conflicts and producing faster realtime translations by dictation of fewer syllables.

R

RAM – Random access memory. The computer's accessible memory for temporary storage and retrieval of programs and data.

RBC – Registered Broadcast Captioner. NVRA's broadcast captioning certification. A realtime test is given to certify speed and accuracy performance meets captioning industry standards.

RCP – Registered CART Provider. NVRA's communications access realtime translations (CART) certification. A realtime test is given to certify speed and accuracy performance meets CART industry standards.

readback – A request that the reporter read aloud from a computer screen a portion of previously dictated text.

realtime – Live conversion of speech-to-text translations.

realtime record – A live and unedited version of a record.

reporting – To transcribe a summarized or verbatim account of proceedings or events.

RVR – Realtime Verbatim Reporter. NVRA's realtime reporting certification. A CVR designation must first be attained before a candidate is allowed to sit for the RVR examination. Text must be produced live through dictation, and silence requirements must be maintained.

S

small speech silencer – A dictation device containing a built-in microphone which covers the mouth and a portion of the chin.

speaker-adaptive – Refers to a speech recognition program developed to adapt its operation to the characteristics of new speakers.

speaker-dependent – Refers to a speech recognition program that recognizes speech based on one human voice.

speaker identifier – User-defined text that identifies a speaker.

speaker-independent – Refers to a speech recognition program that recognizes speech of a large segment of the population (e.g., American English).

speech silencer – A dictation input device designed to prevent the sound of dictation from distracting others in the environment as a recording is being made.

speech-to-text conversion – The process of converting speech into analog signals, analog signals into digital format, then digital format into text.

spelling technique – The part of voice theory language that involves producing letters of the alphabet.

spoken form – The phonetic spelling of how an entry is pronounced in a speech recognition program. This must be given when the written form does not adequately represent its pronunciation.

SR – Speech recognition. Also known as "ASR" for "automatic speech recognition" (and now being used synonymously with "VR" for "voice recognition"). A technology designed to automatically recognize speech input and respond through digitization and algorithm-based programming to transcribe or act upon what a person has said.

SRE – Speech recognition engine. A speech recognition program.

T

top-down processing – The human propensity to distinguish words based on previously known concepts and circumstantial knowledge.

train – To record a pronunciation.

translation – Conversion of audible sounds, signals, or gestures to written text.

trigram – Three-word sequences which are commonly used by a speech recognition program as the basis for statistical analysis of text.

___

U

USB – A peripheral device connection standard that is expected to completely replace serial and parallel ports. Used to connect keyboards, mice, printers, and sound cards.

USB port – An external interface connection point which facilitates communication between a computer and peripheral device.

USB sound card – An external device that converts analog signals into digital form for further processing by a speech recognition engine. (Also known as "USB speech processor" or "USB sound pod.")

___

V

verbatim record – The final, edited version of a record that is produced word for word.

vocabulary – The word bank within a speech recognition engine, containing both base and user-added words.

vocabulary building – The process used to automatically find and add new words to the vocabulary and also for creating a grammatical structure.

voice code – A made-up utterance that does not sound similar to an English word. Used for identifying speakers, producing text, formatting functions, conflict resolution, on-the-fly translations, brief forms, and inclusion of miscellaneous blocks of text and other data.

voice training – The initial process of voice model creation performed by reading aloud stories provided by the speech recognition program so it can learn the way a person speaks. (Also referred to as "general training.")

voice model – A compilation of data pertaining to a specific human voice within a speaker-dependent speech recognition program.

voice writing theory – The culmination of voice codes and their conceptual usages, which together comprise a realtime dictation language.

___

W

written form – The actual spelling of how an entry is written within the vocabulary of a speech recognition program.

INDEX

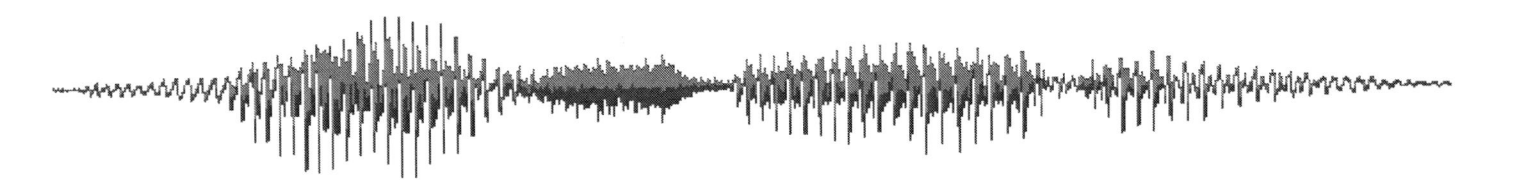

A

accuracy calculation	191
accuracy reinforcement	184-185
adding phrases	73-76, 160-161
adding words	125-126, 158-159
adding word list	158-159, 214
adding vocabulary entries for voice writing theory	129-142
for Dragon-alone use	129-138
for Microsoft Word use	138-140
for CAT program use	141-142
CAT program with a dictionary	141-142
CAT program without a dictionary	142
analyzing documents	176-182
anti-virus software	16, 89, 194
archiving data	193
audio setup	10, 146-149
automatic updates	16, 89

B

back up user	199-200
breathing	24-25
with open-mic headset	24
with speech silencer	24-25
brief forms	76-79, 161, 170-171

C

captioning	5
careers, realtime	3-6
captioning	5
CART	5
court reporting	4
financial call reporting	6
other	6
CART	5
CAT code	12-15, 141-142, 165, 172-173
CAT program	11-15
about	11-15, 215
how it works	12-13
with CAT dictionary	13-14
without CAT dictionary	14-15
vocabulary setup	141-142
with CAT dictionary	141
without CAT dictionary	142
certifications	4-5
closing user	198
computer	
anti-virus software	16, 89, 194
archiving data	193
automatic updates	16, 89
defragmenting	194
maintenance	193-194
setup	85-89
USB connectivity	88-89

conflicts, resolving 64-72, 183-185
    homophones 65-67
    letters 63-64
    other conflicts 72
    recognition problems 183-185
    similar-sounding words 70
    small words 68-70
    undesirable words 68
    words with soft pronunciations 67
    words versus acronyms 71-72
correcting speech recognition errors ... 119, 150-157
court reporting 4

D

defragmenting hard drive 194
deleting
    user 198-199
    vocabulary 208
    vocabulary entries 127
    words by voice (delete command) 81-82
dictation input devices 17-18
    open-mic headset 18
    speech silencer 18
    USB sound card 17
dictation material 189-191
Dictation Mode 117, 138, 143, 151
dictation session
    performing 117-119
    saving 120
    setting to Dictation Mode 117, 151
dictation techniques 23-38
    breathing 24-25
        with open-mic headset 24
        with speech silencer 24-25
    enunciation 29-32
    pace 32-34
        fast speakers 33
        slow speakers 34
    paraphrasing 35-36
    patterns, target dictation 36-38
    punctuating 34
    tone and modulation 26-28
    verbatim 35
    verbatim summarization 35-36
    volume 24-26
        with open-mic headset 24
        with speech silencer 25-26
document analysis 176-182

document preparation 163-174
    for CAT program use 169-174
        CAT program with dictionary 169-171
        CAT program without dictionary 172-174
    for Dragon-alone use 167-168

E

enunciation 29-32
equipment
    for voice realtime writing 15-20
    setup of 85-87
examination lines 49
exporting
    vocabulary 209-210
    word list 212-213
    user files 201-202

F

financial call reporting 6
foot pedal 19
formatting preferences, user 116

H

headset microphone (a.k.a. open-mic headset) 18
homophone conflicts 65-67

I

importing
    vocabulary 210-211
    word list 214
    user files 203-204
improving accuracy 145-185
    correcting speech recognition errors 119, 150-157
    vocabulary building 162-182
    all other techniques 183-185

Index

M
material, dictation 189-191
    from Internet 191
    from SpeedMaster™ CD 189
microphone, room recording 19
modifying formatting properties of
    vocabulary entries 128
    for punctuation marks 130-132
    for speaker identifiers 133-135

N
N-Gram Phrase Extractor 160
numbers 61-62
    producing numbers under 10 as digits 61
    inserting a space between numbers 62

O
on-the-fly translations 79-81
opening (activating) vocabulary 208
opening user 198
open-mic headset 18
    dictating into 24
    proper positioning of 98, 147

P
pace, dictation 32-34
    fast speakers 33
    slow speakers 34
paraphrasing 35-36
parentheticals 52-60
    for captioning 57-59
    for CART 57-59
    for court reporting 52-56
    for financial call reporting 60
patterns, target dictation 36-38
phrases 73-76
    adding 73, 160-161
    pre-punctuated 75
    identifying through N-Gram Phrase
        Extractor 160-161
punctuation 34, 41-42
    effects on speech recognition 34
    pronunciations for 41-42
        single punctuation marks 41-42
        combination punctuation marks 42
    setup in vocabulary 130-132

Q
Q&A markers 50-51
    for court reporting 50
    for financial call reporting 51

R
realtime voice writing, about xv
recognition problems, resolving 183-185
reformatting documents 167-174
    for CAT program use 169-174
    CAT program with dictionary 169-174
    CAT program without dictionary 172-174
    Dragon-alone use 167-168
rename user 198-199
rename vocabulary 208
restore user 199-200

S
setting user formatting preferences 116
setting user options 109-115
saving
    dictation session 120
    user files 157
speaker identifiers 43-48
    for captioning 47
    for CART 46
    for court reporting 44-45
    for financial call reporting 48
    setup in vocabulary 133-135
speech recognition software
    about .... xiv, 7
    how it works 7-9
    tools, terms, and functions 10
speech silencer 18
    dictating into 24-26
    proper positioning of 98, 147
speed-building plan 188
SpeedMaster™ .... xiv, 91, 187-190
    about xiv
    creating a user with 91
    dictation material 189-191
    installation of 188
    operation of 189-190
    voice training story text files 91, 189
spelling technique 63-64
spoken forms 122-123
surge-protection power strip 20, 85-86

225

T
tone and modulation, dictation 26-28

U
USB connectivity 88-89
USB sound card..17, 87
user files
    backing up/restoring.............................. 199-200
    creating new 90-106, 195-196
    deleting 198-199
    opening/closing 198
    renaming... 198-199
    saving 157
user formatting preferences 116
user options.. 109-115

V
verbatim dictation 35
verbatim summarization 35-36
vocabulary
    accessing vocabulary 124
    adding entries... 125-126
    creating a new vocabulary 206-207
    deleting entries 127
    exporting/importing 209-211
    modifying entry formatting properties......... 128
    opening (activating) vocabulary 208
    renaming vocabulary 208
    written- and spoken-form spellings..... 122-123
vocabulary building.. 162-182
    document preparation............................ 163-174
    document analysis................................... 176-182
vocabulary customization 175-176
vocabulary setup... 121-142
    CAT program with a dictionary 141-142
    CAT program without a dictionary 142
    Dragon-alone use..................................... 129-138
    Microsoft Word use 138-140
voice codes
    definition of ....xiii, xv, 39
    brief forms 76-79
    creation methods 40
        ad-hoc combining approach....................40
        doubling-up method 40
        lettering technique 40
    delete commands 81-82

voice codes *(continued)*
    examination lines 49
    homophones 64-67
    insert-space command 62
    numbers under 10 as digits 61
    on-the-fly translations.................................79-81
    parentheticals 52-60
    punctuation marks......................................41-42
    Q&A markers...50-51
    speaker identifiers.......................................43-48
    letters of the alphabet 63-64
voice model, creating 90-108, 195-197
voice training story text files from
SpeedMaster™ CD 91, 189
voice writing, about.. xv
voice writing theory..xv, 39-82
volume, dictation 24-26
    with open-mic headset 24
    with speech silencer 25-26

W
word list
    creating 158-159, 213
    adding 159
    exporting/importing 212-214
written forms 122-123